Topics in Applied Multivariate Analysis

Edited by

DOUGLAS M. HAWKINS
Senior Consultant in Statistics
Council for Scientific and Industrial Research

CAMBRIDGE UNIVERSITY PRESS

CAMBRIDGE

LONDON NEW YORK NEW ROCHELLE

MELBOURNE SYDNEY

CAMBRIDGE UNIVERSITY PRESS
Cambridge, New York, Melbourne, Madrid, Cape Town, Singapore, São Paulo, Delhi

Cambridge University Press
The Edinburgh Building, Cambridge CB2 8RU, UK

Published in the United States of America by Cambridge University Press, New York

www.cambridge.org
Information on this title: www.cambridge.org/9780521243681

First published 1982
This digitally printed version 2008

A catalogue record for this publication is available from the British Library

Library of Congress Catalogue Card Number: 81–15527

ISBN 978-0-521-24368-1 hardback
ISBN 978-0-521-09070-4 paperback

PREFACE

In 1980, the National Research Institute for Mathematical Sciences (NRIMS) of the Council for Scientific and Industrial Research held a highly successful Summer Seminar Series on Extensions of Linear-Quadratic Control Theory. This was followed in February 1981 by one on Applied Multivariate Analysis. The papers presented at that semi= nar formed the penultimate drafts of those in this book.

The choice of Applied Multivariate Analysis as the topic for the second Summer Seminar Series was prompted by the recognition that there is a much wider divergence between theory and practice of mul= tivariate methods than is the case with univariate statistics. As a result of this, multivariate methods are applied widely, but little understood by their users.

The authors of these papers, apart from having considerable prac= tical experience in the application of the techniques they discuss, have also contributed to their theoretical development. Their papers give a distillation of what they regard as the major features of each topic from the viewpoint of applicability, and so may go some way towards bridging the chasm between theory and practice.

Valuable comments on the first drafts and suggestions for improve= ments were made by Mr. J.C. Gower (Chapters 4 and 6), Professor J.N. Darroch (Chapter 3) and Professor J. Aitchison (Chapter 1). The gene= rosity of these reviewers in finding time to go through the manu= script is greatly appreciated.

On behalf of the authors, I should like to acknowledge gratefully the support of NRIMS in subventing and hosting the Summer Seminar Series. The help of colleagues in our various institutions was also invaluable:- in particular, thanks are due to Mrs. M. Russouw (assis= ted by Mrs. A.M. de Villiers) for typing the manuscript; Mesdames P.Colman and A. Myburgh for preparation of graphics, and Mrs. J.S. Galpin for editorial assistance.

Douglas M. Hawkins
NRIMS, CSIR
1981.07.07

CONTRIBUTORS

Prof M W Browne

Dept of Statistics
University of South Africa
P O Box 392
PRETORIA 0001

Dr L P Fatti

National Research Institute
for Mathematical Sciences
C S I R
P O Box 395
PRETORIA 0001

Dr M J Greenacre

Dept of Statistics
University of South Africa
P O Box 392
PRETORIA 0001

Dr D M Hawkins

National Research Institute
for Mathematical Sciences
C S I R
P O Box 395
PRETORIA 0001

Prof G V Kass

Dept of Applied Mathematics
University of the Witwatersrand
Milner Park
JOHANNESBURG 2001

Dr T J v W Kotze

Institute for Biostatistics of
the SA Medical Research Council
P O Box 70
TYGERBERG 7505

Mr M W Muller

National Institute for Personnel
Research
P O Box 32410
BRAAMFONTEIN 2017

Dr E L Raath

National Research Institute
for Mathematical Sciences
C S I R
P O Box 395
PRETORIA 0001

Mrs J A ten Krooden

National Research Institute for
Mathematical Sciences
C S I R
P O Box 395
PRETORIA 0001

Dr L G Underhill

Dept of Statistics
University of Cape Town
RONDEBOSCH 7700

CONTENTS

CONTENTS

DISCRIMINANT ANALYSIS

L. PAUL FATTI, *UNIVERSITY OF THE WITWATERSRAND*

DOUGLAS M. HAWKINS and E. LIEFDE RAATH, *COUNCIL FOR SCIENTIFIC AND INDUSTRIAL RESEARCH*

1. PRELIMINARIES

1.1 General Introduction

Discriminant analysis is concerned with the problem of classify= ing an object of unknown origin into one of two or more distinct groups or populations on the basis of observations made on it. As evi= denced by the examples given below, this problem occurs frequently in various fields as diverse as medicine, anthropology and mining, and the techniques of discriminant analysis have been used successfully in many situations. Computer packages for performing the necessary calculations involved in applying some of the techniques have been readily available for some time, although there are still some seri= ous omissions in most of these packages.

Some examples

1. Haemophilia is a sex-linked genetic disease which is trans= mitted only by females, but whose symptoms are manifest only in males. Under normal medical examination it is impossible to distinguish be= tween females carrying the disease and those not. In order to try and identify female carriers, the levels of a coagulant factor and its related antigen (Factor VIII and Factor VIII RA) in the blood have been suggested as possible discriminators between carriers and non-carriers.

A pilot study was carried out by Gomperts et al (1976) to test how well Factor VIII and its related antigen discriminate between carriers and non-carriers. A sample of 26 white females, of which 11 were known, for genetic reasons, to be carriers and 15 were known to be non-carriers was selected and the Factor VIII and Factor VIII RA levels measured in each subject. Using the linear discriminant func= tion based on the logarithms of the data and the jackknife reclassi= fication procedure to be described in the next section, ten out of the eleven carriers and thirteen out of the fifteen non-carriers were classified correctly. In a parallel study on black females (whose

Factor VIII and Factor VIII RA levels tend to be different from the whites) all ten carriers and fourteen out of fifteen non-carriers were correctly classified. Thus it would appear that carriers of haemophilia can be identified, reasonably reliably, on the basis of their levels of Factor VIII and Factor VIII RA.

2. A common problem occurring in anthropology is that of identi= fying the tribe, race or even sex of a cranium excavated amongst the remains of an ancient civilization (for example De Villiers, 1976). By comparing various measurements (lengths and angles) made on this skull with those made on large numbers of individuals, males and fe= males, from the various tribes at present inhabiting the region, it may be classified to the tribe from which it most probably came. In this problem, we might also be interested in the possibility that it did not originate from any of these tribes, but from another unknown tribe (possibly now extinct). Another possibility is of its occupying some position intermediate between the tribes which could result from intermarriage between members of the different tribes.

A problem arising frequently in this type of application is that of choosing, amongst the large number of possible measurements that can be made on the skull, that subset giving the best discrimination between the various populations.

3. The final example comes from a stratigraphic problem in mining. In the Witwatersrand gold fields the gold bearing reef is one band (the "pay band") in a sedimentary succession, and is usually visually unrecognisable. In badly faulted areas this pay band may fault away, and the miner wishes to know the position in the sedimen= tary succession of the blank band facing him, from which he can de= duce the new position of the pay band.

The trace element geochemistry of a chip of rock from this band provides a means of classification. Given random samples of rock chips from each of the bands, and measurements of the concentrations (in log(parts per million)) of a number of trace elements taken on each chip as well as on the chip of unknown origin, the unknown band has been correctly identified in a high proportion of cases (Hawkins and Rasmussen (1973)).

An interesting feature of this problem is that the rock bands themselves may be considered to be a sample from a "super-population" of rock bands. In different situations, different sets of bands are involved, all drawn randomly from this super-population. The random effects model, discussed later, may well be appropriate here.

There is also another use of discriminant analysis. Multiple regression is generally presented as a method of making a prediction of an unknown variable from a set of predictors, but in many cases one uses it, not to actually make the prediction, but to study the regression coefficients to see how the dependent variable is affected by the predictors. In a closely analogous way, a discriminant function is computed, not to actually classify observations to their source, but to gain a better understanding of what it is that distinguishes the populations Π_i from one another.

1.2 The basic principles of discriminant analysis

Consider the problem of classifying an observation (vector) X into one of k groups or populations $\Pi_1, \Pi_2, \ldots, \Pi_k$ where Π_i is charactezing a probability density function $f_i(X)$. Suppose further that the observation has a prior probability π_i of coming from Π_i, where $\sum_{i=1}^{k} \pi_i = 1$, and that the cost associated with classifying it into Π_i when it has actually come from Π_j is c_{ij}.

Anderson (1958) shows that the rule that minimizes the expected cost of misclassification is to assign the observation X to Π_i if

$$\sum_{\ell=1}^{k} \pi_\ell \; c_{i\ell} \; f_\ell(X) < \sum_{\ell=1}^{k} \pi_\ell \; c_{j\ell} \; f_\ell(X) \quad , \; j=1,\ldots,k; j\neq i \qquad (1.1)$$

where $c_{jj} = 0$, $j = 1,\ldots,k$.

In the situation where the costs of misclassification are all equal, this rule simplifies to: assign X to Π_i if

$$\pi_i \; f_i(X) = \max_{j=1,\ldots,k} \; \pi_j \; f_j(X) \qquad (1.2)$$

Assignment rules (1.1) and (1.2) have been derived considering the discriminant analysis problem from a decision-theoretic viewpoint. Viewing it from a purely probabilistic viewpoint instead, the optimal rule is to assign X to that population Π_i for which the posterior probability is the greatest. Now, using Bayes theorem, the posterior probability of Π_j given X is proportional to $\pi_j \; f_j(X)$, so that the optimal probabilistic rule is also (1.2). So, when the costs of misclassification are all equal, the optimal decision-theoretic and probabilistic classification rules are equivalent.

For the two-group (k=2) case rules (1.1) and (1.2) become,

respectively: Assign X to Π_1 if:

$$\pi_2 c_{12} f_2(X) < \pi_1 c_{21} f_1(X) \qquad (1.3)$$

and to Π_2 otherwise;

and, assign X to Π_1 if:

$$\pi_2 f_2(X) < \pi_1 f_1(X) \qquad (1.4)$$

and to Π_2 otherwise.

In practice, the probability density functions $f_i(X)$ $i=1,\ldots,k$ are seldom known. Usually one assumes that they have some particular parametric form (e.g. a multivariate normal distribution) which de= pends on some unknown parameters. Random samples, called training samples, consisting of observations known to have come from each spe= cific one of these k populations, are then used to construct sample-based classification rules corresponding to (1.1) to (1.4) above.

There are two different methods of setting up such sample-based variants of (1.1) to (1.4). In the *estimative* approach the training samples are used to estimate the unknown parameters using such methods as maximum likelihood. These estimates are then substituted ("plugged in") for the unknown parameters in f_i, and we behave as if the esti= mates were the true unknown values. Provided consistent estimators are used, the estimative approach leads asymptotically, as the train= ing samples become infinitely large, to the correct optimal classi= fication rule. With small training samples, however, the estimative approach has but a tenuous claim to good theoretical properties, though it is within the general class of empirical Bayes procedures.

The more recent *predictive* approach is a fully Bayesian method. This means that the data are regarded as given, and the unknown para= meters as random variables which in due course are integrated out of the model. The steps in this are:

(i) Set up an a priori distribution $g(\theta)$ for the unknown para= meters.

(ii) Construct $f(X,T|\theta)$ - the joint distribution of the unknown to be classified, X, and the training sample, T.

(iii) The joint distribution of X, T and θ is $g(\theta)f(X,T|\theta)$. Integrate θ out of this expression, getting, in due course, the conditional distribution of X given T.

As is generally the case with Bayesian methods, criticism of this model usually centres about the realism or otherwise of the prior distribution $g(\theta)$. This is commonly assumed uninformative, but unless the training sample is extremely small, the exact specification of $g(\theta)$ has little effect on the classification.

The estimative approach consists of setting up the generally very simple algebra for computing the ratio $f_i(X)/f_j(X)$ and then plugging in the estimates of any unknown parameters. In the predictive approach, one must set up the full model corresponding to the particular form of $f_i(X)$ and then integrate out θ. This latter may be no easy feat, and to date predictive procedures have essentially been worked out only for certain models based on underlying normal data.

In the next two sections the classical approach to discriminant analysis, being the estimative approach applied to the multivariate normal distribution, is described. Thereafter, the predictive approach is considered, followed by a section on other approaches to discriminant analysis. In the final section a number of miscellaneous topics are considered.

2. CLASSICAL DISCRIMINANT ANALYSIS

The most common assumption in discriminant analysis is that X is a p-dimensional vector of observations, and that if it comes from Π_i then it follows a multivariate normal distribution with mean vector ξ_i and covariance matrix Σ_i. We will also assume that the costs of misclassification are all equal, so that the decision-theoretic and probabilistic classification rules are the same.

Following the common notation, we will use $N(\xi,\Sigma)$ to denote the p-variate multivariate normal distribution with density

$$f(X) = \frac{1}{\{(2\pi)^p|\Sigma|\}^{\frac{1}{2}}} \exp\{-\tfrac{1}{2}(X - \xi)^T \Sigma^{-1}(X - \xi)\}$$

Later we will also use the notation

$$\text{etr}(A) = \exp\{\text{trace } (A)\}.$$

2.1 Known parameters

In the situation where all the parameters ξ_i, Σ_i and π_i, $i=1,\dots,k$ are known, classification rule (1.2) becomes: assign X to Π_i if:

$$\delta_i^2(X) + \ln|\Sigma_i| - 2\ln\pi_i = \min_{j=1,\dots,k}\{\delta_j^2(X) + \ln|\Sigma_j| - 2\ln\pi_j\} \quad (2.1)$$

where $\delta_j^2(X) = (X - \xi_j)^T \Sigma_j^{-1}(X - \xi_j)$ is the Mahalanobis distance from X to Π_j. Note that, for equal Σ_j and π_j, j=1,...,k, (2.1) is a minimum distance rule.

For two groups (2.1) simplifies to: assign X to Π_1 if

$$Q_{12}(X) \geq \ln \; (\pi_2/\pi_1) \tag{2.2}$$

and to Π_2 otherwise;

where

$$Q_{12}(X) = -\tfrac{1}{2} X^T(\Sigma_1^{-1} - \Sigma_2^{-1})X + X^T(\Sigma_1^{-1}\xi_1 - \Sigma_2^{-1}\xi_2)$$
$$-\tfrac{1}{2}(\xi_1^T \Sigma_1^{-1}\xi_1 - \xi_2^T \Sigma_2^{-1}\xi_2) + \tfrac{1}{2}\ln\,(|\Sigma_2|/|\Sigma_1|) \tag{2.3}$$

is called the quadratic discriminant function. The boundary between the regions in which X would be classified to Π_1 and to Π_2 may, in principle, have any quadratic shape. The more surprising ones include an annulus on which X is allocated to Π_1 while both inside and out= side the annulus it is allocated to Π_2; and a degenerate case in which any X whatever is allocated to Π_1.

2.1.1 *The case of equal covariance matrices*

As elsewhere in statistics, the assumption that the covariance matrices in different groups are equal, i.e. $\Sigma_j = \Sigma$, j=1,...,k sim= plifies matters considerably. Under this assumption, classification rule (2.1) simplifies to: assign X to Π_i if

$$U_i(X) = \max_{j=1,\ldots,k} U_j(X) \tag{2.4}$$

where $U_j(X)$ is the linear function:

$$U_j(X) = X^T \Sigma^{-1}\xi_j - \tfrac{1}{2}\xi_j^T \Sigma^{-1}\xi_j + \ln \pi_j \tag{2.5}$$

For two groups (2.4) simplifies further to: assign X to Π_1 if

$$U_{12}(X)=U_1(X)-U_2(X) = (X-\tfrac{1}{2}(\xi_1+\xi_2))^T\Sigma^{-1}(\xi_1-\xi_2) \geq \ln(\pi_2/\pi_1) \text{ and to } \Pi_2 \text{ otherwise.} \tag{2.6}$$

$U_{12}(X)$ is called the linear discriminant function (LDF).

As a practical matter we might mention that it is seldom wise to compute and report only the LDF. Generally one would wish to guard against, and check for, the possibility that the unknown X does not come from any of the populations sampled. This possibility is easily checked given $\delta_i^2(X)$ (which follows a central χ_p^2 distribution if X comes from Π_i) or $U_i(X)$ (which is normal $(\frac{1}{2}\mu + \ell n\ \pi_i, \mu)$ if X comes from Π_i; $\mu = \xi_i^T\ \Sigma^{-1}\ \xi_i$). It cannot be checked using only the LDF which may give acceptable values even when X is completely incompatible with any of the source populations.

2.1.2 *Probability of misclassification*

The probability of misclassification under any classification rule is a measure of the expected performance of that rule when clas= sifying observations of unknown origin. In order to obtain this, we note that if u_{ij} is the linear discriminant function corresponding to groups Π_i and Π_j, then, if X is from Π_i, u_{ij} is normally distri= buted with mean $\frac{1}{2}\ \delta_{ij}^2$ and variance δ_{ij}^2, whereas if X is from Π_j then it has mean $-\ \frac{1}{2}\ \delta_{ij}^2$ and variance δ_{ij}^2, where

$$\delta_{ij}^2 = (\xi_i - \xi_j)^T\ \Sigma^{-1}(\xi_i - \xi_j) \qquad (2.7)$$

is the Mahalanobis distance between Π_i and Π_j.

For the two-group problem with equal covariance matrices this yields the following probabilities of misclassification using rule (2.4):

$$P_1 = P[\text{Misclassification}|X \text{ from } \Pi_1] = \Phi\left(\frac{\ell n(\pi_2/\pi_1) - \frac{1}{2}\ \delta_{12}^2}{\delta_{12}}\right) \qquad (2.8)$$

and

$$P_2 = P[\text{Misclassification}|X \text{ from } \Pi_2] = \Phi\left(\frac{-\ell n(\pi_2/\pi_1) - \frac{1}{2}\delta_{12}^2}{\delta_{12}}\right) \qquad (2.9)$$

where $\Phi(\cdot)$ is the standard normal distribution function.

The expected probability of misclassification for a randomly chosen observation from Π_1 or Π_2 :

$$P = \pi_1\ P_1 + \pi_2\ P_2$$

is called the error rate of the classification rule. For $\pi_1 = \pi_2$, both misclassification probabilities P_1 and P_2, given in (2.8) and (2.9), are equal, so that the error rate becomes:

$$P = \Phi(-\tfrac{1}{2}\ \delta_{12}) \qquad (2.10)$$

For k > 2 groups, the probabilities of misclassification are expressed in terms of multiple integrals over (k-1)-dimensional nor= mal probability density functions, that can only be evaluated analy= tically in certain special cases. However, using Bonferroni's first inequality, we obtain the following upper bound for the probability of misclassification:

$$P_i = P[\text{misclassification} | X \text{ from } \Pi_i] \le \sum_{\substack{j=1 \\ j\neq i}}^{k} \Phi\left(\frac{\ell n(\pi_j/\pi_i) - \frac{1}{2}\delta_{ij}^2}{\delta_{ij}}\right) \quad (2.11)$$

$$= \sum_{\substack{j=1 \\ j\neq i}}^{k} \Phi(-\tfrac{1}{2}\delta_{ij}) \text{ if } \pi_j=\pi_i, j=1,\ldots,k \quad (2.12)$$

In the case of unequal covariance matrices, no simple expres= sions exist for the probabilities of misclassification, whether there are two or more populations.

2.1.3 *Canonical variables*

In his 1936 paper, Fisher suggested a different approach to the two-group classification problem, which turns out to give the same procedure in terms of $U_{12}(X)$ defined in 2.6: this approach is to seek some scalar linear combination $\alpha^T X$ of the p measurements yield= ing the largest possible Student's t statistic between Π_1 and Π_2. The approach can be generalized to k > 2 populations, but this gene= ralization no longer corresponds to the linear discriminant functions 2.5, and confusion between the two approaches has misled a number of users of discriminant analysis.

The generalization of Fisher's method leads to defining new "canonical variables" Y_i by

$$Y_i = \alpha_j^T X$$

where α_i is the solution of the eigenvector/eigenvalue problem: $(\Sigma_B - \lambda_i\Sigma)\alpha_i = 0$ corresponding to the i-th largest eigenvalue λ_i,

$$\Sigma_B = k^{-1} \sum_{j=1}^{k} (\xi_j - \xi_.)(\xi_j - \xi_.)^T$$

is the between groups covariance matrix, and

$$\xi_. = k^{-1} \sum_{j=1}^{k} \xi_j.$$

These variables have the property that within populations they are independent normal variates with standard deviation 1; between populations they are also independent, and Y_i has variance λ_i, which is a maximum possible. Of the Y_i so defined, only $s = \min(k-1,p)$ can have a nonzero λ_i.

The proportion of the total between groups variance, relative to the within group variance, explained by Y_i is $\lambda_i / \sum_{j=1}^{s} \lambda_j$. The classification rule based on the first r canonical variables, and assuming equal prior probabilities, is: assign X to Π_i if:

$$\sum_{\ell=1}^{r} \{\alpha_\ell^T (X - \xi_i)\}^2 = \min_{j=1,\ldots,k} \sum_{\ell=1}^{r} \{\alpha_\ell^T (X - \xi_j)\}^2 \qquad (2.13)$$

This rule is sub-optimal unless r=s or $\lambda_{r+1} = \ldots = \lambda_s = 0$. For k=2, classification based on the first canonical variable is equivalent to using the linear discriminant function. For $k > 2$ groups, the most common use of the canonical variables is for graphical display using, say, the first two canonical variables.

Another use for the canonical variables when $k \leq p$ is in testing whether X has not come from any of the populations $\Pi_1,\ldots,\Pi_k$, or any combination of them. It is not difficult to show that the last p-k+1 eigenvalues λ_i, i=k,...,p will be identically zero, and that the corresponding canonical variables Y_i, i=k,...,p, will be independently normally distributed with corresponding means $\alpha_i^T \xi.$ and unit variances, no matter which of the k populations X comes from. If, however, X comes from a totally different population Π_* with mean vector ξ_* and covariance Σ then the canonical variables will have means $\alpha_i^T \xi_*$. A test for this can therefore be constructed by computing the sum of squared deviations of the Y_i: $T = \Sigma_k^p (Y_i - \alpha_i^T \xi.)^2$ and comparing it with the chi-squared distribution with p-k+1 degrees of freedom.

Looking at the problem from a geometrical point of view, we see that these last canonical variables are orthogonal to the hyperplane containing all ξ_i. Thus T will have a central χ^2 distribution if the unknown comes from a normal distribution whose mean is coplanar with all ξ_i, and will have a noncentral χ^2 distribution otherwise.

2.2 Unknown parameters

When the parameters of the distributions in the k populations are unknown, the usual procedure in classical discriminant analysis is to estimate them from training samples $\{X_{ij}, j=1,\ldots,N_i\}$ from each of the populations Π_i, i=1,...,k.

Let

$$X_{i.} = N_i^{-1} \sum_{j=1}^{N_i} X_{ij}$$

and

$$S_i = n_i^{-1} \sum_{j=1}^{N_i} (X_{ij} - X_{i.})(X_{ij} - X_{i.})^T$$

where $n_i = N_i - 1$

be the sample mean and covariance matrix corresponding to the train= ing sample from Π_i.

The usual, estimative, approach to discriminant analysis is to replace the parameters in the classification rules given above by their sample estimates (Anderson, 1958). Applying this approach to (2.1) yields the sample-based classification rule: Assign X to Π_i if

$$D_i^2(X) + \ell n|S_i| - 2\,\ell n\,\hat{\pi}_i = \underset{j=1,\ldots,k}{\text{Min}} \{D_j^2(X) + \ell n|S_j| - 2\,\ell n\,\hat{\pi}_j\} \tag{2.14}$$

where

$$D_j^2(X) = (X - X_j)^T S_j^{-1}(X - X_j) \tag{2.15}$$

is the sample-based Mahalanobis distance between X and Π_j, and $\hat{\pi}_j$ is some estimator of π_j.

The corresponding two-group rule is: Assign X to Π_1 if:

$$R_{12}(X) \geq (\hat{\pi}_2/\hat{\pi}_1) \text{ and to } \Pi_2 \text{ otherwise} \tag{2.16}$$

where

$$R_{12}(X) = -\tfrac{1}{2} X^T(S_1^{-1} - S_2^{-1})X + X^T(S_1^{-1} X_{1.} - S_2^{-1} X_{2.}) - \tfrac{1}{2}(X_{1.}^T S_1^{-1} X_{1.} - X_{2.}^T S_2^{-1} X_{2.}) + \tfrac{1}{2}\,\ell n(|S_2|/|S_1|) \tag{2.17}$$

is the sample-based quadratic discriminant function.

Under the assumption that the population covariance matrices in the different populations are equal, the sample-based rule becomes: assign X to Π_i if

$$V_i(X) = \underset{j=1,\ldots,n}{\text{Max}} V_j(X) \tag{2.18}$$

where

$$V_j(X) = X^T S^{-1} X_{j.} - \tfrac{1}{2} X_{j.}^T S^{-1} X_{j.} - \ell n\,\hat{\pi}_j \tag{2.19}$$

and $\quad S = n^{-1} \sum_{j=1}^{k} n_j S_j$, where $n = \sum_{j=1}^{k} n_j = \sum_{j=1}^{k} N_j - k$,

is the pooled sample covariance matrix.

For two groups, the sample-based linear discriminant function is

$$V_{12}(X) = V_1(X) - V_2(X) = (X - \tfrac{1}{2}(X_{1.} + X_{2.}))^T S^{-1}(X_{1.} - X_{2.}) \tag{2.20}$$

and the corresponding classification rule becomes: assign X to Π_1 if

$$V_{12}(X) > \ell n(\hat{\pi}_2/\hat{\pi}_1) \tag{2.21}$$

and to Π_2 otherwise.

There is a formal analogy between the computations of multiple regression and those of two-group discriminant analysis. Using a dummy dependent variable Y taking on the two values $Y = N_2/(N_1+N_2)$ for all observations from Π_1 and $Y = -N_1/(N_1+N_2)$ for all observa= tions from Π_2, it can be shown (Anderson, 1958) that the equation obtained by fitting a regression line to the training sample data, with Y as dependent variable and X as the vector of predictors, is, apart from a constant term, proportional to the linear discriminant function $V_{12}(X)$.

A similar relationship exists between the computations for k-group discriminant analysis and those of multivariate regression with a k-dimensional vector of dummy dependent variables, although the analogy is not quite so straightforward as in the two-group case.

This formal equivalence between discriminant analysis and mul= tiple regression extends also to many of the distributional results in the two areas, and enables useful results to be determined quickly from their regression counterparts. It also has the computational benefit that a minor adaptation of sophisticated regression codes enables them to be used for discriminant analysis also.

2.2.1 *Probabilities of misclassification (Equal Covariance Matrices)*

There are three misclassification probabilities that can be con= sidered:

(i) The optimum probability, for the population-based classi= fication rules (2.4) and (2.6),

(ii) The conditional probability (actual error rate) for the sample-based rules (2.18) and (2.21) based on a particular training sample, and

(iii) The expected (unconditional) probability for the rules (2.18) and (2.21) based on training samples of size N_i from Π_i, $i=1,\ldots,k$.

The optimum probability, given in section 2.1.2, is clearly the smallest that can be achieved.

The conditional probability of misclassification measures the expected performance of a particular sample-based rule. For the two-group problem with equal prior probabilities, it is easy to show that

$$P_i^c = P[\text{misclassification}|X_{1\cdot},X_{2\cdot},S.,X \text{ from } \Pi_i]$$

$$= \Phi\left(\frac{(-1)^i(\xi_i - \tfrac{1}{2}(X_{1.} + X_{2.}))^T S^{-1}(X_{1.} - X_{2.})}{\sqrt{\{(X_{1.} - X_{2.})^T S^{-1}\,\Sigma\, S^{-1}(X_{1.} - X_{2.})\}}}\right) \quad i=1,2 \qquad (2.22)$$

The expected probability, P_i^c is obtained by taking the expected value of P_i^c over the distributions of $X_{i.}$, $i=1,\dots,k$, and S for given sample sizes N_i, $i=1,\dots,k$.

The simplest estimator of P_i^c is obtained by replacing the parameters in it by their sample-based estimators. This yields, for the two-group problem with equal prior probabilities:

$$\hat{P}_1^c = \hat{P}_2^c = \Phi(-\tfrac{1}{2}\, D_{12}) \qquad (2.23)$$

where

$$D_{12}^2 = (X_{1.} - X_{2.})^T S^{-1}(X_{1.} - X_{2.}) \qquad (2.24)$$

is the sample-based Mahalanobis distance between Π_1 and Π_2. This estimator is clearly also obtained by estimating δ_{12}^2 by D_{12}^2 in the optimum probability (2.10).

However, for moderate sample sizes, this estimator is badly biased and gives much too favourable an impression of the true probability of misclassification. Hills (1966) proves that the expected value of (2.23) is less than the optimum probability (2.10), whereas the conditional error rate $P^c = \tfrac{1}{2}(P_1^c + P_2^c)$ is in fact greater than it: i.e. $E[\Phi(-\tfrac{1}{2}\, D_{12})] < \Phi(-\tfrac{1}{2}\delta_{12}) < P^c$.

Dunn and Varady (1966), Lachenbruch and Mickey (1968) and Dunn (1971) show empirically that this bias may indeed be substantial unless the training samples are large.

McLachlan (1974a) gives the following estimator of P^c, with bias of order three with respect to $(N_1^{-1},N_2^{-1},n^{-1})$ where $n = N_1 + N_2 - 2$:

$$\hat{P}_1^c = \Phi(-\tfrac{1}{2}D_{12}) + \phi(\tfrac{1}{2}D_{12})\{N_1^{-1}(p-1)/D_{12} + n^{-1}(\tfrac{D_{12}}{32})(4(4p-1)-D_{12}^2)\} + 0_2 \qquad (2.25)$$

where $\phi(\cdot)$ denotes the standard normal density function, and 0_2 denotes the term of order 2 with respect to $(N_1^{-1},N_2^{-1},n^{-1})$; this is given explicitly by McLachlan (1975a). $\hat{P}_2^c$ may be obtained from (2.25) by replacing N_1 by N_2.

The examples given below highlight the differences between the two estimators (2.23) and (2.25) of P_1^c.

D_{12}^2	6	6	6
p	5	5	10
N_1	10	20	20
N_2	10	20	20
$\hat{P}_1^c$(2.23)	.1103	.1103	.1103
$\hat{P}_1^c$(2.25)	.1972	.1523	.2019

For small sample sizes, relative to the dimension p, the difference be= tween the two estimators can be considerable.

While conditional error rates are of interest in assessing the expected performance of a *particular* discriminant function, the un= conditional or expected error rates, obtained by considering $X_{1.}, X_{2.}$ and S in (2.22) as random variables, are more appropriate when consi= dering the expected performance of the classification rule based on randomly chosen training samples of sizes N_1 and N_2 from Π_1 and Π_2, respectively.

A simple estimator of the expected probability of misclassifi= cation is obtained by replacing δ_{12}^2 by its unbiased estimator,

$$\hat{\delta}_{12}^2 = \left(\frac{n-p-1}{n}\right) D_{12}^2 - \frac{p(N_1+N_2)}{N_1 N_2} \tag{2.26}$$

obtained from the noncentral Hotelling's T^2 distribution, in the opti= mum probability (2.7). This yields

$$\hat{P}_1^e = \hat{P}[\text{Misclassification}|N_1,N_2;\ X \text{ from } \Pi_1]$$

$$= \Phi(-\tfrac{1}{2}\sqrt{\{(n-p-1)D_{12}^2/n - p(N_1+N_2)/N_1N_2\}}) \tag{2.27}$$

An apparent drawback to 2.26 is that it can produce a negative estimate of the Mahalanobis distance. This however will only occur if the two training samples are so close together as to make any classifi= cation a dubious proposition, and use of 2.26 with zero replacing any negative distances is satisfactory (Costanza and Afifi, 1979).

An asymptotic expansion for the distribution of the two-group sample discriminant function V_{12} has been derived by Okamoto (1963). This yields the following expression for the expected probability of mis= classification, in the case of equal prior probabilities, for a randomly chosen member of Π_1:

$$P_1^e = P[\text{Misclassification}|N_1,N_2;\ X \text{ from } \Pi_1]$$

$$= \Phi(-\tfrac{1}{2}\delta_{12})+\phi(\tfrac{1}{2}\delta_{12})\{(2\ N_1\ \delta_{12})^{-1}(\delta_{12}^2/8 + 3(p-1)/2) \quad (2.28)$$

$$+ (2\ N_2\ \delta_{12})^{-1}(\delta_{12}^2/8 - (p-1)/2) + (4n)^{-1}\delta_{12}(p-1)\} + 0_2$$

(Okamoto also gives the terms of order 2 with respect to (N_1^{-1}, N_2^{-1} and n^{-1}).) P_2^e may be obtained from (2.28) by interchanging N_1 and N_2.

The performance of a number of estimators of P_i^c and P_i^e have been studied by Lachenbruch and Mickey (1968) via simulation. Although they make no distinction between these two probabilities, all estima= tors being compared with the conditional probability of misclassification, they find that Okamoto's expression (2.28), with δ_{12}^2 replaced by its almost unbiased estimator (2.26) excluding the last term, performs the best. Estimator (2.27), again excluding the last term, is found to be considerably superior to $\hat{P}_1^c$ given in (2.23). Although they did not in= clude them in their simulations, it is surmised that substituting $\hat{\delta}_{12}^2$ given in (2.26) in Okamoto's expression (2.28), with negative values replaced by zero, and expression (2.27) with the same correction for negative values, will perform better than their counterparts used in the simulation study. Note that (2.27) is equivalent to using the first term only in Okamoto's expression (2.28) with δ_{12}^2 estimated by (2.26).

As an illustration these estimators are evaluated below for the same three combinations of D_{12}^2, p, N_1 and N_2 used earlier.

	D_{12}^2	6	6	6
	p	5	5	10
	N_1	10	20	20
	N_2	10	20	20
$\hat{P}_1^e$	(2.27)	.1931	.1430	.1833
	(2.27) excluding last term	.1587	.1305	.1510
	(2.28) using (2.26)	.2572	.1693	.2476
	(2.28) using (2.26) excluding last term	.2158	.1555	.2081

In a subsequent simulation study McLachlan (1974c) finds the per= formance of his estimator (2.25) of P_1^c and that of Okamoto's estimator (2.28) using the almost unbiased estimator (2.26) of δ_{12}^2 with the last term excluded to be very similar.

An empirical estimator of P_i^c, applicable to any classification rule, is the observed proportion of the training sample from Π_i that is misclassified by this rule. This estimator, the "apparent" error rate, is unfortunately highly biased towards the low side, especially for small training samples. In order to overcome this bias Lachenbruch (1967) proposes a sample reuse (jackknife) procedure, which involves classifying each observation using the classification rule obtained by omitting that observation from the training sample, in turn for all observations from Π_i, and estimating P_i^c by the observed proportion misclassified. In order to overcome the otherwise excessive number of computations involved in obtaining this estimator, Lachenbruch (1967, 1975a) gives convenient computational formulae which require only one matrix inversion in the whole calculation. This procedure is very gene= ral, can be applied to any classification rule with any number of groups, and does not depend on any model assumptions. Lachenbruch and Mickey (1968) show that this jackknife estimator is far superior to the "apparent" error data, but that it is appreciably worse than Oka= moto's estimator in the two-group case. However, when the assumptions of normality and homoscedasticity are violated, the estimators based on these assumptions become unreliable, and Lachenbruch (1975) recom= mends using the jackknife estimator under these circumstances. It must be stressed however, that use of the jackknife estimator in no way justifies the use of the linear or quadratic discriminant functions in any particular application - it merely estimates the error rate of the classification rule, whether or not the rule is appropriate.

Another sample reuse method, the Bootstrap, has been proposed as an alternative empirical method for estimating error rates in dis= criminant analysis. This is a generalization of the jackknife method and involves drawing repeated random samples from the sample distribu= tion functions corresponding to each of the k groups, calculating the discriminant functions and estimating the error rates in each case, and then averaging the error rates over all replications. Efron (1979) shows that the Bootstrap method produces estimators that have similar bias but are less variable than the corresponding Jackknife estimators. It is, however, a moot point whether the enormous amount of extra computa= tions involved in obtaining these estimators is justified by their im= proved performance.

2.2.2 *The random effects model*

A situation sometimes arises where one anticipates a repetitive discriminant analysis of the following form:

In each analysis, k, ξ_i and Σ will be sampled randomly from some super-population. A training sample from each will be drawn and one or more unknowns classified. This entire sequence of sampling the ξ_i and Σ, obtaining training samples, and classifying unknowns will occur re= peatedly.

A good example of such a practical situation is the third open= ing example of our chapter, where a fresh discriminant analysis problem occurs every time the reef faults.

From an initial training sample in this situation, one needs to infer, not how well those k populations may be classified, but how well a future k populations sampled from the same super-population could be classified.

On the assumption that the ξ_i are independent $N(\mu,T)$ variates, Fatti (1979) shows how the characteristics of such future classification problems can be inferred from the eigenvalues of $T\,\Sigma^{-1}$, which in turn may be estimated from a current training sample.

While this situation probably occurs rarely, the importance of being able to infer properties of the classification in the absence of any training data from the actual populations to which classification will be made, sets it apart from the more usual models.

2.2.3 *Biased discriminant analysis*

In paragraph 1.1, we showed how a discriminant analysis could be formulated as a regression problem. It is well known that biased es= timation procedures for multiple regression such as ridge regression can improve (often dramatically) both the estimates of the parameters and the predictions made using the regression equation. It is thus to be suspected that the identical adaptation would also improve a discrimi= nant analysis. That this is indeed the case has been shown by DiPillo (1976, 1977, 1979) and Campbell (1980).

The ridging adaptation consists very simply of replacing S (in the homoscedastic model) or S_i (in the heteroscedastic model) by $S + K$ and $S_i + K$ respectively, and then computing the LDF or QDF using these altered covariance matrices. As in the case of ridge regression, an interesting special choice for K is the matrix ϵ diag S, in which the ridge parameter ϵ may be chosen to control the amount of biasing applied to the classification rule.

This ridging of the LDF and QDF may also be applied to the computation of canonical variates.

Biasing is a relatively new development in the theory of discriminant analysis, and while it may be shown that a ϵ exists which will improve the resultant classification rules, it remains to be seen what order of improvement occurs under what circumstances and how the ϵ giving rise to such improvement can be estimated.

2.3 Robustness of the discriminant function

The most commonly used procedure for discriminating between two populations is the linear discriminant function, which minimizes the total probability of misclassification if sampling is from multivariate normal populations with known means and known equal covariance matrices. If sampling is from multivariate normal distributions with known means and known but unequal covariance matrices, then the quadratic discriminant function minimizes the total probability of misclassification.

To render the above methods optimal requires various assumptions concerning the populations from which the samples have been drawn. The question can now be asked is: How robust are the rules to these assumptions?

In the case of the linear discriminant function the assumptions are:

1. The populations are multivariate normal.
2. The covariance matrices are equal.
3. The a priori probabilities that the sample comes from a given population are known.
4. The means and covariance matrices are known.

If one or more of these conditions are not satisfied, the linear discriminant rule is no longer optimal.

As noted in the preceding section, the a priori probabilities and the parameters of the distribution can be estimated from the data if they are not known, giving fairly satisfactory results provided the training sample is not too small. Thus questions of the robustness of the LDF may be focussed on the first two model assumptions.

If the distribution of the populations is not normal, the linear discriminant function is in general no longer optimal. In some cases it may be possible to improve the performance of the linear discriminant function by the use of transformations. Other times exact distributions such as the multinomial may be desired. A variety of non-normal situa= tions can be envisioned, for example, the variables may be Bernoulli, Poisson, continuous non-normal or the sample populations might be con= taminated distributions.

First let us consider data that are continuous, but follow non-normal distributions. Some simulation studies have looked at the effect of simply applying the LDF or QDF to such data (e.g. Lachenbruch, Sneer= inger and Revo 1973, Clarke, Lachenbruch and Broffitt 1979). Combining these published results with some from our own work, we can make the following generalizations:

(i) If the distributions have lighter tails than the normal, then the LDF (for homoscedastic data) or the QDF (for heteroscedastic data) should perform very adequately.

(ii) If the distributions are heavy tailed and skew, then the LDF and QDF will perform very poorly.

(iii) If the distributions are heavy tailed but symmetric and the training samples are very large, then the QDF may perform reasonably well in terms of *overall* error rate (though not the individual rates). However if the training samples are small, then the heavy distributional tails introduce excessively large sampling errors into the para= meter estimates, and so lead to an unreliable discriminant function.

In no case with non-normal data is an estimator of the error rate based on normal distribution theory appropriate, and one should use the jackknife estimator.

As these remarks imply, the departure from normality to be watched for is a heavy-tailed distribution - especially if it is skew. How to locate such departures is the subject of the final section of this chapter.

2.3.1 *Use of dichotomous variables in the discriminant functions*

One particular case of the use of non-normal variables arises so commonly as to merit a special comment:- this is a dichotomous or binary variable, which takes on only the values 0 and 1. A polytomous variable with c classes may easily be transformed to a vector of c-1 dichotomous variables, and so also comes into the same category of problem. Now there are sufficient situations elsewhere in statistics (for example 2 x k contingency table) in which the correct results are obtained by simply treating binary variables as if they were nor= mal to lead to a reasonable expectation that the same can be done in discriminant analysis. Apart from some situations studied by Krzanowski and given in section 5, so it proves to be. Binary data, despite being non-normal and generally heteroscedastic, often lend themselves well to classification using LDF's (Gilbert 1968).

In the selection of a discrimination procedure with several binary variables however, some care is needed. If the log likelihood ratio for the two populations is plotted against the *number* of vari= ables having value 1, then in some situations the log likelihood ratio does not increase monotonically. This is called a "reversal" and in such cases, Fisher's LDF leads to a significantly greater er= ror rate than classification procedures based on the location model or the full multinomial model (see section 5).

2.3.2 *Robustness to heteroscedasticity*

The other model assumption made when linear discrimination is used is that the source populations all have the same covariance matrix. In view of the very common use of the LDF and much less fre= quent use of the QDF, it is particularly useful to know how the LDF performs when the data are in fact heteroscedastic.

Looking at this question in very general terms, we see that in the homoscedastic model, we need a single covariance matrix of $P = \frac{1}{2} p(p+1)$ elements from all N observations in the training sample. In the heteroscedastic model, we have to estimate k such matrices for a total of kp parameters, the i-th matrix being estimated from N_i observations.

The very success elsewhere of biased estimators (e.g. James-Stein shrinkage, ridge regression, biased discriminant analysis) leads one to suspect that the LDF may, although biased, perform better than the QDF in the presence of a small amount of heterscedasticity. This is indeed true (see for example Marks and Dunn 1974):- if the data are only mildly heteroscedastic, then one obtains better results by using the LDF than by going over to the QDF.

The question of how much heteroscedasticity one can tolerate and still prefer the LDF is not easy to resolve, as it depends on a number of other factors. From mere armchair analysis, one can say that the QDF is less attractive:

(i) if p is large or
(ii) if there are many groups, or
(iii) if any N_i is small

since under any of these conditions, the sampling fluctuations in S_i^{-1} will be much greater than those in S^{-1}.

To give some idea of the practical significance of the degree of this robustness consider the following figures from Marks and Dunn. $N(\xi_1,I)$ and $N(\xi_2,\lambda I)$ are the two populations; a training sample of size 25 is taken from each. At what value of λ are the LDF and QDF equally good? The answers to this are:

p	$(\xi_1 - \xi_2)^T(\xi_1 - \xi_2)$	λ
2	0.75	1.4
2	1.75	1.3
10	0.75	2.5
10	1.75	4.3

Apart from the anticipated increase in this cut-off λ with p, we note that it is also affected by the amount of separation between the population means; the greater this is, the less need there is to take any notice of the heteroscedasticity.

2.3.3 *More robust procedures*

As the preceding sections have shown, the QDF and, more espe= cially the LDF can be applied safely even when there are quite marked departures from model. This of course does not exonerate the user from attempting, by such means as transformation of the data, to en= sure that the model does fit, and, on occasion, it will be found im= possible to bring the tail behaviour of the data into reasonable ag= reement with the normal distribution. Under these circumstances, a sensible procedure may be to either robustify the conventional methods or go to a non-parametric method.

Various methods of trimming the sample means and covariance ma= trices have been developed. In simulation studies done by Broffitt, Clarke and Lachenbruch (1980), involving the normal, log normal and inverse hyperbolic sine normal distributions, it has been found that for normally distributed data, the Huberized discriminant functions performed extremely well, having individual misclassification rate increases over the ordinary discriminant function of less than 1%.

The Huberized discriminant function uses the values

$$D_{ij}^2 = (X_{ij} - X_i^o)^T (S_i^o)^{-1} (X_{ij} - X_i^o)$$

with X_i^o and S_i^o the usual sample estimates of the mean and covariance matrix. A 100α% cut-off value C for D_{ij}^2 is then calculated from the beta distribution since each D_{ij}^2 value is distributed as

$$\frac{(N_i-1)^2}{N_i} \quad \frac{\beta(\frac{p}{2}; N_i-p-1)}{2}$$

when X is normally distributed with dimension p.

If $D_{ij}^2 \le C$, then set $w_{ij} = 1$, otherwise $w_{ij} = \frac{C}{D_{ij}^2}$. The Huberized mean is

$$X_i^H = \left(\sum_{j=1}^{N_i} w_{ij}\right)^{-1} \sum_{j=1}^{N_i} w_{ij} X_{ij}$$

The D_{ij}^2-values are then recomputed, using X_i^H instead of X_i^o. The 100α% cut-off C_s of these D_{ij}^{*2} is found and the covariance matrix is estimated by

$$S_i^H = \left(\sum_{j=1}^{N_i} w_{ij}^*\right)^{-1} \sum_{j=1}^{N_i} w_{ij}^* (X_{ij} - X_i^H)(X_{ij} - X_i^H)^T$$

where $w_{ij}^* = \min(1, \frac{C_s}{D_{ij}^{*2}})$.

If necessary, this process of reweighting and estimation is repeated until convergence is obtained.

Simulations using this procedure have shown that it improves the error rate somewhat when applied to heavy-tailed data, and also reduces the marked imbalance usually found between the error rates of the different source populations.

The above discussion concentrated on the heteroscedastic problem:- the homoscedastic equivalent is easily derived and implemented.

2.3.4 *Discrimination using ranks*

Another possible approach is to use a non-parametric procedure. Let X_{ij} be the j-th observation from population Π_i, $j=1,\ldots,N_i$; $i=1,\ldots,k$. X_{ijm} denotes the p components of X_{ij} where $m=1,\ldots,p$. The m-th components of all X_{ij} are now ranked from the smallest, with rank 1, to the largest, with rank $N = N_1 + \ldots + N_k$. Each component is ranked separately for m=1 to m=p. Then the sample means $X_{i.}(R)$

and sample covariance matrices $S_i(R)$ are computed on the ranks of the observations from each population separately. The new observa= tion to be classified, X_o, is compared, component by component, with all N original observations and each component of X_o is replaced by a rank, a number obtained by linear interpolation between two adja= cent ranks. Then X_o is classified to that population with the lar= gest "estimated density" computed at the vector $r(X_o)$ of ranks of X_o, namely

$$\hat{f}_i(r(X_o)) = (2\pi)^{-\frac{1}{2}p}|S_i(R)|^{-\frac{1}{2}}\exp\{-\tfrac{1}{2}[r(X_o)-X_{i.}(R)]^T S_i^{-1}(R)[r(X_o)-X_{i.}(R)]\} \quad (2.29)$$

Average ranks are used where ties occur.

Simulation studies were performed by Conover and Iman (1980) to compare this rank transformation method with the LDF and QDF as well as with the results of Gessaman and Gessaman (1972), who com= pared the probabilities of misclassification for several types of discrimination methods, using normal distributions with estimated parameters. The results show that the probability of misclassifica= tion using the rank LDF is similar to that for the LDF, and the two methods are virtually equivalent when the populations are normal with equal covariance matrices.

When the covariance matrices differ, either the QDF or rank QDF is better than the LDF or rank LDF, but here again the two are virtually equivalent. When compared with the Gessaman and Gessaman results the rank method again gave satisfactory results, especially for small samples. Overall, for normal populations, it appears that the rank procedures may be used with good results.

The rank transformation method was also compared with the results of the Johnson system of distributions, as described by Lachenbruch et al. (1973). The results can be summarized as follows:

	EQUAL COVARIANCE MATRICES	*UNEQUAL COVARIANCE MATRICES
Normal Distribution	LDF and rank LDF give similar re= sults	QDF and rank QDF give similar re= sults
Non-normal Distributions $N_1 = N_2$	Rank LDF is best	Rank QDF is best
Non-normal Distributions $\frac{N_1}{N_2} = \frac{3}{1}$	QDF is better than rank LDF which in turn is better than LDF	
Non-normal Distributions $N_1 \neq N_2$ Small differences between means	Rank LDF is better than LDF	Rank QDF is better than QDF

*Unequal Covariance Matrices here means that the covariance matrices differ considerably.

When the separation between the populations tends to infinity, so that perfect classification becomes possible, Chanda (1980) shows that the efficiency of the rank LDF relative to its optimum parametric counterpart tends to zero. However, since it too gives correct classi= fications with asymptotic probability 1, this loss of efficiency is not of practical importance.

The conclusion of these simulation studies then is that the rank procedures are not only simple to use and robust with non-normal populations, but also results in the lowest percentage of misclassi= fication when compared with the other procedures examined by Conover and Iman (1980). Furthermore, the rank LDF method is not very sensi= tive to unequal covariance matrices so that, especially in cases where we have reason to believe that the covariances are almost equal, it is much safer to use the rank LDF.

The discussion just given of the robustness of discriminant ana= lysis to departure from model has concentrated on the estimative ap= proach. We are not aware of parallel investigations of the robustness of the predictive approach. However since both approaches depend on the same summary statistics $-(X_{i.}$ and $S_i) -$ it is more than likely that their robustness properties are at least qualitatively similar.

3. VARIABLE SELECTION

In any application of discriminant analysis, some variables will show greater variation between groups, relative to their variation within groups, than other variables. Clearly, these variables will provide better discrimination between groups than the others. The question then arises whether we can drop any of the latter variables from the discriminant analysis without appreciably increasing the er= ror rate. Indeed, in the situation where the classification rules are estimated from training samples, dropping variables may well *decrease* the conditional or expected error rate by reducing the number of para= meters that have to be estimated from the training sample. The ques= tion of discarding variables becomes more complicated when the inter= correlations between variables are taken into account, so that the discriminating power of a particular variable depends on the other variables already in the discriminant function.

Suppose the vector of variables is divided into two sets: $X = (X_{(1)}, X_{(2)})^T$ where $X_{(1)}$ contains q elements and $X_{(2)}$ the remain= ing p-q. The question may then be phrased: Does $X_{(1)}$ provide equally good discrimination as X? Or, equivalently: does $X_{(2)}$ provide any additional discrimination over and above that provided by $X_{(1)}$?

3.1 Selection criteria

Historically, Rao provided the first test for additional dis= crimination in the two-group case. If $D^2_{(q)}$ is the sample Mahalanobis distance between the two groups based on the q variables in $X_{(1)}$ and $D^2_{(p)}$ is the corresponding distance based on all p variables in X, then Rao (1965, 1970) gives the test statistic:

$$F = \left(\frac{n-p+1}{p-q}\right)\left(\frac{N_1N_2}{N_1+N_2}\right)(D^2_{(p)}-D^2_{(q)})/\left(n + \left(\frac{N_1\ N_2}{N_1+N_2}\right)D^2_{(q)}\right) \qquad (3.1)$$

Under the assumption that the q variables in $X_{(1)}$ have been ran= domly selected from the set of p variables, F has an F-distribution with (p-q) and (n-p+1) degrees of freedom under the null hypothesis that the p-q variables in $X_{(2)}$ provide no extra discrimination be= tween the two groups, i.e. that the coefficients of the elements of $X_{(2)}$ in the optimal discriminant function U_{12} are all zero.

For the particular case where q=p-1, i.e. we are testing whether one variable may be dropped without affecting the overall discrimina= ting power, Rao's test statistic becomes:

$$t^2 = (n-p+1)\left(\frac{N_1 N_2}{N_1+N_2}\right)(D^2_{(p)} - D^2_{(p-1)})/(n + (\frac{N_1 N_2}{N_1+N_2})D^2_{(p-1)}) \qquad (3.2)$$

where t has a t-distribution with (n-p+1) degrees of freedom under the null hypothesis.

Kshirsagar (1972) generalizes Rao's criterion to the k-group case. He uses the factorisation of the likelihood ratio statistic:

$$\frac{|A|}{|C|} = \frac{a_{11}\, a_{22.1}\, a_{33.1,2} \cdots\cdots a_{pp.1,2,\ \ldots,p-1}}{c_{11}\, c_{22.1}\, c_{33.1,2} \cdots\cdots c_{pp.1,2,\ \ldots,p-1}} = \prod_{i=1}^{p} W_i \qquad (3.3)$$

where C = A + B is the total sums of squares matrix, $a_{ii.1,2,\ldots,i-1}$ is the first diagonal element in the within-group sums of squares matrix $A_{22.1} = A_{22} - A_{21} A_{11}^{-1} A_{12}$, where A has been partitioned according to the first i-1 and last p-i+1 variables:

$$A = \begin{pmatrix} A_{11} & A_{12} \\ A_{21} & A_{22} \end{pmatrix} ,$$

$c_{ii.1,2,\ldots,i-1}$ is similarly defined, and $W_i = a_{ii.1,2,\ldots,i-1}/c_{ii.1,2,\ldots,i-1}$. The statistic:

$$F_i = (W_i^{-1} - 1)(n - i + 1)/(k-1) \qquad (3.4)$$

where $n = \sum_{i=1}^{k} N_i - k$

has an F-distribution with k-1 and n-q+1 degrees of freedom under the null hypothesis that variable X_q provides no extra discriminating power when the other q-1 variables in $X_{(1)}$ are included in the discriminant function; i.e. that the coefficients of x_q in all k true discriminant functions U_j, j=1,...,k, based on the q variables in $X_{(1)}$, are equal.

Use of the F-criterion in the case of k > 2 groups has been criticised because it emphasizes those groups that are best separated, whereas it is the poorly separated groups that present the problem in discriminant analysis. A method of overcoming this is to consider for inclusion first that variable which, after inclusion, maximizes the minimum between-groups Mahalanobis distance D^2_{ij}, and to include it if it provides significant additional discrimination between the corresponding groups Π_i and Π_j.

Habbema and Hermans (1977) argue that the error rate is a more appropriate criterion for variable selection. Thus, in order to test whether a subset of variables should, or should not be included in the analysis, the discriminant functions are estimated with, and without the subset included, and the error rate is estimated in both cases. The decision whether or not to include the subset is then based on the difference between these two estimated error rates.

McLachlan (1976) derives a criterion, based on the asymptotic distribution of the increase in the conditional error rate, when a subset of variables is deleted from the two-group discriminant function. Given the estimated conditional error rate $\hat{P}^c(p) = \frac{1}{2}(\hat{P}_1^c + \hat{P}_2^c)$, where $\hat{P}_1^c$ and $\hat{P}_2^c$ are given in (2.25), based on all p variables, and $\hat{P}^c(q)$ is the corresponding quantity based on a subset of q variables, then their difference: $\hat{P}^c(p) - \hat{P}^c(q)$ is asymptotically normally distributed with mean $P^c(p) - P^c(q)$, the difference between the corresponding true conditional error rates, and variance:

$$\sigma_{pq}^2 = v(\delta_{12}(p))+v(\delta_{12}(q))-2\{\delta_{12}(q)\phi(\tfrac{1}{2}\delta_{12}(p))/\delta_{12}(p)\phi(\tfrac{1}{2}\delta_{12}(q))\}v(\delta_{12}(q)) \tag{3.5}$$

where $v(y) = \{\phi(\frac{1}{2}y)\}^2(1/4N_1 + 1/4N_2 + y^2/8n)$

Therefore the statistic $\{\hat{P}^c(p) - \hat{P}^c(q)\}/\hat{\sigma}_{pq}$, where $\hat{\sigma}_{pq}$ is the estimator of σ_{pq} obtained by substituting estimates of $\delta_{12}(p)$ and $\delta_{12}(q)$ into (3.5), has approximately a standard normal distribution under the null hypothesis that the p-q variables provide no additional discrimination.

If an empirical estimator, such as the jackknife, is used to estimate the error rates with and without the subset included, then a sign test, based on the differences in individual classifications, can be used as a test for additional discrimination.

It is clear that in the two-group situation the selection rules based on the estimated error rate and those based on the Mahalanobis distance D_{12}^2 between the two groups should produce very similar results. For reasons of computational efficiency, tests based on D_{12}^2 might therefore be preferred, especially if an empirical estimator is used to estimate the error rates.

3.1.1 *Heteroscedastic selection criterion*

Little attention appears to have been paid to variable selection in heteroscedastic discriminant analysis. Letting A_i, $i=1,2,...,k$ be the deviance matrix of the ith population, the likelihood ratio test for identity of all k populations is well-known:-

$$-2 \log LR = \sum_{i=1}^{k} N_i \log\{ \left|\frac{A_i}{N_i}\right| . \left|\frac{C}{N}\right|^{-1} \} \quad (3.6)$$

writing

$$U_{ij} = A_{ijj.1,2,\ldots,j-1}$$

$$U_{Tj} = C_{jj.1,2,\ldots,j-1}$$

for the residual sums of squares on regressed X_j on its predecessors, we find that 3.6 can be written as

$$\sum_{i=1}^{k} N_i \log \sum_{j=1}^{p} \{\frac{U_{ij}}{N_i} . \frac{N}{U_{Tj}}\}$$

$$= \sum_{j=1}^{p} \sum_{i=1}^{k} N_i \log (\frac{U_{ij}}{N_i} . \frac{N}{U_{Tj}}) = \sum_{j=1}^{p} T_j, \text{ say.} \quad (3.7)$$

In this form, the likelihood ratio statistic on using only the first q components is simply $\Sigma_1^q T_j$, and so T_j is clearly recognizable as a measure of the additional heteroscedastic discrimination due to the addition of X_j to its predecessors.

A valuable interpretive aid results from noting that T_j may be decomposed into three terms; one which tests whether the regressions of X_j on $X_1 \ldots, X_{j-1}$ are parallel in the k populations; another which tests whether the residuals about these separate regressions have the same variance; and the homoscedastic statistic for additional discrimination W_j.

3.2 Variable selection procedures

We now describe various procedures for selecting variables to include in the discriminant analysis using the criteria discussed earlier.

The most common method for selecting variables in discriminant analysis is the stepwise procedure. Briefly, this is as follows:

(i) Start with an initial subset of the variables in the discriminant analysis. Possibilities include the empty set and the full set of all the variables.

(ii) Using an appropriate criterion, test each variable in the discriminant functions separately for possible ex= clusion and drop the variable with the smallest non-significant reduction in the criterion. Repeat until no more variables can be removed without causing a sig= nificant reduction in the criterion.

(iii) Test each variable not included in the discriminant functions for possible inclusion, and include that variable which causes the greatest improvement in the criterion, if this improvement is significant. If a variable is included, go back to step (ii), otherwise proceed to step (iv).

(iv) Use the variables included in the discriminant func= tions for classification.

To avoid the possibility of cycling between steps (ii) and (iii), a slightly stricter significance level should be used for the inclusion test than for the elimination test.

The coding, whether one is using the homoscedastic likelihood ratio, the minimum D^2 or the heteroscedastic likelihood ratio, is quite simple, and depends on the "sweep" and "reverse sweep" opera= tors.

The formal inferential problems associated with the procedures are less clear. In order to control the experimentwise probability of including an extraneous variable, Hawkins (1976) suggests that the conservative Bonferroni test be used for introduction of variables and that a variable only be introduced if its significance level is α/(number of unused predictors available for inclusion).

On the other hand, Costanza and Afifi (1979) report their ex= perience that the best classifications result when a very lenient test of size 0.1 to 0.25 is used without the Bonferroni inequality and this suggests that at least in circumstances like those modelled by Costanza and Afifi, the virtual certainty of including an irrele= vant variable is less damaging than the potential omission of a relevant one.

In fact, one can go a stage further. If the objective is to pro= duce correct classifications and one is not concerned about the inter= pretation of the discriminant functions, then there is a strong case for refusing to accept any subset q which differs significantly from the full set of variables. This constraint on the subset will tend to increase the number of variables included above even that of Costanza and Afifi. (It is quite easy to construct examples in which the overall D_p^2 or likelihood ratio is significant, while Costanza and Afifi's approach, and Hawkins' even more so, will successfully eliminate all variables). A possible resolution of this dilemma is to stop elimination if, at any stage, the deleted information as measured by $\Pi_{q+1}^{p} W_i$, becomes significant.

The main practical drawback to the stepwise procedure, as studies by McCabe (1975) and Habbema and Hermans (1977) show, is that it is unlikely that it will yield the optimum subset of variables, especially when there are a large number of variables. Another drawback is that the stepwise procedure only produces one subset, whereas in practice there may be a number of subsets that perform as well or even better, some of which could be more convenient to use from the point of view of, say, measurement costs.

The only means of ensuring that the best subset of variables is found is by exhaustive enumeration, whether explicitly or implicitly. McCabe (1975) describes a computer program for doing so with the minimum amount of computation. Hawkins (1976) suggests that a branch-and-bound algorithm, similar to that proposed by Furnival and Wilson (1974) for the subset problem in multiple regression, would require about 175 $(1.39)^p$ computer operations as compared with the asymptotic number $3p^2 + O(p^2)$ required for the stepwise procedure.

McKay (1976, 1977) derives a simultaneous test procedure (STP) for finding all subsets of the p variables whose discriminatory capabilities are not significantly poorer than those of the whole set, with overall significance level controlled at a certain value. In essence, his procedure requires the investigator to compare the test statistic for additional discrimination of the last p-q variables with the critical value for the corresponding test of whether *all* p variables provide significant discrimination between the groups.

In the two-group case this yields the following acceptance region for the STP of the null hypothesis that a particular subset of q variables discriminates as well as all p variables, with overall significance level α:

$$(n-p+1)\left(\frac{N_1 N_2}{N_1+N_2}\right)(D_p^2-D_q^2)/\left(n + \left(\frac{N_1 N_2}{N_1+N_2}\right)D_q^2\right) < pF_{p,n-p+1;1-\alpha} \qquad (3.8)$$

where $F_{p,n-p+1;\alpha}$ is the $100(1-\alpha)$th percentile of the F-distribu= tion with p and n-p+1 degrees of freedom. (Compare this with Rao's test (3.1) for additional information: the right hand side for that test is $(p-q)\ F_{p-q,n-p+1;1-\alpha}$).

Using the likelihood ratio criterion (Hawkins, 1976; McKay, 1977) the k-group STP becomes: accept the null hypothesis that the subset of q variables discriminates as well as all p variables if:

$$\left|\frac{A_{22.1}}{C_{22.1}}\right| = \prod_{i=q+1}^{p} \frac{a_{ii.1,2,\ldots,i-1}}{c_{ii.1,2,\ldots,i-1}} > U_{p,k-1,n;\alpha} \qquad (3.9)$$

where the right hand side denotes the 100αth percentile of the li= kelihood ratio statistic for testing whether all p variables provide significant discrimination between the k groups. The asymptotic ex= pansions of Box (1949) or Crowther (1980) can be used to evaluate $U_{p,k-1,n;\alpha}$ approximately.

From the practical point of view, one may well wonder whether the different subsets so selected in fact differ in their discrimi= nation:- this is a question of whether the likelihood ratio in fact provides a reliable measure of discrimination.

Given two groups, the probability of misclassification is a monotonically decreasing function of D^2. Thus, provided the normal model appears to fit satisfactorily, we much prefer to work in terms of D^2 rather than use a measure such as a sample reclassification error rate, which is subject to considerably higher random sampling fluctuations.

If there are three more groups, then the error rate is no lon= ger a function of the likelihood ratio, and a satisfactory value of the latter criterion may well correspond to a very poor value of the former. One solution is to carry out a subset selection using esti= mated misclassification probabilities as criterion (Habbema and Her= mans, 1977). This however involves a substantial computational load. A computationally lighter criterion is to gauge a subset according to the minimum intergroup Mahalanobis distance it gives rise to. Reasoning that when misclassifications occur, they will generally be to the nearby populations, we conclude that this criterion should be very successful in identifying subsets with low error rates. It is very easily programmed, and may be grafted on to a stepwise regres= sion or discriminant analysis code. To date however, there appears to be little practical experience with it.

4. PREDICTIVE DISCRIMINATION

4.1 Introduction

Sometimes it is of interest to an investigator to assess in some way the relative odds or probability that a multivariate observation Z belongs to one of k multivariate normal populations, $\Pi_i(i=1,\ldots,k)$. This problem has been thoroughly investigated by Geisser (1964). If the prior probability that Z belongs to Π_i is π_i, $\sum_{i=1}^{k} \pi_i = 1$, and the parameters are known, it is relatively easy to compute the likelihood that Z belongs to Π_i and then to combine this with the prior probability, using Bayes' theorem, to obtain the posterior probability that Z belongs to Π_i. When π_i and the parameters of Π_i are known, this procedure is equivalent to the classical discrimination procedure.

When the parameters of Π_i are unknown, a prior probability density, reflecting our ignorance, can be assigned to the parameters of Π_i, whereafter a solution can be obtained for the posterior probability that the observation belongs to Π_i, assuming π_i is known. The estimative procedure however, "plugs" in estimates of the parameters which causes the posterior probability to be biased.

Within the Bayesian framework the case of unknown prior probabilities π_i is much easier to handle since a prior distribution can be assumed for the π_i, and they can then be integrated out.

4.2 Some results in predictive discrimination

4.2.1 *General theory*

Suppose we have k populations Π_i, i=1,...,k each specified by a continuous density $f(\cdot|\theta_i,\psi_i)$, where θ_i is the set of distinct unknown parameters of Π_i and ψ_i is the set of distinct known parameters of Π_i. For each of the populations we have data X_i based on N_i independent observations. Let Z be a new observation to be assigned to one of the k populations, with prior probability π_i of belonging to Π_i, $\sum_{i=1}^{k} \pi_i = 1$.

The predictive density of Z on the hypothesis that it was obtained from Π_i is

$$f(Z|X,\psi,\Pi_i) = \int f(Z|\theta_i,\psi_i,\Pi_i)p(\theta|X,\psi)d\theta \tag{4.1}$$

where $p(\theta|X,\psi)$ is the posterior density and is given by

$$p(\theta|X,\psi) \propto L(X|\theta,\psi)g(\theta|\psi)$$

with $L(X|\theta,\psi) = \prod_{i=1}^{k} L(X_i,|\theta_i,\psi_i)$, the joint likelihood and $g(\theta|\psi)$ the joint prior density of θ for known ψ, $\theta = \bigcup_{i=1}^{k} \theta_i$, $\psi = \bigcup_{i=1}^{k} \psi_i$.

The posterior probability that Z belongs to Π_i is then given by

$$P(Z \in \Pi_i|X,\psi,\Pi_i) \propto \pi_i\ f(Z|X,\psi,\Pi_i) \tag{4.2}$$

For classification purposes Z is then assigned to that population for which this expression is a maximum.

If we let $P(\Pi_j|\Pi_i)$ be the predictive probability that Z has been classified as belonging to Π_j when in fact it belongs to Π_i, we have a predictive probability of misclassification defined by

$$\sum_{i=1}^{k} P(\Pi_i^{\,c}|\Pi_i) = 1 - \sum_{i=1}^{k} P(\Pi_i|\Pi_i) \tag{4.3}$$

where $P(\Pi_i^{\,c}|\Pi_i) = \pi_i\ (1 - \int_{R_i} f(Z|X,\psi,\Pi_i)dZ)$ (4.4)

$$P(\Pi_i|\Pi_i) = \pi_i \int_{R_i} f(Z|X,\psi,\Pi_i)dZ \tag{4.5}$$

where R_i is the classification region to Π_i.

If we want to classify jointly n independent observations $Z_1,\ldots,Z_n$, each having prior probability π_i of belonging to Π_i, we compute the joint predictive density

$$f(Z_1,\ldots,Z_n|X,\psi,\Pi_{i_1},\ldots,\Pi_{i_n}) = \int P(\theta|\psi,X) \prod_{j=1}^{n} f(Z_j|\theta_{i_j},\psi_{i_j},\Pi_{i_j})d\theta \tag{4.6}$$

and from this the joint posterior probability

$$P(Z_1 \in \Pi_{i_1},\ldots,Z_n \in \Pi_{i_n}|X,\psi,\Pi\} \propto (\prod_{j=1}^{n} \pi_{i_j})f(Z_1,\ldots,Z_n|X,\psi,\Pi_{i_1},\ldots,\Pi_{i_n}) \tag{4.7}$$

4.2.2 *Multivariate normal discrimination*

In this case it is assumed that an observation $Z^T = (Z_1,\ldots,Z_p)$ is observed with known prior probability π_i of belonging to Π_i which is $N(\xi_i,\Sigma_i)$ with $\xi_i = (\xi_{1i},\ldots,\xi_{pi})$

and $\Sigma_i = (\sigma_{uti})$, $i=1,\ldots,k$; u, $t=1,\ldots,p$

Define $X_{i.} = \frac{1}{N_i} \sum_{j=1}^{N_i} X_{ij}$

$$(N_i-1)S_i = \sum_{j=1}^{N_i} (X_{ij}-X_{i.})(X_{ij}-X_{i.})^T = A_i, \qquad S = \frac{1}{N-k} \sum_{i=1}^{k} A_i,$$

$$N = \sum_{i=1}^{k} N_i$$

When Σ_i and ξ_i are both known, the posterior density function is identical to the discriminant function in the classical case for known parameters. The two other situations most commonly found in practice are the following:

(a) Σ_i unknown, ξ_i unknown

Here we have prior density $g(\xi_i,\Sigma_i)d\xi_i \; d\Sigma_i^{-1} \propto |\Sigma_i|^{\frac{1}{2}(p+1)} d\xi_i \; d\Sigma_i^{-1}$ and therefore

$$f(Z|X_{i.},S_i,\Pi_i) \propto \left(\frac{N_i}{N_i+1}\right)^{\frac{1}{2}p} \frac{\Gamma(\frac{1}{2}N_i)}{\Gamma(\frac{1}{2}(N_i-p))} |A_i|^{-\frac{1}{2}} \left[1 + \frac{N_i(X_{i.}-Z)^T S_i^{-1}(X_{i.}-Z)}{N_{i-1}^2}\right]^{-\frac{1}{2}N_i} \tag{4.8}$$

(b) $\Sigma_i = \Sigma$ but unknown, ξ_i unknown.

The prior density is $g(\xi_i,\Sigma)d\xi_i \; d\Sigma^{-1} \propto |\Sigma|^{\frac{1}{2}(p+1)} d\xi_i \; d\Sigma^{-1}$ so that

$$f(Z|X_{i.},S,\Pi_i) \propto \left(\frac{N_i}{N_i+1}\right)^{\frac{1}{2}p} \left[1 + \frac{N_i(X_{i.}-Z)^T S^{-1}(X_{i.}-Z)}{(N_i+1)(N-k)}\right]^{-\frac{1}{2}(N-k+1)} \tag{4.9}$$

These two cases are the predictive equivalents of the heterosec= astic and homoscedastic cases respectively. Note however that the predictive classification rule, which allocates to Π_j rather than Π_i if $f(Z|X_{i.},S,\Pi_i) < f(Z|X_{j.},S,\Pi_j)$ reduces to a linear function of Z only in exceptional circumstances, $N_i \equiv N_j$, and is otherwise nonlinear even in the homoscedastic case.

4.3 Proportional covariance matrices

In the derivation of Geisser's result for different and unknown covariance matrices, it is assumed a priori that the Σ_i are mutually independent with diffuse prior distributions, which implies that one expects them to differ considerably. In practice this is not always the case, since one would expect that measurement of the same charac= teristic in different populations would give rise to at least similar even if not identical covariance matrices. Therefore a model has been developed for the case where $\Sigma_i = a_i\Sigma, \Sigma$ has a diffuse prior distribution and the $\{a_i\}$ follow an exchangeable Dirichlet prior distribution with parameter α, i.e.

$$f(a_1,a_2,\ldots,a_k) \propto \begin{cases} \prod_{i=1}^{k} a_i^{\alpha-1} & \Sigma a_i = 1 \\ 0 & \text{elsewhere} \end{cases} \tag{4.10}$$

so that the prior distribution is

$$f(a,\xi,\Sigma) \propto |\Sigma|^{-\frac{1}{2}(p+1)} \left[\prod_{i=1}^{k} a_i^{\alpha-1}\right] \tag{4.11}$$

After integration over ξ and Σ, we have for the case of multivariate normal populations that

$$f(Z,a|\text{data},\Pi_i) \propto \prod_{j=1}^{k} \{|S_j|^{\frac{1}{2}(N_j-p-2)} a_j^{\alpha-\frac{1}{2}pN_j-1}\} a_i^{-\frac{1}{2}p}$$
$$\cdot\left|\sum_{j=1}^{k} a_j^{-1} S_j + \frac{N_i}{N_i^2-1} a_i^{-1}(Z-X_{i.})(Z-X_{i.})^T\right|^{-\frac{1}{2}(N+1)} \tag{4.12}$$

For special cases this expression can now be integrated out over the a_j but in general a modal estimator of the a_j must be substituted.

4.4 The random effects model

This model was discussed earlier in the context of classical discriminant analysis. In the predictive approach, use of this model is equivalent to assuming an informative normal $N(\mu,T)$ prior distribution for the population means $\xi_1,\xi_2,\ldots,\xi_k$. Considering the univariate ($p=1$) case, and using the following general form of diffuse prior distribution for the common variance σ^2 and the mean and variance μ and τ^2 of the ξ_i:

$$g(\sigma^2,\mu,\tau^2) \propto \sigma^{-\nu_1} \tau^{-\nu_2} d\sigma^2 \, d\mu \, d\tau^2$$

yields the following expression for the predictive density:

$$f(Z|\nu_1,\nu_2,\text{data},\Pi_i) \propto (A_3^*)^{-\frac{1}{2}(N+\nu_1+\nu_2-5)} F(\tfrac{1}{2}(2-\nu_2),$$
$$\tfrac{1}{2}(N+\nu_1+\nu_2-5);\ \tfrac{1}{2}(k-1);\ A_1^*/A_3^*) \text{ for } \nu_2 < 2 \tag{4.13}$$

where A_1^* is the between-groups sum of squares and A_3^* is the total sum of squares of the data in the training sample, with the observation Z of unknown origin included in the sample from Π_i, and $F(\alpha,\beta;\gamma;x)$ is the hypergeometric function (see, for example, Johnson and Kotz (1969)).

It is interesting that the parameter ν_2 cannot assume its usual value $\nu_2 = 2$ in the diffuse prior distribution. Using the values $\nu_1 = 2$ and $\nu_2 = 1$ in the predictive density (4.13) results in the following expression for the posterior probability of Π_i:

$$P[\Pi_i|Z,\text{data } \pi] \propto \pi_i(A_3^*)^{-\frac{1}{2}(N-2)}F(\tfrac{1}{2},\tfrac{1}{2}(N-2);\ \tfrac{1}{2}(k-1);A_1^*/A_3^*) \qquad (4.14)$$

In the multivariate $(p > 1)$ case the predictive density is expressed in terms of the hypergeometric function with matrix argument, for which no simple expression exists.

4.5 Predictive discrimination with an informative prior for the covariance matrix

This model makes provision for the case where one has no prior knowledge of what the covariance matrices are, but has reason to believe that they are related. Since the covariance matrices are un= known, a diffuse prior for the Σ_i has to be used. Let ξ_i have a dif= fuse prior and Σ_i be distributed a priori according to a $W^{-1}(\Gamma,\nu)$ distribution, where Γ has a diffuse prior. Since the ξ_i and Σ_i are distributed independently, we thus have

$$f(\xi_i,\Sigma_i) \propto |\Gamma|^{\frac{1}{2}(\nu-p-1)}|\Sigma_i|^{-\frac{1}{2}\nu}\exp\{-\tfrac{1}{2}\text{tr}\Sigma_i^{-1}\Gamma\} \qquad (4.15)$$

The Γ and ν are hyperparameters, where ν will usually be spe= cified a priori. In certain special cases, Γ may also be known or spe= cified, for example, Γ may be diagonal or the identity matrix.

If we assume that Z comes from population Π_r with prior proba= bility π_r, we have, after integration with respect to ξ_i, and Σ_i, that

$$f(Z,\Gamma|\text{data}) \propto |\Gamma|^{\frac{1}{2}[k(\nu-p-1)-p-1]}\prod_{i=1}^{k}|S_i|^{\frac{1}{2}(n_i-p-1)}|\Gamma + B_i|^{-\frac{1}{2}(n_i^*+\nu-p-1)} \qquad (4.16)$$

where

$$n_i^* = \begin{cases} n_i & i\neq r \\ n_i+1 & i=r \end{cases}, \quad n_i = N_i-1$$

$$B_i = \begin{cases} n_iS_i \\ n_iS_i + \dfrac{N_i}{N_i+1}(Z-X_{i.})(Z-X_{i.})^T & i\neq r \end{cases} \qquad (4.17)$$

In some cases this equation can be evaluated analytically, but in the general case an estimate of Γ, based on the S_i, must be used. A simple classical estimate uses the fact that

$$E[S_i|\Sigma_i] = \Sigma_i$$

while $E[\Sigma_i] = \frac{\Gamma}{\nu-2p-2}$, $\nu > 2p + 2$

and thus

$$E[S_i|\Gamma] = \frac{1}{\nu-2p-2}\Gamma$$

so that $(\nu-2p-2)S_i = G_i$ provides an unbiased estimator of Γ. A pooled estimator could be obtained from

$$\hat{\Gamma} = \frac{\sum_{i=1}^{k} w_i G_i}{\sum_{i=1}^{k} w_i}$$

where the w_i are any system of weights not summing to zero, for exam= ple $w_i = (1 - N_i^{-1})$.

Since the accuracy of this estimator of Γ depends primarily on the magnitude of k and secondarily on that of the N_i, the error in= duced by using this estimator for Γ will be small for large k. If k is small but the N_i are large, Γ will be estimated rather inaccurately but since the role played by Γ in the predictive density in this case is relatively minor, the error will again be small. Only in the case of N_i and k small, will the error be large.

This model provides a useful intermediate case between the homo= scedastic model in which $\Sigma_i = \Sigma$ for all i, and the general heteroscedas= tic case. We may note in passing that this pooling of the different S_i is easily and naturally done within the predictive Bayesian context, but is not at all natural within the estimative context.

The method is, however, related both conceptually and computa= tionally to the ridge-type biased discriminant analysis procedures of DiPillo (1976, 1977, 1979). The latter procedures, by analogy with ridge regression, may be considered as including in the covariance matrix dummy observations from $N(0,kI)$, and these have an obvious Bayesian interpretation in terms of prior information on Σ_i.

4.6 Comparison of the predictive and classical approaches to discriminant analysis

In the previous sections, we have discussed the two main ap= proaches to the problem of discriminant analysis namely the predictive method and the classical (or estimative) method. It is generally agreed by the protagonists of both approaches that there is very little or no difference between the classification rule produced by the two methods: for example the homoscedastic models yield identical classification rules if the N_i are equal, and differ markedly only if the N_i are simultaneously small and markedly unequal.

The crux of the debate is the posterior probabilities, which can differ by several orders of magnitude between the two methods, as illustrated by Aitchison, Habbema and Kay (1977). On this point, we consider that the predictive method is correct, and the estimative simply wrong. While the estimative method may be patched up to reduce this inferiority (Moran and Murphy, 1979), we can see no reason to prefer it to a method which gives the correct result ab initio.

The difference between the methods can be illustrated for the case of a multivariate normal distribution. Suppose we have N_i observations from population Π_i, with $X_{i.}$ and S_i the sample mean and covariance, then the density function for the estimative method is given by

$$r(X|\Pi_i,\text{data}) = \frac{1}{\{(2\pi)^p\ |S_i|\}^{\frac{1}{2}}} \exp\{-\tfrac{1}{2}(X-X_{i.})^T S_i^{-1}(X-X_{i.})\} \qquad (4.18)$$

while for the predictive method, using a diffuse prior for the parameters, we have a multivariate t distribution:

$$q(X|\Pi_i,\text{data}) = \frac{\Gamma(\frac{1}{2}N_i)}{\pi^{\frac{1}{2}p}\Gamma(\frac{1}{2}(N_i-p))} \left|\frac{N_i-1}{N_i} S_i\right|^{-\frac{1}{2}} \{1 + (X+X_{i.})^T \frac{N_i}{N_i^2-1} S_i^{-1}(X-X_{i.})\}^{-\frac{1}{2}N_i} \qquad (4.19)$$

The distributional result (4.19) is respectable in classical as well as Bayesian theory:- it consists simply of the distribution defining the tolerance region for a future observation X given an initial sample from the population Π_i. This distribution may be derived from conventional distributional methods.

It is easy to see that as $N_i \to \infty$, (4.18) and (4.19) become identical. Thus in large samples, they will give very much the same posterior probability for each Π_i. For small N_i, however, this is very far from being the case.

Aitchison (1975) considered the relative merits of the predictive and estimative methods as methods of density function estimation by defining an appropriate measure of the relative overall closeness of $q(X|\Pi_i,Z)$ and $r(X|\Pi_i,Z)$, based on data Z, to the true $p(X|\Pi_i,\theta)$. This measure, based on an information divergence measure of Kullback and

Leibler (1951) is defined by

$$\int_Z p(Z|\theta)dZ \int_X p(X|\Pi_i,\theta) \log \frac{q(X|\Pi_i,Z)}{r(X|\Pi_i,Z)} dX \tag{4.20}$$

For $p(X|\Pi_i,\theta)$ multivariate normal, Aitchison (1975) shows that this measure is independent of θ and positive, indicating greater overall closeness of the predictive density function estimator to the true density function.

For simulation studies, Aitchison, Habbema and Kay (1977) con= sidered two p-dimensional multivariate normal populations and data sets that are small in relation to the parameter dimension. They com= pared four methods for each of the situations, namely the predictive method for equal covariance matrices, P_e, the predictive method for unequal covariance matrices, P_u, the estimative method for equal covariance matrices, E_e, and the estimative method for unequal co= variance matrices, E_u. Thus E_e is the LDF and E_u the QDF. The results of these studies can be summarized as follows:

For equal covariance matrices, P_e is far superior to the other methods. For unequal covariance matrices the P_u method is slightly better than the P_e method, which in turn is better than the E_e method. E_u performed far worse than the others in both situations.

The predictive method also gives less extreme asymptotic values for the posterior probability in the case of equal prior probabili= ties and for unequal priors except when r(X) and q(X) are on opposite sides of 0.5 and q(X) is further away from 0.5 than r(X). (McLachlan, 1979).

There are cases in which one only wants a mechanical classifi= cation rule, and then the choice between the predictive and the esti= mative approach is of little import. We believe however, that a cau= tious practitioner of discriminant analysis should study the posterior probabilities of all Π_i, and make his classification with the degree of tentativeness that their spread dictates, and for this purpose the usual estimative calculations can be hopelessly misleading.

5. OTHER APPROACHES TO DISCRIMINANT ANALYSIS

In this section a number of approaches to discriminant analysis that do not assume multivariate normality of the observation vectors will be discussed. Logistic discriminant analysis is based on the assumption that the posterior probabilities of each of the groups, given a particular observation, has a logistic form, and does not make any specific assumptions about the distributions associated with the groups. On the other hand, both the location model and the multinomial discriminant analysis procedure are based on specific distributional assumptions, although in the latter case they are very broad. Finally, the discriminant analysis procedure based on density function estimation via the kernel method and those based on the rank transformation method are all truly nonparametric.

5.1 Logistic discriminant analysis

In classical discriminant analysis, the parameters in the distributions corresponding to each population are first estimated and then used, together with the observation to be classified, to estimate the posterior probabilities of each of the populations via Bayes' formula:

$$\hat{P}[\Pi_i|X] \propto \pi_i \ \hat{f}_i(X) \quad , \quad i=1,\ldots,k$$

where $\hat{f}_i(X)$ is the density function corresponding to Π_i, with its parameters replaced by their estimates. X is then assigned to that population with highest posterior probability. (See (1.2).) In contrast, logistic discriminant analysis is based on the assumption that the posterior probabilities have the linear logistic form:

$$P[\Pi_i|X] = p_{ix} = \exp(\alpha_{io} + \alpha_{i1} X_1 + \ldots + \alpha_{ip} X_p) p_{kx} \qquad (5.1)$$
$$i=1,\ldots,k-1$$

and $$P[\Pi_k|X] = p_{kx} = 1/\{1 + \sum_{i=1}^{k-1} \exp(\alpha_{io} + \alpha_{i1} X_1 + \ldots + \alpha_{ip} X_p)\} \qquad (5.2)$$

where $X = (X_1, X_2, \ldots, X_p)^T$,

so that the parameters in the posterior probabilities may be estimated directly.

Cox (1966) and Day and Kerridge (1967) note that the posterior probabilities from a number of distributions used in discriminant analysis have the logistic form. These include:

(i) the multivariate normal distribution with equal dispersion matrices,

(ii) the multivariate Bernoulli distribution with independent components,

(iii) the multivariate Bernoulli distribution where the variables follow a log-linear model with equal second- and higher-order effects in all k groups (see Chapter 3), and

(iv) A combination of distributions (i) and (iii).

Giving training samples of sizes N_i from each of the k popula= tions Π_i, i=1,...,k, the maximum likelihood equations for the (p+1)(k-1) coefficients α_{ij}, j=0,...,p; i=1,...,k-1 are:

$$\sum_{\forall x} (N_{ix} - N_x\, p_{ix}) x_j = 0 \qquad j=0,\ldots,p;\ i=1,\ldots,k-1 \tag{5.3}$$

where N_{ix} is the frequency that the vector $x = (x_o, x_1, \ldots, x_p)$ occurs in the training sample from Π_i,

$x_o = 1$

and $N_x = \sum_{i=1}^{k} N_{ix}$.

While equations (5.3) were originally derived assuming that the training sample was randomly selected from the mixture of groups Π_i, i=1,...,k, as they occur in the whole population, Anderson (1972) shows that when the Π_i are sampled separately these equations still hold, except that the estimates of α_{io} are modified as follows:

$$\hat{\alpha}^*_{io} = \hat{\alpha}_{io} + \log_e (\pi_i\, N_k/N) - \log_e(\pi_k\, N_i/N) \tag{5.4}$$

where $N = \sum_{i=1}^{k} N_i$,

$\hat{\alpha}_{io}$ is the estimate obtained from (5.3), and $\hat{\alpha}^*_{io}$ is the modified es= timate. In practice π_i is replaced by an estimate, $\hat{\pi}_i$. Note that if $\hat{\pi}_i = N_i/N$, i=1,...,k, then $\hat{\alpha}^*_{io} = \hat{\alpha}_{io}$.

Equations (5.3) can be solved iteratively for the $\hat{\alpha}_{ij}$ using the Newton-Raphson procedure. Anderson (1972) states that starting values of zero for all parameters usually work well, although the usual li= near discriminant function coefficients could also be used.

Unfortunately, the maximum likelihood estimators are not unique when there is a complete separation of groups in the training sample, i.e. when a linear discriminant function can be found that classifies all N sample points correctly. However, in this case any solution to equations (5.3) will give complete separation of groups in the train= ing samples so that, whereas the estimates of the α_{ij} may not be re= liable, any solution will tend to give good discrimination. Anderson (1974) also gives a heuristic solution to the estimation problem ari= sing in the two-group problem when a binary variable is constant in all observations in the training sample from one of the groups.

Inspection of the form of the posterior probabilities 5.1 and 5.2 shows that the classification rule allocating X to Π_i or Π_k de= pends on a discriminant function of the X, namely $\alpha_{io} + \alpha_i X$ which is linear in X. This function differs from the LDF, however, in its estimation. The LDF is derived indirectly from the mean vectors and covariance matrix of the data, while the logistic DF is calibrated directly from the initial training sample and its allocations, and is not dependent on a normality assumption for the validity of the estimator used.

The logistic discriminant function is thus clearly applicable to a far broader class of distributions than the classical linear discriminant function and, in particular, it can often handle the commonly occurring situation (see below) where there are both binary and continuous variables. It is, however, generally not as efficient as the classification rule (1.2) based on the appropriate distribu= tion (O'Neill, 1980). Efron (1975) shows that the asymptotic error rate of the classification rule based on the logistic discriminant function is typically between 1½ times and twice as high as that of the linear discriminant function when the assumptions of normality and homoscedasticity are valid.

On the other hand, when the normality and homoscedasticity as= sumptions are violated, especially when some of the variables are binary, Press and Wilson (1978) show that the logistic discriminant function will outperform the linear discriminant function, although generally not by a large amount. Since it is not determined by second order moments, it will, as an added bonus, have less sampling error with heavy-tailed data.

The applicability of logistic discriminant analysis can be broadened by considering a quadratic logistic discriminant function, but at the expense of greatly increasing the number of parameters. Anderson (1975) proposes an approximation to this function that con= siderably reduces the number of parameters to be estimated.

5.2 Binary and continuous variables - the location model

Looking at the discriminant problem in the light of the loglinear model (Chapter 3), we see that it is a particular property of the multivariate normal distribution that the pairwise interrelationships among variables are sufficient to determine all higher order inter= relationships. In the homoscedastic model this property extends notion= ally to include among the variables the population membership variable.

In the heteroscedastic model, it is necessary to use also third order interrelationships between the population membership variable and pairs of variables X.

This parsimony of parametrization does not necessarily extend to distributions other than the normal, and in particular, to vec= tors of mixed binary and interval measurements.

Krzanowski (1975, 1976), generalizing an idea first proposed by Chang and Afifi (1974), proposes a model for the two-group situa= tion where q of the variables are binary and the remaining p-q are conditionally normal with mean vector $\xi_i^{(m)}$ corresponding to the mth cell in the multinomial representation of the binary variables in Π_i, ($m=1,\ldots,2^q$; $i=1,2$), and common covariance matrix Σ in all cells of both populations. This "location" model is often appropriate in medical applications where some of the classification variables are symptoms of a presence/absence nature while others, such as weight and blood pressure, are continuous.

If $X_{(1)} = (X_1,X_2,\ldots,X_q)^T$ is the vector of binary (0,1) variables and $X_{(2)} = (X_{q+1},\ldots,X_p)$ the vector of normally distributed variables, then Krzanowski shows that the optimal rule that classifies X into the population with maximum posterior probability, is to assign $X = (X_{(1)}^T,X_{(2)}^T)^T$ to Π_1 if

$$(X_{(2)} - \tfrac{1}{2}(\xi_1^{(m)} + \xi_2^{(m)}))^T\Sigma^{-1}(\xi_1^{(m)} - \xi_2^{(m)}) \geq \ell n(\pi_{2m}/\pi_{1m}) \quad (5.5)$$

and to Π_2 otherwise,

where $m = 1 + \sum_{j=1}^{q} X_j\, 2^{j-1}$ and π_{im} is the prior probability of an obser= vation from Π_i falling in cell m. Writing

$$\xi_i^{(m)} = \nu_i + \Sigma\,\alpha_{ji}\, X_j^{(m)} + \Sigma\,\Sigma\,\beta_{jki}\, X_j^{(m)}\, X_k^{(m)} + \ldots$$
$$\ldots + \delta_{12}\quad X_1^{(m)}\, X_2^{(m)},\ldots,X_q^{(m)}\ \ldots \quad (5.6)$$

the classification rule (5.5) may be reduced to a polynomial discriminant function.

In order to overcome the problem of the large number of parameters that need to be estimated before classification rule (5.5) can be applied in practice, a second-order approximation to the full model is proposed, whereby the parameters $\xi_i^{(m)}$ are expressed in the form:

$$\xi_i^{(m)} = \nu_i + \sum_{j=1}^{q} \alpha_{j,i}\, X_j^{(m)} + \sum_{j<k}\Sigma\, \beta_{jk,i}\, X_j^{(m)}\, X_k^{(m)}$$

where $x_j^{(m)}$ is the value of X_j (0 or 1) in cell m. A corresponding second-order log-linear model is proposed as an approximation for the probabilities π_{im}. Given training samples from the two groups in the proportions that they occur in the population, maximum likelihood estimators of the $\xi_i^{(m)}$ in (5.6) can be obtained by multiple regression and of the π_{im} by iterative scaling. (See chapter 4 or Plackett (1974)). Particular success with the location model can follow from the discovery of further simplifications in 5.6 enabling its reduction to fewer parameters. In particular, if $\beta \equiv 0$ and $\alpha_{j1} \equiv \alpha_{j2}$, then linear discrimination will be satisfactory. If the $\beta_{jk1} \equiv \beta_{jk2}$ then the quadratic discriminant function will be satisfactory. The latter situation also extends to data in which $X_{(1)}|X_{(2)}$ is not homoscedastic.

5.3 Multinomial discriminant analysis

Suppose that X has a Multinomial distribution with s possible outcomes, or cells, and that if X comes from Π_i the probability that it falls in cell ℓ is $p_{i\ell}$, where $\sum_{\ell=1}^{s} p_{i\ell} = 1$. Classification rule (1.2) then becomes: Assign X to Π_i if it falls in any cell ℓ such that

$$\pi_i \, p_{i\ell} = \max_{j=1,\ldots,k} \pi_j \, p_{j\ell} \tag{5.7}$$

For the two-group case this simplifies to: Assign X to Π_1 if it falls in any cell ℓ such that

$$\pi_1 \, p_{1\ell} > \pi_2 \, p_{2\ell} \tag{5.8}$$

and to Π_2 otherwise.

The error rate for this rule is clearly:

$$P[\text{misclassification}] = \sum_{i=1}^{k} \pi_i \sum_{\ell \in \bar{S}_i} p_{i\ell} \tag{5.9}$$

where S_i is the set of indices $\{\ell | \pi_i p_{i\ell} = \max_{j=1,\ldots,k} \pi_j p_{j\ell}\}$ i.e. S_i represents the set of cells for which observations are assigned to Π_i, and $\bar{S}_i$ is the complement of S_i.

One way in which the multinomial distribution arises in practice is when the original data vector consists of p variables and the j th variable has been categorised into m_j classes, j=1,...,p, so that $s = \sum_{j=1}^{p} m_j$. In particular, if each of the variables is binary, then the number of cells is $s = 2^p$.

There are two main shortcomings to the multinomial classifica= tion procedure:

(a) A situation can arise where observations are assigned to Π_i whatever their outcome. Specifically, if

$$\pi_i \, p_{i\ell} = \max_{j=1,\ldots,k} \pi_j \, p_{j\ell}$$

for all $\ell=1,\ldots,s$, then observations will always be assigned to Π_i.

(b) When the $p_{i\ell}$ are unknown and s is large, then there is a large number of parameters to estimate. Several approaches have been suggested for estimating the $p_{i\ell}$:

(i) The full multinomial approach. Here $p_{i\ell}$ is estimated by:

$$\hat{p}_{i\ell} = N_{i\ell}/N_i \qquad (5.10)$$

where there are $N_{i,\ell}$ observations in the training sample from Π_i falling in cell ℓ and $N_i = \sum_{\ell=1}^{s} N_{i\ell}$. Clearly, if s is large N_i has to be very large for the estimators to be reliable. Particular problems arise if any of the $N_{i\ell}$ are zero, as it is usually unreasonable to suppose that any $p_{i\ell}$ is zero.

(ii) A lower order approach. Moore (1973) considers the Bahadur reparameterisation of the multinomial probabilities in the case where X is a p-dimensional vector of binary variables, and proposes a first or second order approximation so as to reduce the number of parameters to be estimated. The first order model assumes the p binary variables to be independent, requiring the estimation of only p parameters, whereas the second order model also requires the estimation of p(p-1)/2 correlation coefficients between all pairs of variables.

On the basis of a simulation study of the two-group case, Moore (1973) concludes that in the situation where no reversals in sign of corresponding correlation coefficients occur between the two groups, then for moderate sample sizes the first order model performs better than the second order or full model. In this case the linear discrimi= nant function, ignoring the binary nature of the variables, performs almost as well as the first order multinomial model. On the other hand, when reversals in sign do occur, then the first order model and the linear discriminant function perform significantly worse than the full multinomial model. Use of the second order model is not recommended.

A procedure for variable selection has been proposed by Gold= stein and Rabinowitz (1975) for two-group discriminant analysis. They also remark on the generally poor performance of the multinomial classification rule.

The overall conclusion is that multinomial discriminant analy= sis should only be used when there is evidence of "reversals" (i.e. heteroscedasticity) between the groups and that large training samples are needed. Even in this case the quadratic logistic discriminant analysis procedure proposed by Anderson (1975) might be preferred.

5.4 Discriminant analysis by the kernel method

This is a non-parametric approach to discriminant analysis in which the probability density functions $f_i(X)$, $i=1,\ldots,k$, in classi= fication rules (1.1) to (1.4) are estimated directly from the train= ing samples by the kernel function method. Suppose that $K(y|x,\lambda)$ is a function (kernel function) centered about x, where λ represents a smoothing or spread parameter, such that

$$\int_{-\infty}^{\infty} K(y|x,\lambda)dy = 1$$

Given a training sample $X_{i1},\ldots,X_{iN_i}$ from Π_i, the corresponding proba= bility density function $f_i(X)$ at the point $X = x$ is estimated by

$$\hat{f}_i(x) = N_i^{-1} \sum_{j=1}^{N_i} K(x|X_{ij},\lambda) \tag{5.11}$$

Usually $K(y|x,\lambda)$ is a non-negative function, so that it can be consi= dered as a density function itself, centered about x with λ some mea= sure of spread of the distribution. For example, if

$$\begin{aligned} K(y|x,\lambda) &= 1/2\lambda \quad \text{for } |y-x| \le \lambda \\ &= 0 \quad \text{elsewhere} \end{aligned}$$

then $\hat{f}_i(x)$ is proportional to the number of observations from the train= ing sample falling in the class interval $[x-\lambda,x+\lambda]$.

Given a kernel function $K(\cdot)$ with k values $\lambda_1,\lambda_2,\ldots,\lambda_k$ (possi= bly all equal) of the smoothing parameter λ, training samples from each of the k populations and an observation X of unknown origin, then X is assigned to one of the k groups according to one of the rules (1.1) to (1.4), with $f_i(X)$, $i=1,\ldots,k$, replaced by

$$\hat{f}_i(X) = N_i^{-1} \sum_{j=1}^{N_i} K(X|X_{ij},\lambda_i) \tag{5.12}$$

Two obvious problems arise with the use of this approach. They are:

(a) the choice of kernel function, and

(b) the choice of smoothing parameters $\lambda_1,\lambda_2,\ldots,\lambda_k$.

For p-dimensional continuous data the most common kernel used is the spherical normal density function:

$$K(y|x,\lambda) = (2\Pi\lambda)^{-\frac{1}{2}p} \exp\{-\tfrac{1}{2}(y-x)^T(y-x)/\lambda\} \tag{5.13}$$

Note that this kernel is not invariant under any but the most trivial transformations of the data vectors. Clearly this kernel per= forms best if the data have components of equal variance and no cor= relation, and may perform almost arbitrarily poorly as one departs from this ideal. Thus it is desirable at least to standardize all components so that they have the same spread. Better yet, one should multistandardize so that the components have the same spread, and also no correlation between components. Unfortunately it is far from clear how one should ideally perform either of these transformations on actual data.

The choice of smoothing parameter is important. As λ decreases the kernel function becomes more peaked, so that a narrower range of sample points around x is included in the density estimate. The re= sulting density estimate will be more responsive to local variations in the underlying density function but its sampling variability will be high. On the other hand, increasing λ smooths the resulting den= sity estimate and reduces its sampling variability, but increases its bias. In the limit, as λ tends to infinity, the estimated density function will be uniform.

A spherical Cauchy kernel, with slightly heavier tails, has been proposed as an alternative to (5.13), but van Ness and Simpson (1976) show that it gives very similar results.

When the data come from a p-dimensional multivariate binary dis= tribution, Aitchison and Aitken (1976) propose the kernel:

$$K(y|x,\lambda) = \lambda^{p-D^2(x,y)}(1-\lambda)^{D^2(x,y)} \tag{5.14}$$

where $D^2(x,y) = (x-y)^T(x-y)$ and x and y denote p-dimensional vectors of binary (0,1) elements.

This can also be written as:

$$K(y|x,\lambda) = \prod_{i=1}^{p} \lambda^{1-Z_i} (1-\lambda)^{Z_i} \quad (5.15)$$

where Z_i = 1 if the ith elements of x and y are equal
= 0 otherwise

When $\lambda = 1$, $K(y|x,1) = 1$ if $y = x$
= 0 if not,

so that the density at y is estimated by the relative frequency of y occurring in the sample, for all 2^p possible values of y.

If $\lambda = \frac{1}{2}$, $K(y|x,\frac{1}{2}) = (\frac{1}{2})^p$ for all y and x, so that the esti= mated density is uniform. More useful values of λ will clearly lie between these two extremes.

Kernel (5.14) may be extended to handle categorical data with more than two categories for both situations where the categories have a natural ordering and where they do not. (Aitchison and Aitken, 1976; van der Merwe and Kotze, 1980).

For situations where some of the data elements are continuous and others binary, a kernel function consisting of the product of the two functions (5.13) and (5.14), corresponding to the continuous and binary variables, respectively, is proposed by Aitchison and Aitken (1976). It is worth stressing that the factorisation of all the kernel functions mentioned above into separate factors corres= ponding to each of the p variables, in no way implies that the vari= ables themselves are independent.

Coming to the choice of smoothing parameter, the maximum likeli= hood estimator of λ_i, is that value that maximizes the "pseudo-likeli= hood" function

$$L(\lambda_i|TS) = \prod_{j=1}^{N_i} \hat{f}_i(X_{ij}|TS,\lambda_i) \quad (5.16)$$

where $\hat{f}_i(X_{ij}|TS,\lambda_i)$ is the kernel function estimator (5.12) of $f_i(X_{ij})$ when the training sample TS and smoothing parameter λ_i are used. How= ever, it can be shown that this procedure leads to the oversharp values $\lambda_i = 0$ with the spherical normal kernel (5.13) and $\lambda_i = 1$ with kernel (5.14), respectively. Habbema et al (1974) propose the follow= ing jackknife procedure for getting around this problem: find the value of λ_i that maximizes the function

$$L^*(\lambda_i|TS) = \prod_{j=1}^{N_i} \hat{f}(X_{ij}|TS-X_{ij},\lambda_i) \quad (5.17)$$

where $TS-X_{ij}$ denotes the training sample with observation X_{ij} removed. Aitchison and Aitken (1976) show that with this choice of λ_i the kernel function estimator $\hat{f}_i(X)$ converges in probability to $f_i(X)$ in the multivariate binary case.

The performances of the kernel function and linear discriminant function methods in the two-group case have been compared by Van Ness and Simpson (1976) by means of a simulation study. Using simulated data from $N_p(0,I)$ and $N_p(\mu,I)$ distributions for Π_1 and Π_2, respectively, where $\mu = (\delta_{12}^2, 0, \ldots, 0)^T$, they come to the conclusion that the kernel method using (5.13) outperforms the LDF for all but very small dimensions. However, since their simulation used covariance matrix I, the situation most favourable to the spherical normal kernal, their conclusion cannot be supposed to have any very great generality. So while the kernel method has been shown to perform well under ideal conditions, it remains unclear how well it will perform under everyday conditions with imperfect information about the scaling necessary to obtain approximate sphericity in the data.

Thus, as in other statistical procedures, the non-parametric kernel function method should not be used when available parametric methods are appropriate. However, when the assumptions underlying these methods do not hold, the kernel function method, with its minimal number of assumptions, may be a viable competitor, in spite of the arbitrariness in the choice of kernel function and the fact that classification is based only on data in the neighbourhood of the observation X of unknown origin, rather than on the entire training sample, as in parametric procedures.

5.5 Two methods based on ranks

These two methods use the ranking operation at different stages in the classification procedure.

The first, the rank transformation method, is applicable when all the variables are continuous. Rank each variable in all the observations in the training samples from the smallest, with rank 1, to the largest, with rank $N = \sum_{i=1}^{k} N_i$, separately for each of the p variables. The normal-based linear or quadratic discriminant function procedure is then applied to the data vectors with each of their elements replaced by their corresponding ranks. The observation to be classified is replaced by a vector of "ranks", where the "rank" of the ℓth variable is obtained by linear interpolation between the ranks of the two adjacent values in the training sample.

As discussed in section 2.3.4, this method is of very general applicability, assuming little more than that the regression of any component on any other is monotonic, and has quite high efficiency when applied to normal data.

The second method, applicable to the two-group situation, is based on the following result: If $D(\cdot)$ is any p-variable function computed from the training samples in such a way that the observa= tions within each of the groups is treated symmetrically, and if the $D(X_{ij})$, for $j=1$ to N_i, are ranked from 1 to N_i, then the vector $(R_1,R_2,\ldots,R_{N_i})$, where $R_j = \text{rank}(D(X_{ij}))$, has a uniform distribution over the set of all permutations of the integers 1 to N_i. The linear and quadratic discriminant functions are both functions that treat the observations within each of the training samples symmetrically.

Using this result together with the linear discriminant func= tion, Broffitt et al (1976) construct a classification rule based on the relative ranks of the linear discriminant function of X (the observation of unknown origin) assuming, in turn, that X has come from Π_1 and then from Π_2.

In principle, their procedure may be used with any function $D(\cdot)$ that separates the two populations adequately, as long as it satis= fies the above mentioned criteria. Randles et al (1978) propose two other rules, one based on the quadratic discriminant function (2.17) and the other on a weighted average of the linear and quadratic dis= criminant functions.

One of the main features of this rank method is that, apart from being distribution free, it produces misclassification probabili= ties that are effectively the same for both groups, in contrast with most parametric procedures where the two probabilities can be marked= ly different.

6. MISCELLANEOUS TOPICS

In this section, a number of topics associated with discriminant analysis are briefly mentioned.

6.1 Sequential discrimination

There are two types of sequential procedure in discriminant ana= lysis. The first is applicable if it is possible to obtain several independent observations from the individual to be classified. If we preassign a maximum probability of misclassification for each popu= lation, then, in principle, we can use a sequential test to decide whether to assign an individual to one of the populations or whether to make another observation on the individual. In the two-group case with normal data and equal covariance matrices, and with maximum misclassification probabilities P_1 and P_2 for Π_1 and Π_2 respectively, the classification rule, based on the sequential probability ratio test, is: If after m observations on the individual, where X. is their mean,

$$m(X.-\tfrac{1}{2}(\xi_1+\xi_2))^T\Sigma^{-1}(\xi_1-\xi_2) \begin{cases} \geq \ell n(1-P_1)/P_2 & \text{assign it to } \Pi_1 \\ \leq \ell n\, P_1/(1-P_2) & \text{assign it to } \Pi_2 \end{cases}$$

otherwise take another observation. In the practical situation, where ξ_1, ξ_2 and Σ are unknown, an approximate rule may be obtained by re= placing them by their training sample estimates.

The second sequential procedure applies to the variables ob= served on an individual. Here one assumes that variables are obser= ved sequentially, presumably in increasing order of cost, and after each observation a decision is made whether to assign the individual to one of the populations, or whether to observe the next variable. For the two-group normal case with equal covariance matrices, and P_1 and P_2 defined as before, Mallows (1953) proposes the following sequential rule, where $U_{12}(X^{(1)})$ denotes the population-based linear discriminant function, defined in (2.6), based on the vector $X^{(1)}$ of the first q observations observed: If

$$U_{12}(X^{(1)}) \begin{cases} \geq \ell n(1 - P_1)/P_2 & \text{assign the individual to } \Pi_1 \\ \leq \ell n\, P_1/(1 - P_2) & \text{assign the individual to } \Pi_2 \end{cases}$$

otherwise observe variable X_{q+1}. In the practical situation, $U_{12}(X^{(1)})$ will be replaced by the corresponding sample-based function $V_{12}(X^{(1)})$ defined in (2.20).

A decision-theoretic sequential rule, based on both the costs of misclassification and of observing the various variables and using a dynamic programming approach, is proposed by Hora (1980).

Kendall and Stuart (1966) propose the following simple sequen= tial rule for discriminating between two populations: First divide the range of variation of each variable into three mutually exclusive regions on the basis of the training sample, such that the first re= gion contains only observations from Π_1, the second only from Π_2 and the third from both. Then, starting with the first variable, assign the observation of unknown origin to Π_1 or Π_2 depending on whether it falls in the first or second region , respectively. If it falls in the third region, repeat the procedure with the second variable and continue, if necessary, until the observation is classified.

It is evident that this method makes no use of the joint distribution of the variables. A refinement of the procedure is to examine all variables, including those previously employed, at each step. If any observations remain unclassified at the end, then the joint frequency distributions of all pairs of variables can be used in the same manner to try and classify them. This is proposed by Richards (1972) and can, in principle, be extended, with increasing complexity, to the joint frequency distributions of all sets of three variables.

In this procedure, as well as in the other sequential procedures, there is the possibility that an individual remains unclassified at the end, after all variables (or observations) have been used.

6.2 Discriminant analysis using unclassified observations

Unclassified observations, if sampled randomly from the mixture of populations $\Pi_i, i=1,\ldots,k$ contain information about the distributions associated with each of the Π_i as well as the mixing proportions $\pi_i, i=1,\ldots,k$.

In the two-group situation with normal populations and equal covariance matrices Day (1969) has derived the maximum likelihood equations for the parameters ξ_1, ξ_2, Σ and $\pi_1 (\pi_2 = 1 - \pi_1)$ which may be solved by iterative techniques for the corresponding maximum likelihood estimators.

The efficiency of the discriminant analysis procedure obtained by substituting these estimators into the linear discriminant function, has been investigated by Ganesalingham and McLachlan (1978) in the univariate case. They show that the asymptotic efficiency of this procedure, relative to the usual procedure using classified observations, as measured by the ratio of the first order terms in the asymptotic expansions of their respective expected error rates, is low when the square root of the Mahalanobis distance δ_{12} between Π_1 and Π_2 is less than 2 but improves as δ_{12} is increased. For widely separated populations with $\delta_{12} \geq 4$, the asymptotic relative efficiency is greater than 50% unless one of the prior probabilities π_1 or π_2 is very small (< .05).

A generalization of this procedure, that incorporates both classified and unclassified observations, is given by O'Neill (1978), who shows that in the statistically interesting range $2.5 \leq \delta_{12} \leq 4$

of between-group distances, the information in an unclassified obser= vation varies from approximately one-fifth to two-thirds of that of a classified observation. This can be of considerable importance when the cost of obtaining unclassified observations is much lower than that for classified observations.

An empirical procedure for incorporating a large number of un= classified observations into the linear discriminant function is pro= posed by McLachlan (1975b). The procedure is as follows: First esti= mate the linear discriminant function on the basis of the observations of known origin only, and use this to classify the other observations. Now re-estimate the function using as training sample from each Π_i those observations known to have come from it as well as those that have been assigned to it. This procedure can be repeated by re-classi= fying the observations of unknown origin using the new function, and then repeating the estimation procedure with the re-classified obser= vations. This procedure can clearly be repeated any number of times.

A moment's thought shows up the strong connection between this algorithm (a member of the general EM class, see p.320) and the itera= tive reallocation method used in cluster analysis, which is discussed in chapter 6 of this volume. The only difference, in fact, is that there is a good starting point for the iterative reallocation, and that the classifications of the classified observations in the train= ing sample are never altered. On the strength of this analogy alone, one would expect good performance of the approach so long as the sepa= ration between the populations is not too small, and in fact McLachlan (1975b) shows that this procedure is asymptotically optimal as the number of unclassified observations tends to infinity and the initial training sample of observations of known origin is moderately large. A simulation study shows that this procedure leads to an improvement in the error rate when the size of the sample of unclassified obser= vations is finite, but at least twice that of the initial training sample. The greatest improvement comes after the first iteration, subsequent iterations being of negligible value.

When the number of unclassified observations is small relative to the initial training sample, McLachlan (1977b) proposes that the mean vectors only be re-estimated using the unclassified observations in the above manner.

The analogy of McLachlan's approach to an iterative reallocation suggests the adaptation of another method from cluster analysis, namely the iterative EM estimation of the mixture model comprising the Π_i and the normal populations. This would have the advantage over McLach= lan's approach of providing consistent estimates of the population means and covariance matrices. It would appear however that this approach has not yet been evaluated in practice.

6.3 Missing values in discriminant analysis

There are several methods of handling incomplete data vectors in the training samples, ranging from ignoring them to estimating the missing values, by some means and then including them in the estimated classification rule. Chan, Gilman and Dunn (1976) as well as two earlier studies by Chan and Dunn (1972, 1974) use simulation experiments to compare a number of these methods under a variety of conditions, used in conjunction with the normal-based classification rule (2.18) with equal covariance matrices. As a result of these studies, they recommend the following method:

First estimate the mean vectors and common covariance matrix from the training samples, with all missing values replaced by their corresponding group means. Then, using regression equations based on the estimated mean vectors and covariance matrix, estimate the mis= sing values in each data vector using the observed variables as pre= dictors. Now use the completed training sample to estimate the dis= criminant functions.

A closely related rule is to estimate the regression equations using complete data vectors only, although this may be inefficient if a large proportion of the training samples have missing values.

A drawback of the above procedures is that they do not produce consistent estimators of the covariance matrix, and hence of the dis= criminant functions. Little (1978) proposes an adjustment to the es= timated covariance matrix, which may be applied iteratively, and which produces consistent estimators.

6.4 Testing for unsampled source populations

In addition to the technique described in section 2.1.3 for testing whether the observation X of unknown origin has not come from any of the sampled populations $\Pi_1,\ldots,\Pi_k$, a test can be performed using Hotelling's T^2 statistic when the data is normal. Suppose that X has been assigned to Π_i. Then a test of the null hypothesis that X has actually come from Π_i can be based on

$$T^2 = (X-X_{i.})^T S^{-1}(X-X_{i.})N_i/(1+N_i) \tag{6.1}$$

Under the null hypothesis $T^2(n-p+1)/pn$ has an F-distribution with p and n-p+1 degrees of freedom. For the case when the covariance matrices are unequal, S is replaced by S_i in (6.1), and then $T^2(n_i-p+1)/pn_i$ has an F-distribution with p and n_i-p+1 degrees of freedom under the null hypothesis.

In practice, even when one does not really expect to have un= knowns which do not come from any of the sampled populations, one would be wise to recognize it as a possible departure from model, and to test for it by routinely computing T^2 for each source population. If all such T^2 are significantly large, then it may be concluded that X does not come from any of the sampled populations.

This provides a strong motivation for carrying out discriminant analysis using the Mahalanobis distance of the unknown from each popu= lation rather than the mathematically equivalent linear or quadratic discriminant functions:- the former approach can easily test for un= sampled source populations while the latter cannot.

6.5 Constrained discriminant analysis

In situations where the consequences of misallocating an obser= vation from one population is much more serious than it is for an observation from another, such as might be the case in medical diag= nosis, the decision-theoretic approach is to specify different costs of misallocation, which leads to classification rules of the form (1.1). Alternatively, one might want to specify upper bounds on some or all of the probabilities of misclassification.

Anderson (1969) considers the problem of finding the optimum classification rule for the general k-group problem when there are upper bounds on the various probabilities of misclassification. This rule, which includes the possibility of leaving an observation un= classified, can be found analytically when k=2, but it requires an approximate, iterative method of solution to find the rule when $k > 2$.

The above method is based on the assumption that the parameters of each of the populations are known. For the two-group case with unknown normal distribution and equal covariance matrices, Anderson (1973a,b) derives a cutoff point C_1, based on the asymptotic distri= bution of the "Studentised" linear discriminant function

$$V^*_{12}(X) = (V_{12}(X) - \tfrac{1}{2} D^2_{12})/D_{12} \tag{6.2}$$

such that the rule obtained by assigning X to Π_1 when $V^*_{12}(X) > C_1$ has a specified value (to $0(n^{-2})$) for the expected probability of mis= classification. McLachlan (1977a) finds cutoff points C^*_1 and C^*_2 such that the classification rule:

$$\begin{array}{ll} \text{Assign X to } \Pi_1 & \text{if } V^*_{12}(X) > C^*_1 \\ \text{Assign X to } \Pi_1 & \text{if } V^*_{12}(X) < C^*_2 \\ \text{Do not classify X} & \text{if } C^*_2 \leq V^*_{12}(X) \leq C^*_1 \quad , \end{array} \tag{6.3}$$

has conditional probabilities of misclassification which, with cer= tain confidence, have specified upper bounds.

6.6 Equal mean discriminant analysis

In the normal, two-group situation with unequal covariance ma= trices but equal mean vectors, the quadratic discriminant function (2.3) simplifies to:

$$Q_{12}(X) = \tfrac{1}{2}\{\ell n(|\Sigma_2|/|\Sigma_1|) - \tfrac{1}{2}(X-\xi)^T(\Sigma_1^{-1}-\Sigma_2^{-1})(X-\xi)\} \tag{6.4}$$

where ξ is the common mean vector. When the parameters are unknown, they may be replaced by their training sample estimators.

A practical example where this model is applicable is in the discrimination between monozygotic and like-sexed dizygotic pairs of twins. The vector of differences between the measurements of various anatomical features on each pair will have zero mean for both mono= zygotic and dizygotic twins, but the variances will tend to be much smaller in the former case than in the latter. Thus the quadratic discriminant function (6.4), with common mean vector $\xi = 0$, will be appropriate.

Lachenbruch (1975b) proposes a classification rule based on the absolute value $|X|$, as an alternative to that based on the quadratic discriminant function (6.4), when $\xi = 0$, which he claims is more ro= bust against long-tailed contamination of the data.

7. TESTING FOR MODEL FIT

The question of the adequacy of the model assumptions is one that is strangely neglected by many users of discriminant analysis, despite its being fairly easy to assess.

The model assumptions in the usual predictive and estimative parametric discriminant analysis are that the data are independent and normally distributed; furthermore if the analysis is a homoscedas=

tic one, the model requires that any heteroscedasticity not be too severe. (As noted earlier, if there is a mild degree of heteroscedas= ticity, it is generally preferable to ignore it and use a homoscedas= tic procedure).

The first requirement, of mutual independence of the training sample observations, will generally be self-evidently true, and we will say no more about checking it.

The second requirement is normality of distribution of the ob= servations from each of the populations. In principle, one might test this in the same way as one tests for univariate normality, for ex= ample using a χ^2 test of the observed frequencies in the cells of a tabulation; or by a test based on the empiric distribution function of the data. However, tests of the former type are unworkable for multivariate data:- as the number of dimensions increases, so does the number of cells in the tabulation, and soon insuperable practi= cal problems of empty cells arise; in more than three or four dimen= sions, it is also difficult to find the expected frequencies of the cells unless the data have been multistandardized (e.g. Dempster, 1969) to make the components uncorrelated. Approaches based on the empiric distribution function are also less attractive for multi= variate data than for univariate:- they suffer from both computa= tional and distributional problems.

The standard methods of testing for multivariate normality therefore proceed in a slightly indirect way. Multivariate normality of data has many implications. For example:

(i) That the regression of any component on any other set of components is linear;

(ii) That the conditional distribution of any subvector of the data given any other subvector is homoscedastic;

(iii) That the quadratic form $(X-\xi)^T\Sigma^{-1}(X-\xi)$ follows a χ^2 dis= tribution.

Thus a test for multivariate normality may be set up by testing whether the data support one or several of these implications of mul= tivariate normality. If they do not, then normality is disproved; if they support several such implications, then normality may be regar= ded as an increasingly acceptable model for the data.

In this indirect process of testing, it is important to con= centrate on properties that matter for the purpose at hand. For ex= ample the iris data discussed later are recorded to one decimal only, and one of the measurements consists of low integer multiples of 0.1; normality, which implies continuity and the possibility of negative data values, thus cannot hold; nevertheless, for the purposes of discriminant analysis, the data are quite adequately approximated by a multivariate normal distribution. Similarly, binary variables, which are most emphatically not normally distributed can frequently be handled almost optimally by discriminant analysis based on normality assumptions. As is well-known, such variables have an asymptotic χ^2 distribution for their quadratic form:- thus they pass the quadratic form test, but are not normal, yet are likely despite this to be amenable to normal-based discriminant analysis.

Cox and Small (1978) discuss tests of multivariate normality based on properties (i) and (ii), and Krzanowski's location model provides a good example of a situation in which non-linearity or non-parallelism of regression is the crucial indicator of whether or not linear or quadratic discriminant analysis may be used. In general, however, parametric discriminant analysis is based on the differences between certain quadratic forms of the type (iii), and provided these quadratic forms have the needed χ^2 distribution to an adequate degree of approximation, the parametric discriminant analysis will generally prove acceptable. We shall thus concentrate on testing the quadratic forms for a fit to their normal theory distribution. Of course, if the data consist of a mixture of continuous and binary variables, then Krzanowski's "reversal" test should certainly be applied as well.

Provided the samples were sufficiently large, a powerful method of doing this would be to discretize the interval variables, append to each vector its source population, and subject the resulting vector to an analysis using the log-linear model. Reversals of a very generalized form could be detected in the form of significant third and higher order interactions involving the source population variable. Generally however so detailed a check will not be necessary.

If one is to concentrate on verifying that the quadratic forms do approximate a χ^2 distribution adequately, then a natural question that follows is what sort of departures from model have the most serious consequences; it is against such departures that the procedure should guard most carefully.

Here, there is general consensus that heavy-tailed and especially skew departures from model are most damaging. The damage is of two types:

(i) If the distribution is heavy-tailed, then the usual dis= criminant procedures, which make use of first and second moments, will be non-robust, and so excessively sensitive to a few extreme observations. If this is suspected to be the case, then a more robust procedure such as huberizing or outlier detection and elimination is indicated.

(ii) With a non-normal, and especially any skewed distribution for the X_i, the theoretically optimal classification rule may not be capable of a reasonable approximation by a li= near or quadratic discrimination rule (see for example Lachenbruch et al 1973) and so even a stable, robust me= thod of analysis must fail, since the classification rule being estimated is the wrong one.

We thus favour testing for model fit by looking for evidence of multivariate heavy-tailed behaviour.

One intuitively attractive pair of statistics for this test are the multivariate coefficients of skewness and kurtosis (Mardia 1974), defined as follows. Suppose for the moment that we have a single sample $X_1,\ldots,X_N$ believed to be from a $N(\xi,\Sigma)$ distribution. Define the sample mean and covariance matrix:-

$$X. = \frac{1}{N}\sum_{i=1}^{N} X_i$$

$$S = \frac{1}{N-1}\sum_{i=1}^{N} (X_i - X.)(X_i - X.)^T$$

Then the skewness and kurtosis are defined by:

$$b_1 = \frac{1}{N^2}\sum_{i=1}^{N}\sum_{j=1}^{N} \{(X_i - X.)^T S^{-1}(X_j - X.)\}^3 \qquad (7.1)$$

$$b_2 = \frac{1}{N}\sum_{i=1}^{N} \{(X_i - X.)^T S^{-1}(X_i - X.)\}^2. \qquad (7.2)$$

Approximate (asymptotic) distribution theory shows that, under a normal model

$$Nb_1/6 \sim \chi^2_\nu$$

with

$$\nu = p(p+1)(p+2)/6$$

and

$$b_2 \sim N\{p(p+2),\ 8p(p+2)/N\}$$

Note that in contrast with the scalar case, the skewness is not signed, and so one can only test whether skewness is present or absent; it cannot be further classified into "left" and "right" skewness.

The actual situation is slightly more complex than this, since one's calibration data also cover more than one population. Thus the model is

$$X_{ij} \sim N(\xi_i, \Sigma_i) \quad j=1,\ldots,N_i;\ i=1,\ldots,k.$$

Provided all N_i are sufficiently large then b_1 and b_2 may be computed separately for the data from each source, yielding, say, $b_{1i}, b_{2i}, i=1,\ldots,k.$

These statistics may easily be combined to give composite sta= tistics for skewness and kurtosis, for example:

$$\sum_{i=1}^{k} N_i\, b_{1i}/6 \sim \chi^2_{k\nu} \tag{7.3}$$

$$\sum_{i=1}^{k} N_i\, b_{2i} \sim N\{N\, p(p+2),\ 8Np(p+2)\}\quad N = \sum_{i=1}^{k} N_i \tag{7.4}$$

Alternatively for inferential purposes, the statistics could be combined using Fisher's method, which is more powerful but not as easy to apply. If one had prior grounds to believe that all Σ_i were equal, it would be convenient to use the pooled estimate of the common Σ in computing these skewness and kurtosis measures for each source. It is not clear, however, what effect this procedure would have on the distribution theory needed for testing the model fit, and so it is common practice to compute b_{1i} and b_{2i} in isolation for each source population.

The standard test for homoscedasticity is Bartlett's test. It is sensitive to the assumption of normality and gives excessive Type I errors with heavy-tailed data. This is not a serious drawback in the present context since heavy tails in the data also invalidate the parametric discriminant analysis. It is wise, however, to apply Bart= lett's test only after the check on the distribution.

Let S_i denote the sample covariance matrix of the i th popula= tion, and define

$$S = \sum_{i=1}^{k} (N_i-1)S_i/(N-k)$$

$$N = \sum_{i=1}^{k} N_i$$

Bartlett's statistic is then

$$G = -\sum_{i=1}^{k} \tfrac{1}{2}(N_i-1) \log \frac{|S_i|}{|S|} \qquad (7.5)$$

The null distribution of G can be approximated very closely in terms of χ^2 distributions (see for example Press 1972 p.177), so that this test, like the skewness and kurtosis ones, is easily ap= plied.

A great convenience of G is that it can be decomposed into terms which can be interpreted to tell us something about the nature of any heteroscedasticity found.

One obvious decomposition is that of the definition of G. The determinant $|S_i|$ is the generalized variance of the i th source, and so the term

$$-\tfrac{1}{2}(N_i-1) \log \frac{|S_i|}{|S|}$$

may be interpreted directly, as a standardized measure of how the i th source's generalized variance compares with the pooled generalized variance from all sources, large negative terms corresponding to highly variable sources, and large positive terms to more constant sources.

The other decomposition rests on the fact that the determinant $|S_i|$ may be decomposed into a product of partial variances:

$$|S_i| = s_{i11} . s_{i22.1}\, s_{i33.1,2} \cdots s_{ipp.1,2,\ldots,p-1}$$

as can $|S|$, and so G may be written as

$$G = \sum_{j=1}^{p} T_j \qquad (7.6)$$

where

$$T_j = -\sum_{i=1}^{k} \tfrac{1}{2}(N_i-1) \log\{\frac{s_{ijj.1,2,\ldots,j-1}}{s_{jj.1,2,\ldots,j-1}}\}$$

This decomposition focusses on the individual variables in the data vector; T_j is a measure of the additional heteroscedasticity due to the j th component of the data conditionally on the values of the first j-1 components. Under the normal model, the T_j are independent and asymptotically distributed as χ^2 with j(k-1) degrees of freedom.

Note that G is not affected if the variables in the vectors X are permuted; but the decomposition into T_j terms is; this fact can be used in a fashion rather like stepwise regression to separate the variables into a set that are homoscedastic, and another that are conditionally (but not necessarily marginally) heteroscedastic.

As an illustration of these tests for normality and homosce= dasticity, we shall join the long list of re-analysts of Fisher's (1936) iris data, conveniently tabulated in Kendall (1975). The data cover k=3 iris species with sample sizes N_i = 50 and p=4 measure= ments.

The skewness and kurtosis statistics are as follows. For test= ing purposes, we also define

$$C_i = N_i \, b_{1i}/6 \mathrel{\dot\sim} \chi^2_{24}$$

$$Z_i = \frac{b_{2i} - p(p+2)}{\{8p(p+2)/N_i\}^{\frac{1}{2}}} \mathrel{\dot\sim} N(0,1)$$

i	b_{1i}	C_i	b_{2i}	Z_i
1	2.90	24.2	25.49	0.76
2	2.84	23.7	21.97	1.04
3	2.97	24.7	23.3	-0.36

Mere inspection is enough to show that there is no evidence of either skewness or departure from normal kurtosis. The data are thus not heavy-tailed, and so amenable to normal-theory discriminant analysis.

Turning now to the question of scedasticity, we find G = 146.7, which is very highly significant (its asymptotic distribution would be a χ^2_{20}, and there is clearly no need to compute the more accurate approximation to the fractiles of G to discover that the value ob= served is hugely significant).

If G is reduced into the components for the three source popu= lations, we get

$$G = 152.33 + 44.87 + (-50.54)$$

showing a clear and marked decrease in generalized variance on passing from population 1 to 2 to 3.

In reducing G into components T_j, we ordered the variables by selecting at each stage that candidate variable making a minimum con= tribution to G. This gave:

Variable	T_j	j(k-1)
2	2.11	3
1	34.26	6
3	47.61	9
4	62.68	12

thus demonstrating that while variable 2 is marginally homoscedastic, the remaining variables are conditionally highly heteroscedastic.

To try to understand the data better, let us look at the means and standard deviations of each variable in each population.

		Variable			
Source		1	2	3	4
1	Mean	5.01	3.43	1.46	0.25
	s.d.	0.35	0.38	0.17	0.11
2	Mean	5.93	2.77	4.26	1.33
	s.d.	0.51	0.31	0.47	0.20
3	Mean	6.59	2.97	5.52	2.03
	s.d.	0.64	0.32	0.55	0.27

Going down the columns, it looks as though each standard de= viation is roughly proportional to the square root of the mean.This suggests trying a square root transformation of the data. Making this transformation and repeating the analysis, we find that norma= lity remains a fully acceptable hypothesis, while G is reduced to 78.44.

This latter value of G, while still highly significant, is considerably lower than that for the original data, and one might be tempted to analyse the square root transformed data using a homo= scedastic normal model. The reduction of G into its components by source yields:

$$G = 30.91 + 48.61 + (-1.08)$$

showing that sources 1 and 2 have a lower generalized variance than the pooled matrix, while source 3 is about equal.

The reduction by variables gives:

Variable	T_j	j(k-1)
2	0.54	3
4	14.49	6
1	28.00	9
3	35.42	12

showing that there is still conditional heteroscedasticity of variables 4, 1 and 3. A quick check of the marginal standard deviations of these variables shows that this is not due to marginal heteroscedasticity, and so must be tied up with differences in correlations.

A completely different approach to testing for model fit is given by Hawkins (1981). This provides a simultaneous check on normality and homoscedasticity, and is especially good for outlier detection.

Define $$V_{ij} = (X_{ij}-X_{i.})^T S^{-1}(X_{ij}-X_{i.}) \quad (7.7)$$

where $X_{i.}$ is the mean vector of the i th source, and S the pooled covariance matrix.

Let $$F_{ij} = \frac{(N-k-p)n_i \, V_{ij}}{p\{(n_i-1)(N-k)-n_i \, V_{ij}\}} \quad (7.8)$$

and let A_{ij} be the tail area to the right of F_{ij} under an F-distribution with p and N-k-p degrees of freedom.

If the data are normal and homoscedastic, then each A_{ij} follows an exact uniform distribution, and by checking the A_{ij} for uniformity, one can infer whether the model holds well enough. Departures from model tend to have the following effects:

(i) Heavy tails in the data lead to a U shaped distribution for the A_{ij} with an excess of values near 0 and near 1.

(ii) Heterscedasticity generally causes one population to have an excess of A_{ij} near 1, and another an excess near 0.

(iii) As a particular case of (i), if an A_{ij} is excessively close to 0, then X_{ij} is an outlier; A_{ij} is equivalent to the optimal test statistic for a single outlier (Hawkins 1980 p.107).

Much may be learned from a detailed study, including plotting of the distribution function, of the A_{ij}. As a quick summary statistic however, one can merely compute the Anderson-Darling statistic for the A_{ij} from each source and for the totality of A_{ij}, and the first two components of these Anderson-Darling statistics.

If the normal homoscedastic model is satisfied, then the components are approximately N(0,1). The first component measures the tendency of the A_{ij} to cluster near either 0 or 1, while the second measures the tendency of the A_{ij} to have a U shaped distribution.

This approach is less powerful against homoscedasticity than the skewness/kurtosis/Bartlett procedure outlined above. It does have two advantages however.

(i) It uses only the V_{ij}, which are normally computed anyway as part of the discriminant analysis. It thus requires no additional complicated coding or computa= tion.

(ii) It can be applied even when some (or all) N_i do not exceed p.

Applying this test procedure to the Fisher iris data, we get

Source	Anderson-Darling	Component 1	Component 2
1	12.45	-4.62	-2.87
2	0.63	-0.97	0.05
3	6.04	2.89	-2.23
Totality	3.12	-1.56	-2.92

The 0.1% fractile of the Anderson-Darling statistic is 3.86. Thus sources 1 and 3 show a very highly significant departure from model. Looking at the components this is clearly due in large part to component 1, on which source 1 has an excess of A_{ij} near 1, and source 3 an excess near 0. The significant U shaped behaviour suggested by component 2 should be deferred until the homoscedasti= city problem has been resolved. Making the square root transforma= tion produces

Source	Anderson-Darling	Component 1	Component 2
1	0.52	0.37	-1.03
2	0.92	-1.23	-0.36
3	1.94	-1.10	-2.17
Totality	1.67	-1.13	-2.06

This shows that the normal homoscedastic model now fits much better, apart from a mild U shape tendency in source 3 and in the totality of A_{ij}. On closer inspection, we find that source 3 includes one A_{ij} value of 0.0012 (indicating at the 6% level that the corres= ponding observation is an outlier), and another of 0.0023. Thus there is an indication that the residual departure shown up by the Bartlett statistic may be due to a few outliers in the data.

REFERENCES

AHMED, S.W. and LACHENBRUCH, P.A. (1977). Discriminant analysis when scale contamination is present in the initial sample. In Van Ryzin, J. (Ed.) Classification and clustering. Proceed. Advanced Sem 1976. Academic Press, Inc.

AITCHISON, J. (1975). Goodness of prediction fit. *Biometrika*, 62, 547-554.

AITCHISON, J. and AITKEN, C.G.G. (1976); Multivariate binary dis= crimination by the kernel method. *Biometrika*, 63, 413-420.

AITCHISON, J., HABBEMA, J.D.F. and KAY, J.W. (1977). A critical comparison of two methods of statistical discrimination. *Applied Statist.*, 26, 15-25.

ANDERSON, J.A. (1969). Constrained discrimination between k popu= lations. *J.Roy. Statist. Soc. B.*, 31, 123-139.

ANDERSON, J.A. (1972). Separate sample logistic discrimination. *Biometrika*, 59, 19-36.

ANDERSON, J.A. (1974). Diagnosis by logistic discriminant func= tion: further practical problems and results. *Applied Sta= tist.*, 23, 397-404.

ANDERSON, J.A. (1975). Quadratic logistic discrimination. *Biome= trika*, 62, 149-154.

ANDERSON, T.W. (1958). An introduction to multivariate statisti= cal analysis. Wiley, New York.

ANDERSON, T.W. (1973a). An asymptotic expansion of the distribu= tion of the studentized classification statistic W'. *Ann. Statist.*, 1, 964-972.

ANDERSON, T.W. (1973b). Asymptotic evaluation of the probabilities of misclassification by linear discriminant functions. Dis= criminant analysis and applications, T. Cacoullos, ed., Academic Press, New York, 17-35.

BOX, G.E.P. (1949). A general distribution theory for a class of likelihood criteria. *Biometrika*, 36, 317-346.

BROFFITT, B., CLARKE, W.R. and LACHENBRUCH, P.A. (1980). The Effect of Huberizing and Trimming on the Quadratic Discriminant Function. *Commun. Statist.*, A9(1), 13-25.

BROFFITT, J.D., RANDLES, R.H. and HOGG, R.V. (1976). Distribution-free partial discriminant analysis. *J. Amer. Statist. Assoc.*, 71, 934-939.

CAMPBELL, N.A. (1980). Shrunken estimators in discriminant and ca= nonical variate analysis. *Applied Statist.*, 29, 1, 5-14.

CHAN, L.S. and DUNN, O.J. (1972). The treatment of missing values in discriminant analysis - 1. The sampling experiment. *J. Amer. Statist. Assoc.*, 67, 473-477.

CHAN, L.S. and DUNN, O.J. (1974). A note on the asymptotic aspect of the treatment of missing values in discriminant analysis. *J. Amer. Statist. Assoc.*, 69, 672-673.

CHAN, L.S., GILMAN, J.A. and DUNN, O.J. (1976). Alternative ap= proaches to missing values in discriminant analysis. *J.Amer. Statist. Assoc.*, 71, 842-844.

CHANDA, K.C. (1980). Asymptotic properties of classification rules based on Wilcoxon-type statistics. *J. Amer. Statist. Assoc.*, 75, 726-8.

CHANG,P.C.and AFIFI, A.A. (1974): Classification based on dichoto= mous and continuous variables. *Journal of the American Sta= tistical Association*, 69, 336-339.

CLARKE, W.R., LACHENBRUCH, P.A. and BROFFITT, B. (1979). How non-normality affects the Quadratic Discriminant Function. *Commun. Statist.* A8(13), 1285-1301.

CONOVER, W.J. and IMAN, R.L. (1980). The rank transformation as a method of discrimination with some examples. *Commun. Statist. Theor.Meth.*, A9(5), 465-487.

COSTANZA, M.C. and AFIFI, A.A. (1979). Comparison of stopping rules in forward stepwise discriminant analysis. *J.Amer. Statist. Assoc.*, 74, 777-785.

COX, D.R. (1966). Some procedures associated with the logistic re= sponse curve. Research Papers in Statistics, Festschrift for J. Neyman, F.N. David Ed., Wiley, London, 55-71.

COX, D.R. and SMALL, N.J.H. (1978). Testing multivariate normality. *Biometrika*, 65, 263-272.

CROWTHER, N.A.S. (1980) Approxmate Distributions for the Box Class of Likelihood Criteria. *S.Af. Statist. J.*, 14, 1, 61-71.

DAY, N.E. (1969). Estimating the components of a mixture of normal distributions. *Biometrika*, 56, 463-474.

DAY, N.E. and KERRIDGE, D.F. (1967). A general maximum likelihood discriminant. *Biometrics.*, 23, 313-323.

DEMPSTER, A.P. (1969). Elements of Continuous Multivariate Analysis. Addison-Wesley, Reading.

DE ROUEN, T.A. and SARMA, Y.R. (1975) G_1-minimax procedures for the case of prior distributions in discriminant analysis. *Biometrika*, 62, 403-406.

DE VILLIERS, H. (1976). A second adult human mandible from Border cave, Inquraruma District, Kwazulu, South Africa. *S. Afr. J. Sci.*, 72, 212-215.

DiPILLO, P.J. (1976). The application of bias to discriminant ana= lysis. *Commun. Statist. Theor.Meth.*, A5(9), 843-854.

DiPILLO, P.J. (1977). Further applications of bias to discriminant analysis. *Commun. Statist-Theor. Meth.*, A6(10), 933-943.

DiPILLO, P.J. (1979). Biased discriminant analysis: Evaluation of the optimum probability of misclassification. *Commun. Sta= tist.-Theor. Meth.*, A8(14), 1447-1457.

DUNN, O.J. (1971). Some expected values for probabilities of cor= rect classification in discriminant analysis. *Technometrics*, 13, 345-353.

DUNN, O.J. and VARADY, P.D. (1966). Probabilities of correct clas= sification in discriminant analysis. *Biometrics*, 22, 908-924.

EFRON, B. (1975). The efficiency of logistic-regression compared to normal discriminant analysis. *J. Amer. Statist. Assoc.*, 70, 892-898.

EFRON, B. (1979). Bootstrap methods: another look at the jackknife. *Ann. Statist.*, 7, 1-26.

FATTI, L.P. (1979). The Random Effects Model in Discriminant Analy= sis. Unpubl. Ph.D. Thesis, University of the Witwatersrand, South Africa.

FISHER, R.A. (1936). The use of multiple measurement in taxonomic problems. *Ann. Eugen.*, 7, 179-188.

FURNIVAL, G.M. and WILSON, R.W. (1974). Regression by leaps and bounds. *Technometrics*, 16, 499-511.

GANESALINGHAM, S. and McLACHLAN, G.J. (1978). The efficiency of a linear discriminant function based on unclassified initial samples. *Biometrika*, 65, 658-662.

GEISSER, S. (1964). Posterior odds for multivariate normal classi= fications. *J. R. Statist. Soc.*, B, 26, 69-76.

GEISSER, S. (1966). Predictive discrimination. In Krishnaiah, P.R. (Ed.) Multivariate Analysis. Proc. Int. Sym. Dayton, Ohio, Academic Press, New York.

GESSAMAN, M.P. and GESSAMAN, P.H. (1972). A comparison of some Mul= tivariate Discrimination Procedures. *J. Amer. Statist. Assoc.*, 67, 468-477.

GILBERT, E.S. (1968). On Discrimination using Qualitative Variables. *J. Amer. Statist. Assoc.*, 63, 1399-1412.

GOLDSTEIN, M. and RABINOWITZ, M. (1975). Selection of variates for the two-group multinomial classification problem. *J. Amer. Statist. Assoc.*, 70, 776-781.

GOMPERTS, E.D., FATTI, L.P., VAN DER WALT, J.D., FEESEY, M., HARTMAN, E. and SNELL, R.J. (1976). Factor VIII and factor VIII related antigen in normal South African blacks and a black carrier group. *Thrombosis Research* 9, 293-299.

HABBEMA, J.D.F. and HERMANS, J. (1977). Selection of variables in discriminant analysis by F-statistic and error rate. *Techno= metrics*, 19, 487-493.

HABBEMA, J.D.F., HERMANS,J.and VAN DEN BROEK, K. (1974): A stepwise discriminant analysis program using density estimation. *Compstat 1974, Proceedings in Computational Statistics*, Wien, Physica Verlag, 101-110.

HAWKINS, D.M. (1976). The subset problem in multivariate analysis of variance. *J. Roy. Statist. Soc.*, B, 38, 132-139.

HAWKINS, D.M. (1980). Identification of Outliers. Chapman and Hall, London.

HAWKINS, D.M. (1981). A new test for multivariate normality and homoscedasticity. *Technometrics*, 23, 105-110.

HAWKINS, D.M. and RASMUSSEN, S.E. (1973). Use of discriminant ana= lysis for classification of strata in sedimentary successions. *J. Math. Geology*, 5, 163-177.

HILLS, M. (1966). Allocation rules and their error rates. *J. Roy. Statist. Soc.*, B, 28, 1-31.

HORA, S.C. (1980). Sequential discrimination. *Commun. Statist.- Theor. Meth.*, A9(9), 905-916.

JOHNSON, N.L. and KOTZ, S. (1969). Distributions in statistics: discrete distributions. Houghton Miffin, Boston.

KENDALL, M.G. (1975). Multivariate Analysis. Griffin, London.

KENDALL, M.G. and STUART, A. (1966). The advanced theory of statis= tics, volume 3, Griffin, London.

KOFFLER, S.L. and PENFIELD, D.A. (1979). Nonparametric Discrimina= tion Procedures for nonnormal distributions. *J. Statist. Comp. Simul.*, 8, 281-299.

KRZANOWSKI, W.J. (1975). Discrimination and classification using both binary and continuous variables. *J.Amer. Statist. Assoc.*, 70, 782-790.

KRZANOWSKI, W.J. (1976). Canonical representation of the location model for discrimination or classification. *J. Amer. Statist. Assoc.*, 71, 845-848.

KRZANOWSKI, W.J. (1977). The performance of Fisher's linear discri= minant function under non-optimal conditions. *Technometrics*, 19, 191-200.

KRZANOWSKI, W.J. (1979). Some linear transformation for mixtures of binary and continuous variables, with particular reference to linear discriminant analysis. *Biometrika*, 66, 33-39.

KSHIRSAGAR, A.M. (1972). Multivariate analysis. Marcel Dekker, Inc., New York.

KULLBACK, S. and LEIBLER, R.A. (1951). On information and sufficiency. *Ann. Math. Statistics*, 22, 79-86.

LACHENBRUCH, P.A. (1967). An almost unbiased method of obtaining confidence intervals for the probability of misclassification in discriminant analysis. *Biometrics*, 23, 639-645.

LACHENBRUCH, P.A. (1975a). Discriminant Analysis. Hafner Press, New York.

LACHENBRUCH, P.A. (1975b). Zero-mean difference discrimination and the absolute linear discriminant function. *Biometrika*, 62, 397-401.

LACHENBRUCH, P.A. and MICKEY, M.R. (1968). Estimation of error rates in discriminant analysis. *Technometrics*, 10, 1-11.

LACHENBRUCH, P.A., SNEERINGER, C. and REVO, L.T. (1973). Robustness of the linear and quadratic discriminant functions to certain types of non-normality. *Comm. Statist.*, 1, 39-56.

LITTLE, R.J.A. (1978). Consistent regression methods for discrimi= nant analysis with incomplete data. *J. Amer. Statist. Assoc.*, 73, 319-322.

MALLOWS, C.L. (1953). Sequential discrimination. *Sankhyā*, 12, 321.

MARDIA, K.V. (1974). Applications of some measures of multivariate skewness and kurtosis in testing normality and robustness studies. *Sankhya*, B, 36, 115-128.

MARKS, S. and DUNN, O.J. (1974). Discriminant Functions when Cova= riance Matrices are unequal. *J. Amer. Statist. Assoc.*, 69 (346), 555-559.

McCABE, G.P. (1975). Computations for variable selection in discrimi= nant analysis. *Technometrics*, 17, 103-109.

McKAY, R.J. (1976). Simultaneous procedures in discriminant analy= sis involving two groups. *Technometrics*, 18, 47-53.

McKAY, R.J. (1977). Simultaneous procedures for variable selection in multiple discriminant analysis. *Biometrika*, 64, 283-290.

McLACHLAN, G.J. (1974a). An asymptotic unbiased technique for esti= mating the error rates in discriminant analysis. *Biometrics*, 30, 239-249.

McLACHLAN, G.J. (1974b). The asymptotic distributions of the condi= tional error rate and risk in discriminant analysis. *Biometrika*, 61, 131-135.

McLACHLAN, G.J. (1974c). Estimation of the errors of misclassifica= tion on the criterion of asymptotic mean square error. *Technometrics*, 16, 255-260.

McLACHLAN, G.J. (1975a). Confidence intervals for the conditional probability of misallocation in discriminant analysis. *Biometrics*, 31, 161-167.

McLACHLAN, G.J. (1975b). Iterative reclassification procedure for constructing an asymptotically optimal rule of allocation in discriminant analysis. *J. Amer. Statist. Assoc.*, 70, 365-369.

McLACHLAN, G.J. (1976). A criterion for selecting variables for the linear discriminant function. *Biometrics*, 32, 529-534.

McLACHLAN, G.J. (1977a). Constrained sample discrimination with the studentized classification statistic W. *Commun. Statist.*, A6, 575-583.

McLACHLAN, G.J. (1977b). Estimating the linear discriminant func= tion from initial samples containing a small number of unclas= sified observations. *J. Amer. Statist. Assoc.*, 72, 403-406.

McLACHLAN, G.J. (1979). A comparison of the estimative and predic= tive methods of estimating posterior probabilities. *Commun. Statist.* A 8(9), 919-929.

MOORE, D.H. (1973). Evaluation of five discrimination procedures for binary variables. *J. Amer. Statist. Assoc.*, 68, 399-404.

MORAN, M.A. and MURPHY, B.J. (1979). A closer look at two alterna= tive methods of statistical discrimination. *Appl. Statist.*, 28, no.3, 223-232.

OKAMOTO, M. (1963). An asymptotic expansion for the distribution of the linear discriminant function. *Ann. Math. Statist.*, 34, 1286-1301.

O'NEILL, T.J. (1978). Normal discrimination with unclassified ob= servations. *J. Amer. Statist. Assoc.*, 73, 821-826.

O'NEILL, T.J. (1980). The general distribution of the error rate of a classification procedure with application to logistic re= gression discrimination. *J. Amer. Statist. Assoc.*, 75, 154-160.

PLACKETT, R.L. (1974). The Analysis of categorical data. Griffin, London.

PRESS, S.J. (1972). Applied multivariate analysis. Holt, Rinehart and Winston, Inc., New York.

PRESS, S.J. and WILSON, S. (1978). Choosing between logistic regres= sion and discriminant analysis. *J. Amer. Statist. Assoc.*, 73, 699-705.

RAO, C.R. (1965). Linear Statistical Inference and its Applications. Wiley, New York.

RAO, C.R. (1970). Inference on discriminant function coefficients. Essays in Probability and Statistics, Bose, R.C. et al., Eds., University of North Carolina.

RANDLES, R.H., BROFFITT, J.D., RAMBERG, J.S. and HOGG, R.V. (1978). Discriminant analysis based on ranks. *J. Amer. Statist. Assoc.*, 73, 379-384.

RICHARDS, L.E. (1972). Refinement and extension of distribution-free discriminant analysis. *Appl. Statist.*, 21, 174-176.

VAN DER MERWE, C.A. and KOTZÉ, T.J.v.W. (1980). Bekendstelling van 'n verdelingsvrye klassifikasieprosedure gebaseer op die skat= ting van waarskynlikheidsverdelings deur gebruik te maak van die sleutelfunksie metode (kernel function method). Tech. Re= port, no.3, Institute for Biostatistics, P.O.Box 70, Tyger= berg, South Africa.

VAN NESS, J.W. and SIMPSON, C. (1976). On the effects of dimension in discriminant analysis. *Technometrics*, 18, 175-187.

COVARIANCE STRUCTURES

MICHAEL W. BROWNE, *UNIVERSITY OF SOUTH AFRICA*

1. INTRODUCTION

Structural models for covariance matrices are used when studying relationships between variables and are employed predominantly in the social sciences. The best known of these is the factor analysis model but recently there has been rapid development of extensions and alter= natives. Most of this work has appeared in psychometric journals. Text= books on applied multivariate analysis are recently including chapters on factor analysis. They tend, however, to be out of date and give little attention to modern developments such as efficient computational methods in maximum likelihood factor analysis, effective methods for oblique rotation, methods for obtaining standard errors of factor load= ings and extensions of the factor analysis model.

This paper will present a personal view of structural models for covariance matrices arising from continuous variates. The lack of ro= bustness of statistical tests involving the assumption of multivariate normality will receive some attention and alternative approaches will be examined . Modern developments will be considered but no attempt will be made to provide an exhaustive review. Readers who require addi= tional information are referred to the review article by Bentler (1980) and to the bibliographies in Harman (1976) and Mulaik (1972).

The paper is divided into two main parts. Part I is concerned with general technical background common to all structural models for covariance matrices. The manner in which covariance structures arise will be examined and general procedures for estimating parameters and comparing the adequacy of alternative models will be dealt with. Part II is concerned with some specific covariance structures.

PART I

GENERAL TECHNICAL BACKGROUND

1.1 Genesis of structural models for covariance matrices

Let $X : N \times p$ represent a data matrix whose rows are independently and identically distributed with mean ξ and covariance matrix Σ. We shall assume throughout that N is substantially greater than p. The usual unbiased estimators of ξ and Σ are

$$\bar{x} = N^{-1} X^T 1 \tag{1.1.1}$$

and

$$S = (N-1)^{-1}(X^T - \bar{x}\, 1^T)(X - 1\,\bar{x}^T) \tag{1.1.2}$$

$$= (N-1)^{-1} X^T(I - 1\, N^{-1}\, 1^T)X \tag{1.1.3}$$

where 1 is a column vector with unit elements and I is the identity matrix. In many applications data matrices of this type arise from repeated measurements made on persons. It will be convenient, therefore, to regard a row of X as representing a series of measurements on a person, and a column of X as the values a variable assumes over different persons.

In general, a covariance structure is a model where the elements of the covariance matrix $\Sigma : p \times p$ are regarded as functions of q parameters, γ_i, $i=1,\ldots,q$, say, where $q \le \frac{1}{2}p(p+1)$, the number of nonduplicated elements of the symmetric matrix Σ. Thus Σ is a matrix valued function of a $q \times 1$ vector variable γ contained in some specified domain G.

It is of interest to ask what considerations lead to the development of structural models for covariance matrices in practice. There are two main types of model which need not be mutually exclusive. The first arises when knowledge of the measurements made leads to hypotheses either of equality of certain variances, covariances or correlations or of specified values, generally zero, for these. Thus the covariance matrix Σ, or the correlation matrix P, has a specified pattern. A simple example of a hypothesis giving rise to a patterned covariance matrix is the sphericity hypothesis where all variances are equal and all covariances or correlations are zero:

$$\Sigma = \psi I\,. \tag{1.1.4}$$

Here $q = 1$ and the parameter vector γ consists of the single parameter ψ contained in the domain

$$G = \{\psi;\ 0 < \psi < \infty\}\quad .$$

The second type of covariance structure arises when the observed variates are regarded as functions of certain hypothetical unobservable variates, commonly known as "latent variates". If the vector x^T represents a typical row of X, we have

$$x = f(y) \tag{1.1.5}$$

where $f(\cdot)$ is a p x 1 vector valued function with the vector variate y as argument. Certain assumptions concerning the distribution of the latent vector variate y then lead to some structural form

$$\Sigma = \Sigma(\gamma) \tag{1.1.6}$$

for the covariance matrix of the observable vector variate, x. Here we shall be concerned only with the case where $f(\cdot)$ is a continuous function and x is continuous. There has, however, been considerable progress with latent variate models for discrete observable variates (cf. e.g. Muthén, 1978; Bartholomew, 1980).

The best known latent variate model is the factor analysis model. Consider a set of p linear minimum mean square error regression equations of the observable variates in x on a set z of m unobservable variates or "common factors" where

$$\mathcal{E}(z) = 0 : m \times 1 \tag{1.1.7}$$

$$\mathrm{Cov}(z,z^T) = \Phi : m \times m \tag{1.1.8}$$

and m is specified a priori.

We then have

$$x = \xi + \Lambda z + u \tag{1.1.9}$$

where Λ: p x m represents the regression weights of the observable variates in x on the common factors z and the p elements of u are residual variates or unique factors. Since (1.1.8) represents p linear minimum mean square error regression equations it follows that

$$\mathcal{E}(u) = 0 : p \times 1 \quad , \tag{1.1.10}$$

$$\mathrm{Cov}(z,u^T) = 0 : m \times p \quad . \tag{1.1.11}$$

Now let

$$\mathrm{Cov}(u,u^T) = \Psi : p \times p \tag{1.1.12}$$

It then follows from (1.1.9) and (1.1.11) that Σ is of the form

$$\Sigma = \mathrm{Cov}(x,x^T) = \Lambda\Phi\Lambda^T + \Psi \tag{1.1.13}$$

At this point (1.1.13) is tautological and, given any positive definite Σ and any positive integer m it is always possible to choose a Λ, Φ and

Ψ such that (1.1.13) is satisfied. It becomes a model if the assump= tion is made that Ψ is diagonal,

$$\Psi = D_\psi . \qquad (1.1.14)$$

This implies that the residual variates are uncorrelated so that all linear dependences between the observed variates are accounted for by the factors. Then Σ satisfies the structural model

$$\Sigma = \Lambda\Phi\Lambda^T + D_\psi \qquad (1.1.15)$$

which is falsifiable under certain conditions. Thus (1.1.9) corresponds to (1.1.5) and (1.1.15) corresponds to (1.1.6) with the parameter vec= tor γ consisting of elements of Λ, Φ and D_ψ .

It is important to remember that the model involving latent va= riates, (1.1.5) or (1.1.9), is a sufficient condition for the covariance structure (1.1.6) or (1.1.15) but is not a necessary condition. Even if (1.1.6) or (1.1.15) holds one cannot infer that a process correspon= ding to (1.1.5) or (1.1.9) generated the data (see Section 1.3).

In some cases a latent variable model can give rise to a simple pattern for the covariance matrix so that the two types of covariance structure we have described are not mutually exclusive. Suppose that there is a single factor, all regression weights are equal and all re= sidual variances are equal in (1.1.9). Then we can express Σ in the form

$$\Sigma = 1\, \varphi\, 1^T + \psi I \qquad (1.1.16)$$

so that all diagonal elements of Σ are equal and all non-diagonal ele= ments of Σ are equal. Thus Σ simultaneously is the consequence of a la= tent variate model and has a pattern structure.

1.2 Invariance of covariance structures under changes of scale

Frequently the scale in which a quantity is measured is comple= tely arbitrary. It is therefore of interest to consider whether or not a model is affected by linear transformations of the data. Replacement of x by $Ax + \beta$, where A is any nonsingular p x p matrix and β is any p x 1 vector, involves replacement of Σ by $A \Sigma A^T$. Since the covariance matrix is unaffected by β, all covariance structures are invariant un= der changes of location. We shall say that a *covariance structure* $\Sigma = \Sigma(\gamma)$ *is invariant under a certain class of linear transformations* if, given any γ contained in the parameter space G and any nonsingular p x p matrix A belonging to that class, there exists a γ^* contained in

G such that

$$\Sigma(\gamma^*) = A\ \Sigma(\gamma)A^T \tag{1.2.1}$$

In practical applications most linear transformations involve only change of scale. We shall say that a covariance structure is *in= variant under changes of scale* if it is invariant under linear trans= formations with A belonging to the class of all p x p diagonal matri= ces

$$A = D_\alpha . \tag{1.2.2}$$

For example a restricted factor analysis covariance structure (1.1.15) is invariant under changes of scale if no elements of Λ or diagonal elements of D_ψ have elements restricted to have specified nonzero values or to be equal to other parameters. A covariance structure of this type is not invariant under changes of scale if elements of Λ in the same column are specified to be equal, diagonal elements of D_ψ are specified to be equal or elements of Λ or D_ψ are specified to have nonzero values. In particular the covariance structure in (1.1.16) is not invariant under changes of scale.

In a great many cases, such as in factor analysis, it is the structure of the correlation matrix rather than that of the covariance matrix which is of primary interest. Such structures are, however, treated as covariance structures because of the fact that the distri= bution of the sample covariance matrix is simpler than that of the sam= ple correlation matrix. This can be done when the rescaling of variables to have unit variances does not affect the covariance structure. Clearly, any covariance structure that is invariant under changes of scale implies a similar structure for the correlation matrix P. If $\Sigma(\gamma)$ is invariant under changes of scale then there exists a γ^* such that

$$P = \Sigma(\gamma^*) .$$

The converse is not necessarily true however. Consider the di= rect product structure (Swain, 1975b)

$$\Sigma = \Sigma_m \otimes \Sigma_t \tag{1.2.3}$$

where Σ_m is of order p_m x p_m, Σ_t is of order p_t x p_t and $p = p_m p_t$. Then the correlation matrix has the same structural form

$$P = P_m \otimes P_t$$

but the model (1.2.3) is not invariant under change of scale. The model is no longer satisfied if Σ is replaced by $D_\alpha \Sigma\ D_\alpha$ and D_α is *not* of the form

$$D_\alpha = D_{\alpha_m} \otimes D_{\alpha_t} \quad .$$

Some models for correlation matrices,

$$P = P(\gamma_\rho) \quad , \tag{1.2.4}$$

such as models for patterned correlation matrices (Section II.3) need modification before being treated as covariance structures as they are not invariant under changes of scale. Such models may be written as covariance structures of the form

$$\Sigma(\gamma) = D_\zeta \ P(\gamma_\rho) D_\zeta \tag{1.2.5}$$

where the diagonal elements of the diagonal matrix D_ζ are standard deviations and γ consists of the elements of γ_ρ and the diagonal elements of D_ζ. The covariance structure $\Sigma(\gamma)$ then is invariant under changes of scale.

A class of structures which are invariant under a particular type of change of scale will be of interest subsequently because it results in some simplification in the form of a likelihood ratio rest statistic and this in turn leads to other simplifications. This is the class of structures which are *invariant under a constant scaling factor* where A is of the form

$$A = \alpha I. \tag{1.2.6}$$

If $\Sigma(\gamma)$ is a member of this class, then given any $\gamma \in G$ and any positive scalar α^2 there exists a $\gamma^* \in G$ such that

$$\Sigma(\gamma^*) = \alpha^2 \Sigma(\gamma) \tag{1.2.7}$$

Clearly any structure that is invariant under changes of scale is also invariant under a constant scaling factor but the converse is not true. For example a factor analysis model with elements in some columns of Λ specified to be equal or diagonal elements of D_ψ specified to be equal will be invariant under a constant scaling factor but will not be invariant under changes of scale.

1.3 Estimability of parameters

Given a sample covariance matrix S, we wish to decide whether the model

$$\Sigma = \Sigma(\gamma) \tag{1.3.1}$$

is a reasonable one for the population covariance matrix, and, if so, obtain an estimate of the parameter vector γ. Before this can be done it is essential to investigate whether or not γ can in fact be estima= ted. In common statistical terminology a parameter is said to be esti= mable only if it has a linear unbiased estimator. Here we shall depart from this restrictive terminology and refer to a parameter vector, γ, as estimable if it has a consistent estimate. This is motivated by the opinion that consistency is the minimal property a statistic should have to be regarded as an estimator of a particular parameter.

A parameter vector, γ, can have a consistent estimate, and there= fore in our terminology be estimable, only if it is identified in the following sense. A parameter γ will be said to be identified in a para= meter space G if

$$\Sigma(\gamma) = \Sigma(\ddot{\gamma}) \quad ,$$

with $\ddot{\gamma} \in G$, implies that $\ddot{\gamma} = \gamma$. (Strictly this definition assumes that G is closed and bounded but this point will not be considered further as it is of little practical consequence.)

For example consider the unrestricted factor analysis model, with $\Phi = I$,

$$\Sigma = \Lambda\Lambda^T + D_\psi \tag{1.3.2}$$

where no assumptions are made about the elements of Λ and the diagonal elements of D_ψ are assumed to be nonnegative, ($G = \{\Lambda, D_\psi : \Lambda{:}p \times m \text{ real}, D_\psi : p \times p \text{ diagonal and nonnegative definite}\}$). Then Λ is not identified, since if

$$\ddot{\Lambda} = \Lambda\ \Theta \tag{1.3.3}$$

and $\Theta : m \times m$ is orthogonal then

$$\Sigma = \ddot{\Lambda}\ \ddot{\Lambda}^T + D_\psi \tag{1.3.4}$$

Consequently Λ is not estimable. What is done in such a situation is to restrict the parameter space by imposing additional restrictions and obtain an estimable function $\ddot{\Lambda}$ of Λ. Let $\ddot{\Lambda}$ be defined by (1.3.3) where Θ is chosen such that

$$\ddot{\Lambda}^T\ D_\psi^{-1}\ \ddot{\Lambda} \text{ is diagonal} \tag{1.3.5}$$

and all column sums of $\ddot{\Lambda}$ are positive. Then $\ddot{\Lambda}$ is uniquely determined and $\ddot{\Lambda}$ and D_ψ will in general be estimable (although there are other mild conditions to be met (Anderson & Rubin, 1956, Section 5)).

We note in passing that the identification conditions in (1.3.5) are convenient for mathematical purposes since they can be shown to determine $\ddot{\Lambda}$ uniquely but they do not usually lead to interpretable re= sults. What is usually done in practice is to define Θ as that ortho= gonal matrix which maximises a scalar valued function $\varphi(\ddot{\Lambda})$ of $\ddot{\Lambda}$ such as the Varimax function (Kaiser, 1958). While the results are conve= nient for the purpose of interpretation it is difficult to establish whether $\varphi(\ddot{\Lambda})$ will have a unique maximum and therefore identify $\ddot{\Lambda}$. This is implicitly assumed or hoped to be the case.

When a covariance structure is a consequence of a latent variate model of the form of (1.1.5) it is important to bear in mind that the latent vector variable, y, is still a random vector even after an ob= servation on x has been made (cf. Bartholomew, 1981). The posterior distribution of y after x has been observed is usually not concentra= ted at a single point. Consequently an infinity of possible values of the latent variable y can give rise to the same value of the observed vari= able x. In the factor analysis model this has given rise to much acri= monious discussion (see (Steiger, 1979) for historical background) with much confusion caused by the treatment of the random variable y as if it were a mathematical variable. A clear perspective of the situation is given in Bartholomew (1981).

1.4 Covariance structures as regression models

Since S is an unbiased estimator of Σ we have

$$S = \Sigma(\gamma) + E \tag{1.4.1}$$

where the expected value of the symmetric matrix

$$E = S - \Sigma(\gamma) \tag{1.4.2}$$

is a null matrix.

Insight may be gained by expressing (1.4.1) in the form of a gene= ral nonlinear regression model which includes linear regression models as special cases. Let vecps(S) represent a $\frac{1}{2}p(p+1) \times 1$ vector formed from the nonduplicated elements of the symmetric matrix S (cf. Nel, 1980, p.39) and let

$$s^* = \text{vecps}(S) \quad , \tag{1.4.3}$$

$$\sigma^*(\gamma) = \text{vecps}(\Sigma(\gamma)) \quad , \tag{1.4.4}$$

$$e^* = \text{vecps}(E) \quad , \tag{1.4.5}$$

$$p^* = \tfrac{1}{2}p(p+1) \quad . \tag{1.4.6}$$

Then (1.4.1) may be expressed in the form

$$s^* = \sigma^*(\gamma) + e^* \qquad (1.4.7)$$

where $\sigma^*(\gamma)$ is a $p^* \times 1$ vector valued function of the $q \times 1$ vector γ,

$$\mathcal{E}(e^*) = 0 \qquad (1.4.8)$$

and

$$\mathrm{Cov}(e^*, e^{*T}) = \Upsilon^* \qquad (1.4.9)$$

where Υ^* is a $p^* \times p^*$ nonnegative definite matrix which converges to the null matrix as the sample size N tends to infinity:

i.e. $$\lim_{N\to\infty} (\Upsilon^*) = 0 \ . \qquad (1.4.10)$$

Examination of (1.4.7) shows certain similarities between a co= variance structure and a general nonlinear regression model (cf. e.g. Gallant, 1975). There are two main differences. Firstly, in a regres= sion model p^* represents the number of observations and increases as the sample size is increased whereas in a covariance structure p^* is a function of the number, p, of measurements and remains constant as the sample size N increases. Secondly, in a regression model Υ^* is of the form,

$$\Upsilon^* = \upsilon U \ , \qquad (1.4.11)$$

where U is a known matrix (often an identity matrix) so that any speci= fic diagonal element is not affected by an increase in sample size, whereas in a covariance structure the diagonal elements of Υ^* decrease as N is increased. In summary, in a covariance structure model an in= crease in the sample size N does not affect the number of observations in the regression equation (1.4.7) but does reduce the error variance for each observation. There is a close parallel between a covariance structure model (1.4.7) and a type of econometric model considered by Malinvaud (1970, Chapter 9, §5).

The similarity of a covariance structure model to a regression model suggests the application of methods employed for fitting regression models to the analysis of covariance structures.

1.5 Estimation and fit

Let $\hat{\gamma}$ represent an estimate of the parameter vector γ and let

$$\hat{\Sigma} = \Sigma(\hat{\gamma}) \cdot \qquad (1.5.1)$$

The estimation methods considered here all choose $\hat{\gamma}$ in such a manner as to minimise a "discrepancy function" $F(S;\hat{\Sigma})$. This is a scalar valued function which gives an indication of the discrepancy between the sample covariance matrix S and the fitted matrix $\hat{\Sigma}$ and has the fol=

lowing properties:

(i) $F(S;\hat{\Sigma}) \geq 0$;

(ii) $F(S;\hat{\Sigma}) = 0$ if and only if $\hat{\Sigma} = S$,

(iii) $F(S;\hat{\Sigma})$ is continuous in S and $\hat{\Sigma}$

In recent literature on covariance structures it has been common practice to refer to such a function as a "loss function". This departs from the definition of a loss function in the Kendall & Buckland (1971) dictionary of statistical terms since both S and $\hat{\Sigma}$ are known quantities and no unknown true quantity is involved. Consequently the term "discrepancy function" is preferred here.

The asymptotic distribution of an estimator, $\hat{\gamma}$, will depend on the particular discrepancy function minimised. If γ is identified and $\Sigma(\hat{\gamma})$ is continuous, however, all estimators, $\hat{\gamma}$, obtained by minimising discrepancy functions $F(S;\Sigma(\hat{\gamma}))$ are consistent (Browne, 1981).

One class of discrepancy functions is suggested by the similarity of a covariance structure model and a general nonlinear regression model. This is the class of generalised least squares discrepancy functions of the form

$$F(S;\hat{\Sigma}|U) = (s^* - \hat{\sigma}^*)^T U^{-1}(s^* - \hat{\sigma}^*) \qquad (1.5.2)$$

where $\hat{\sigma}^* = \text{vecps}(\hat{\Sigma})$ and the $p^* \times p^*$ symmetric matrix U^{-1} must be positive definite for requirements (i) and (ii) of a discrepancy function to be met. The corresponding estimators will be best generalised least squares (BGLS) estimators in the sense of having minimum asymptotic variances within the class of generalised least squares estimators if (but not necessarily only if) U converges in probability to any constant multiple of $\lim_{n\to\infty} n\, \Upsilon^*$ where

$$n = N - 1 . \qquad (1.5.3)$$

It is convenient to choose this multiple to be one so that

$$\underset{n\to\infty}{\text{plim}}\, U = \lim_{n\to\infty} \Upsilon \qquad (1.5.4)$$

where

$$\Upsilon = n\, \Upsilon^* , \qquad (1.5.5a)$$

$$= n\, \text{Cov}(s^*, s^{*T}) , \qquad (1.5.5b)$$

$$= n\, \text{Cov}(e^*, e^{*T}) . \qquad (1.5.5c)$$

The elements of Υ will depend on the kurtosis of the distribution of x and additional notation is required. Let σ_{ij} represent a typical element of Σ, let

$$\sigma_{ijk\ell} = \mathcal{E}(x_i - \xi_i)(x_j - \xi_j)(x_k - \xi_k)(x_\ell - \xi_\ell) \tag{1.5.6}$$

represent a typical multivariate fourth order moment about the mean and let

$$\kappa_{ijk\ell} = \sigma_{ijk\ell} - \sigma_{ij}\sigma_{k\ell} - \sigma_{ik}\sigma_{j\ell} - \sigma_{i\ell}\sigma_{jk} \tag{1.5.7}$$

represent a typical fourth order cumulant. The coefficient of excess kurtosis for the marginal distribution of x_i then is

$$\gamma_{2(i)} = \frac{\kappa_{iiii}}{\sigma_{ii}^2} \tag{1.5.8a}$$

$$= \frac{\sigma_{iiii}}{\sigma_{ii}^2} - 3 \tag{1.5.8b}$$

while the Mardia (1970,1974) multivariate coefficient of excess kurtosis is

$$\gamma_{2,p} = \sum_{i=1}^{p} \sum_{j=1}^{p} \sum_{k=1}^{p} \sum_{\ell=1}^{p} \sigma^{ij} \sigma^{k\ell} \kappa_{ijk\ell} \tag{1.5.9a}$$

$$= \sum_{i=1}^{p} \sum_{j=1}^{p} \sum_{k=1}^{p} \sum_{\ell=1}^{p} \sigma^{ij} \sigma^{k\ell} \sigma_{ijk\ell} - p(p+2) \tag{1.5.9b}$$

$$= \mathcal{E}\{(x - \xi)^T\Sigma^{-1}(x - \xi)\}^2 - p(p+2) \tag{1.5.9c}$$

where σ^{ij} is a typical element of Σ^{-1}. If x_i has a normal distribution then $\gamma_{2(i)} = 0$ and if x has a p-variate normal distribution then all fourth order cumulants $\kappa_{ijk\ell}$ are zero with the result that $\gamma_{2,p} = 0$.

The elements of Υ may now be expressed in the form (cf. Kendall & Stuart, 1969, p.321)

$$\upsilon_{ij,k\ell} = n\ \mathrm{Cov}(s_{ij}, s_{k\ell})$$

$$= \sigma_{ik}\sigma_{j\ell} + \sigma_{i\ell}\sigma_{jk} + \frac{n}{N}\kappa_{ijk\ell} \tag{1.5.10}$$

and in particular

$$\upsilon_{ii,ii} = n\ \mathrm{Var}(s_{ii}) = 2\sigma_{ii}^2 + \frac{n}{N}\kappa_{iiii}$$

$$= \sigma_{ii}^2(2 + \frac{n}{N}\gamma_{2(i)}). \tag{1.5.11}$$

Examination of (1.5.11) shows how the variance of a diagonal element of S increases if the distribution of the observed variable is leptokurtic with a positive excess kurtosis coefficient and decreases if it is platykurtic with a negative excess kurtosis coefficient. Similarly the covariances of elements of S in (1.5.10) depend on the values of the fourth order cumulants of the multivariate distribution of x.

If all fourth order cumulants are zero, so that $\gamma_{2,p}$ is zero, (1.5.10) becomes

$$\upsilon_{ij,k\ell} = \sigma_{ik}\sigma_{j\ell} + \sigma_{i\ell}\sigma_{jk} \tag{1.5.12}$$

so that the variances and covariances of elements of S are functions of the expected values of elements of S. This is the case when x has a multivariate normal distribution which is the situation assumed in the development of almost all available methods for the analysis of covariance structures.

Such a situation leads to much algebraic simplification in the form of any discrepancy function yielding BGLS estimators. This is accompanied by consequent simplifications in the evaluation of its derivatives which are required by iterative optimisation algorithms for obtaining estimates. If (1.5.12) is satisfied, a typical element of T^{-1} is given by

$$\upsilon^{ij,k\ell} = \frac{(2 - \delta_{ij})(2 - \delta_{k\ell})}{4} (\sigma^{ik}\sigma^{j\ell} + \sigma^{i\ell}\sigma^{jk}) \tag{1.5.13}$$

where δ_{ij} is Kronecker's delta function. Thus the elements of the $p^* \times p^*$ inverse matrix T^{-1} are simple functions of the elements of the $p \times p$ inverse matrix Σ^{-1}. If V is any consistent estimator of Σ,

$$\text{i.e.}\quad \underset{n\to\infty}{\text{plim}}\, V = \Sigma \quad , \tag{1.5.14}$$

and the elements of V are substituted for elements of Σ in (1.5.12) the $p^* \times p^*$ matrix U with elements of the form

$$u_{ij,k\ell} = v_{ik}\, v_{j\ell} + v_{i\ell}v_{jk} \tag{1.5.15}$$

will be a consistent estimator of T so that (1.5.4) is satisfied. It follows (cf. 1.5.13) that U^{-1} will have elements of the form

$$u^{ij,k\ell} = \frac{(2 - \delta_{ij})(2 - \delta_{k\ell})}{4} (v^{ik}\, v^{j\ell} + v^{i\ell}\, v^{jk}) \tag{1.5.16}$$

and it can be shown (Browne, 1974) that the discrepancy function (1.5.2) can be expressed as

$$F(S;\hat{\Sigma}|U) = F(S;\hat{\Sigma}|V) = \tfrac{1}{2}\text{tr}[(S - \hat{\Sigma})V^{-1}]^2 \ . \tag{1.5.17}$$

Formula (1.5.17) is algebraically equivalent to (1.5.2) with U defined by (1.5.15) but is far more efficient computationally involving matri= ces of order $p \times p$ instead of $p^* \times p^*$.

Thus the property of no excess kurtosis results in considerable simplification of a discrepancy function yielding BGLS estimators. One possible choice for V which satisfies (1.5.14) is

$$V = S \tag{1.5.18}$$

Another possible choice for V is

$$V = \hat{\Sigma}_{MWL} = \Sigma(\hat{\gamma}_{MWL}) \tag{1.5.19}$$

where $\hat{\gamma}_{MWL}$ is the estimator which maximises the Wishart likelihood function for S which is appropriate if x has a multivariate normal distribution. This MWL estimator may be obtained by minimizing the discrepancy function (cf. e.g. Jöreskog, 1969)

$$F_{MWL}(S,\hat{\Sigma}) = \log|\hat{\Sigma}| - \log|S| + \mathrm{tr}[S\hat{\Sigma}^{-1}] - p. \tag{1.5.20}$$

What is interesting is that the choice of V given by (1.5.19) in the discrepancy function (1.5.17) will usually yield $\hat{\gamma}_{MWL}$ (cf. Browne,1974, Proposition 6). In other words $F(S;\hat{\Sigma}|V = \hat{\Sigma}_{MWL})$ and $F_{MWL}(S,\hat{\Sigma})$ will usu= ally (and almost certainly if N is sufficiently large) have their ab= solute minima at the same point. The MWL estimator may then be regar= ded as a member of the asymptotically equivalent class of generalised least squares estimators with minimum asymptotic variances under the assumption of no kurtosis. Another overlapping but not completely equivalent class of discrepancy functions yielding estimators with minimum asymptotic variances under the assumption of no kurtosis has been proposed, and carefully investigated, by Swain (1975a, 1975b).

The estimators obtained by minimising discrepancy functions $F(S;\hat{\Sigma}|V)$ of the form of (1.5.17) with V consistent for Σ will not in general yield best generalised least squares estimators when $\eta_{2,p}$ is nonzero. These estimators will, on the other hand, still be consistent and the amounts by which the minimum attainable asymptotic variances are exceeded may be unimportant.

Other more serious difficulties are involved, however. When $\gamma_{2,p}$ is nonzero some of the fourth order cumulants $\kappa_{ijk\ell}$ in (1.5.10) will also be nonzero. Use of a discrepancy function of the form of (1.5.17) therefore implies misspecification of the coefficient matrix U in (1.5.2) in the sense that (1.5.4) is not satisfied. This can result in an incorrect test statistic and incorrect standard errors for the estimators. Difficulties of this nature are discussed in Sections 1.6 and 1.7.

In practice situations where there is appreciable kurtosis do occur. It is reassuring therefore that there exists a class of distri= butions in which misspecification of U by use of (1.5.15) when $\gamma_{2,p}$

is nonzero still results in BGLS estimators. Furthermore simple corrections for the test statistic and standard errors are available.

This class is the class of elliptical distributions which is sometimes employed in robustness studies (cf. Devlin, Gnadadesikan & Kettenring, 1976; Muirhead & Waternaux, 1980). If all second moments are finite the density function is of the form (Kelker, 1970)

$$f(x) = c|\Sigma|^{-\frac{1}{2}} g(k(x-\xi)^T\Sigma^{-1}(x-\xi)) \qquad (1.5.21)$$

where c and k are constants, ξ is the mean vector, Σ is the covariance matrix and $g(\cdot)$ is a nonnegative function. The class of elliptical distributions includes the multivariate normal distribution, the multivariate t-distribution and certain contaminated normal distributions (e.g. Muirhead & Waternaux, 1980, p.33). It is convenient for the investigation of statistical procedures involving the sample covariance matrix as it includes both leptokurtic and platykurtic distributions.

If x has an elliptical distribution with mean ξ and covariance matrix Σ, then the marginal distributions of the x_i are symmetric with coefficients of relative kurtosis

$$\eta_{2(i)} = \eta \quad , \quad i = 1,\ldots,p, \qquad (1.5.22)$$

where the coefficient of relative kurtosis of a variate is defined by

$$\eta_{2(i)} = \sigma_{iiii}/3\sigma_{ii}^2 \qquad (1.5.23a)$$

$$= (\gamma_{2(i)} + 3)/3\sigma_{ii}^2 \qquad (1.5.23b)$$

Thus a normal distribution has a relative kurtosis of 1, a platykurtic distribution has a relative kurtosis between 0 and 1, and a leptokurtic distribution has a relative kurtosis greater than 1. Similarly if x has an elliptical distribution the multivariate coefficient of relative kurtosis is

$$\eta_{2,p} = \eta \qquad (1.5.24)$$

where the multivariate coefficient of relative kurtosis of a p x 1 vector variate is defined by (cf. (1.5.9))

$$\eta_{2,p} = \sum_i \sum_j \sum_k \sum_\ell \sigma^{ij}\sigma^{k\ell}\sigma_{ijk\ell}\big/p(p+2) \qquad (1.5.25a)$$

$$= \mathcal{E}\{(x-\xi)^T\Sigma^{-1}(x-\xi)\}^2\big/p(p+2) \qquad (1.5.25b)$$

$$= \{\gamma_{2,p} + p(p+2)\}\big/p(p+2) \quad . \qquad (1.5.25c)$$

When x has an elliptical distribution the elements of Υ are given by

$$\upsilon_{ijk\ell} = n\,\mathrm{Cov}(s_{ij}, s_{k\ell})$$

$$= \eta\{\sigma_{ik}\,\sigma_{j\ell} + \sigma_{i\ell}\,\sigma_{jk}\} + \frac{n}{N}(\eta - 1)\sigma_{ij}\,\sigma_{k\ell}. \qquad (1.5.26)$$

Substitution of elements of the consistent estimator $\hat{\Sigma}_{MWL}$ for the elements of Σ and a consistent estimate $\hat{\eta}$ (cf. 1.7.3) for η in (1.5.26) will yield a U satisfying (1.5.4). A surprising and very convenient result (Browne, 1981) now follows. If the covariance structure $\Sigma(\gamma)$ is invariant under a constant scaling factor (cf. Section 1.2) and if U is defined in the manner described here then $\hat{\gamma}_{MWL}$ minimizes $F(S;\hat{\Sigma}|U)$ in (1.5.2). Consequently $\hat{\gamma}_{MWL}$ is a BGLS estimator not only when x has a multivariate normal distribution but also when x has any elliptical distribution, provided that $\Sigma(\gamma)$ is invariant under a constant scaling factor. More important in practice is the fact that corrections for the test statistic (Section 1.7) and standard errors (Section 1.6) are available.

Any estimator which minimises (1.5.17) is asymptotically equivalent to $\hat{\gamma}_{MWL}$ when (1.5.14) holds. This also applies to estimators of the Swain (1975a, 1975b) family. Consequently all such estimators have the same convenient asymptotic properties as $\hat{\gamma}_{MWL}$ when x has an elliptical distribution.

It is possible to choose U in such a manner that (1.5.4) is always satisfied provided only that all eighth order moments of the distribution of x are finite. BGLS estimators can then be obtained. Substitution of (1.5.7) into (1.5.10) shows that the matrix

$$\bar{\Upsilon} = \lim_{n\to\infty} \Upsilon \qquad (1.5.27a)$$

has typical elements

$$\bar{\upsilon}_{ijk\ell} = \lim_{n\to\infty} \upsilon_{ijk\ell} = \sigma_{ijk\ell} - \sigma_{ij}\sigma_{k\ell}\,. \qquad (1.5.27b)$$

Consequently (cf. (1.1.1),(1.1.2)), if

$$w_{ijk\ell} = N^{-1} \sum_{r=1}^{N} (x_{ri}-\bar{x}_i)(x_{rj}-\bar{x}_j)(x_{rk}-\bar{x}_k)(x_{r\ell}-\bar{x}_\ell) \qquad (1.5.28)$$

$$w_{ij} = N^{-1} \sum_{r=1}^{N} (x_{ri}-\bar{x}_i)(x_{rj}-\bar{x}_j) \qquad (1.5.29a)$$

$$= \frac{n}{N}\, s_{ij} \qquad (1.5.29b)$$

the matrix U with typical element

$$u_{ij,k\ell} = w_{ijk\ell} - w_{ij}w_{k\ell} \tag{1.5.30}$$

will satisfy (1.5.4) provided that $\mathcal{E}w^2_{ijk\ell}$ is finite. It is easily seen that (1.5.30) represents a sample covariance between product variables of the form $(x_{ri}-\bar{x}_i)(x_{rj}-\bar{x}_j)$ and $(x_{rk}-\bar{x}_k)(x_{r\ell}-\bar{x}_\ell)$ with sample means w_{ij} and $w_{k\ell}$. The matrix U defined by (1.5.30) will then be positive definite with probability 1 provided that Υ is positive definite and N is greater than p*. It will however be a biased estimator of Υ.

At the expense of some additional computation an unbiased esti= mator of Υ may be obtained (Browne, 1981) from

$$u_{ij,k\ell} = \frac{N(N-1)}{(N-2)(N-3)}(w_{ijk\ell} - w_{ij}w_{k\ell})$$

$$- \frac{N}{(N-2)(N-3)}(w_{ik}w_{j\ell} + w_{i\ell}w_{jk} - 2n^{-1}w_{ij}w_{k\ell}). \tag{1.5.31}$$

While the matrix U defined by (1.5.31) could possibly not be positive definite it is unlikely that this would be the case in any reasonable application with N considerably greater than p* since the second term in (1.5.31) is of order 1/N.

Thus BGLS estimates with a corresponding test of fit and stan= dard errors are therefore available under a mild assumption about the distribution of x. The discrepancy function given in (1.5.2) with U defined either by (1.5.30) or by (1.5.31) is minimised. Such a proce= dure is not free from shortcomings however. The matrix U involved is of order p* x p* and becomes very large as the number of variables p becomes even moderately large. For example, if p = 20 then U is of order 110 x 110. This leads to difficulties because of requirements of much computer storage and a great deal of computation results from operations involving the large matrix U. Furthermore, because of the fourth order sample moments involved a large sample will be required to estimate Υ with any accuracy.

We shall now consider the effect on the discrepancy functions of applying linear transformations to the data. Suppose that X is replaced by XA^T where A is a nonsingular matrix belonging to a certain class (e.g. all nonsingular diagonal matrices). The minimum values of the discrepancy functions (1.5.2) with (1.5.30) or (1.5.31), (1.5.17) with (1.5.18) and (1.5.20) will not be affected provided that the covariance structure model $\Sigma(\gamma)$ is invariant under that particular class of trans= formations (c.f. Section 1.2). Furthermore if $\hat{\gamma}$ yields the minimum $F(S;\Sigma(\hat{\gamma}))$ of the discrepancy function before the transformation there will exist a $\hat{\gamma}^*$ such that

$$F(S;\Sigma(\hat{\gamma})) = F(A\ S\ A^T;\ \Sigma(\hat{\gamma}^*)). \qquad (1.5.32)$$

In particular if A is diagonal one can then say that the estimation processes considered here are scale free (see also Krane & McDonald, 1978) whenever the covariance structure model is invariant under changes of scale (cf. Section 1.2).

It sometimes happens that there are missing observations in X. If the elements of S or U are calculated using different subsets of the original sample the resulting matrices may not be positive definite. Consequently the functions we have just considered will no longer necessarily be bounded below by zero and can no longer be used as discrepancy functions. A possible discrepancy function to be used when there are missing data (or when the estimated covariance matrix S is indefinite for some other reason) is (1.5.2) with U diagonal and

$$u_{ij,ij} = s_{ii}s_{jj} \quad . \qquad (1.5.33)$$

Provided that the diagonal elements of S are always strictly positive, the resulting weighted least squares estimates will be consistent and the estimation process will be scale free. For models which do not impose constraints on diagonal elements of Σ, such as the factor ana= lysis model (1.1.15) with no constraints on D_ψ, use of (1.5.2) with (1.5.32) is equivalent to defining

$$V = \text{Diag}[S] \qquad (1.5.34)$$

in (1.5.17).

In factor analysis the estimator which is obtained by minimising (1.5.17) with (1.5.34), or equivalently, (1.5.2) with (1.5.17) is known as the MINRES estimator (Harman & Jones, 1966). It has commonly been employed in situations with complete data but it is not clear what ad= vantages it would have over the estimators we considered earlier in this case.

Up to this point we have assumed that the population covariance matrix has a particular structure $\Sigma = \Sigma(\gamma)$ and that an estimate of the parameter vector γ is required. Now we consider the situation where we have not been able to obtain a good fit and a misspecification of the model is suspected. Discrepancy functions of the type we have considered for estimation purposes sometimes make the detection of outlying elements of S difficult since the method of generalised least squares can result in a preponderance of residuals of the same sign in $(S - \Sigma(\hat{\gamma}))$. Further= more large deviations from the model of some elements of S can give rise to apparently large residuals corresponding to other elements.

Difficulties of this type can be circumvented by regarding a covariance structure model as a general regression model (1.4.7) and applying a robust regression procedure (Mosteller & Tukey, 1977, Chapter 14) which is insensitive to outliers. One possibility is iteratively reweighted regression using biweights (Mosteller & Tukey, 1977, Section 14h). On each iteration the criterion function

$$F^*(S;\hat{\Sigma}) = \sum_j \sum_{i\leq j} u^{ij,ij}\{s_{ij} - \sigma_{ij}(\hat{\gamma})\}^2 \qquad (1.5.35)$$

is reduced (minimised if $\sigma^*(\gamma)$ is linear), where the weights $u^{ij,ij}$ are functions of the residuals obtained during the previous iteration:-

$$u^{ij,ij} = \{\text{Max}[0,1 - \{(s_{ij} - \hat{\sigma}_{ij})/ck\sqrt{s_{ii}\ s_{jj}}\}^2]\}^2/s_{ii}\ s_{jj} \qquad (1.5.36)$$

where k is the median of the absolute scaled residuals, $|(s_{ij}-\hat{\sigma}_{ij})/\sqrt{s_{ii}s_{jj}}|$, and c is some positive constant (e.g. c=6). Thus maximum weights are given to small residuals and the weights decrease as residuals increase in absolute value. Large residuals receive no weight at all.
A good approximation to initiate the iterative procedure is necessary otherwise convergence of the weights (1.5.36) may not occur.

This robust regression approach is suggested here for investigating deviations from covariance structure models rather than for obtaining estimates of population parameters. The whole approach is defined in terms of an iterative algorithm and it is not clear what function, if any, is minimised overall. We note that the function employed at each stage, $F^*(S;\hat{\Sigma})$, does not meet requirement ii) for a discrepancy function since $F^*(S;\hat{\Sigma}) = 0$ does not imply that $S = \hat{\Sigma}$. Consequently biweighted least squares estimators of parameters in covariance structure models need not be consistent.

Yates (1979) has proposed an interesting iteratively reweighted least squares procedure in exploratory factor analysis in which weights are applied in such a manner as to encourage unifactorial residual clusters.

In some cases, the parameter vector, γ, is required to satisfy r additional equations

$$h(\gamma) = 0 \qquad (1.5.37)$$

where $h(\gamma)$ is a r x 1 vector valued function of γ. Some of these equations will represent conditions imposed to ensure identification of γ (cf.e.g. (1.3.5)) and others will represent restrictions on the parameters. In such a situation the discrepancy function $F(S;\hat{\Sigma})$ or if appropriate the function $F^*(S;\hat{\Sigma})$, is minimised subject to (1.5.37) being satisfied.

1.6 Asymptotic distributions of estimates

It was pointed out in Section 1.5 that any estimator, $\hat{\gamma}$, which is obtained by minimising a discrepancy function is consistent pro= vided that γ is identified and $\Sigma(\gamma)$ is continuous. The limiting dis= tribution of $n^{\frac{1}{2}}(\hat{\gamma}-\gamma)$ then is multivariate normal with the null vector as expected value. We shall now examine the asymptotic covariance matrix of $n^{\frac{1}{2}}(\hat{\gamma}-\gamma)$ where $\hat{\gamma}$ is any GLS estimator, and not necessarily a "best" GLS estimator. The results provided will therefore be rele= vant to cases where MWL estimators, or other asymptotically equivalent estimators, are misapplied to kurtose distributions as well as to the widely used MINRES estimator (cf. Section 1.5) whose asymptotic dis= tribution is not well known.

Consider initially the situation where no conditions of the form of (1.5.37) are applied. Let the $p^* \times q$ matrix Δ^* represent the par= tial derivatives of elements of Σ with respect to the elements of γ. Thus,

$$\Delta^* = \frac{\partial\sigma^*}{\partial\gamma^T} \tag{1.6.1}$$

in notation specified by Nel (1980, Section 3.1). We assume that the matrix U of (1.5.2) converges in probability to a limit $\bar{U}$,

$$\operatorname*{plim}_{n\to\infty} U = \bar{U} \tag{1.6.2}$$

and that $\bar{\Upsilon}$ in (1.5.27) exists.

Then the limiting covariance matrix of

$$\hat{\delta}_{\gamma} = n^{\frac{1}{2}}(\hat{\gamma}-\gamma) \tag{1.6.3}$$

is (Browne, 1981)

$$\mathrm{LCov}(\hat{\delta}_{\gamma},\hat{\delta}_{\gamma}^T) = \{\Theta(\bar{U}^{-1})\}^{-1}\{\Theta(\bar{U}^{-1}\bar{\Upsilon}\bar{U}^{-1})\}\{\Theta(\bar{U}^{-1})\}^{-1} \tag{1.6.4}$$

where $\Theta(Z)$ is a $q \times q$ matrix valued function of a $p^* \times p^*$ nonsingular symmetric matrix Z defined by

$$\Theta(Z) = \Delta^{*T} Z \Delta^* \quad . \tag{1.6.5}$$

In (1.6.4) we have the covariance matrix for any GLS estimator obtained by minimising a discrepancy function of the form of (1.5.2) with a specific choice for U, when x has any distribution with finite fourth order moments so that $\bar{\Upsilon}$ defined by (1.5.27) exists. When $\hat{\gamma}$ is a BGLS estimator so that (cf. 1.5.4)

$$\bar{U} = \bar{\Upsilon} , \tag{1.6.6}$$

(1.6.5) simplifies considerably and becomes

$$\mathrm{LCov}(\hat{\delta}_\gamma, \hat{\delta}_\gamma) = \{\Theta(\bar{T}^{-1})\}^{-1} \tag{1.6.7a}$$

$$= \{\Delta^{*T}\ \bar{T}^{-1}\Delta^*\}^{-1}\ . \tag{1.6.7b}$$

It is easily shown (cf. Browne, 1974, Proposition 3) that the diagonal elements of (1.6.7) are not greater than those of (1.6.4). Thus the standard errors of BGLS estimates are not only smaller than those of arbitrary GLS estimators but are more easily computed. In practice estimates of these standard errors are obtained by replacing Δ^* and $\bar{T}$ by estimates.

It was seen in Section 1.5 that $\bar{T} = T$ assumes a special form (1.5.12) when the distribution of x has no excess kurtosis. This leads to an algebraic simplification of the matrix valued function $\Theta(\cdot)$ in (1.6.4) and (1.6.7). Let Δ represent the $p^2 \times q$ matrix of derivatives of all elements, including duplicated elements, of Σ with respect to the elements of γ. Thus Δ is formed by duplicating rows of Δ^*. If (cf. (1.5.13);(1.5.16)) the Z in (1.6.5) is a $p^* \times p^*$ matrix of the form

$$z_{ij,k\ell} = \frac{(2-\delta_{ij})(2-\delta_{k\ell})}{4}(t_{ik}t_{j\ell} + t_{i\ell}t_{jk}) \tag{1.6.8}$$

where T is a p x p matrix then

$$\Theta(Z) = \Phi(T)$$
$$= \tfrac{1}{2}\Delta^T(T \otimes T)\Delta \tag{1.6.9}$$

where $T \otimes T$ is a Kronecker product. One can then express (1.6.4) in the form

$$\mathrm{LCov}(\hat{\delta}_\gamma, \hat{\delta}_\gamma^T) = \{\Phi(\bar{V}^{-1})\}^{-1}\{\Phi(\bar{V}^{-1}\Sigma\bar{V}^{-1})\}\{\Phi(\bar{V}^{-1})\}^{-1} \tag{1.6.10}$$

where (cf. (1.5.17))

$$\bar{V} = \operatorname*{plim}_{n\to\infty} V. \tag{1.6.11}$$

If $\hat{\gamma}$ is a BGLS estimator then (cf. 1.6.7)

$$\mathrm{LCov}(\hat{\delta}_\gamma, \hat{\delta}_\gamma^T) = \{\Phi(\Sigma^{-1})\}^{-1}. \tag{1.6.12}$$

In particular (1.6.12) gives the asymptotic covariance for the MWL es= timator under normality assumptions, and the information matrix then is $n\Phi(\Sigma^{-1})$. The asymptotic covariance matrix for arbitrary GLS esti= mators such as MINRES estimators is given by (1.6.10) when $\eta_{2,p}$ is unity.

The form of the function $\Phi(T)$ in (1.6.9) is mathematically simple but numerically inefficient since each element requires of the order of p^4 multiplications. A form which requires of the order of p^3 mul= tiplications for the $k\ell$-th element of $\Phi(T)$ is

$$[\Phi(T)]_{k\ell} = \tfrac{1}{2}\mathrm{tr}\left[T \frac{\partial\Sigma}{\partial\gamma_k} T \frac{\partial\Sigma}{\partial\gamma_\ell}\right] \qquad (1.6.13)$$

where $\frac{\partial\Sigma}{\partial\gamma_k}$ is the symmetric p x p matrix whose columns are stacked to form the k-th column of Δ. One may then employ (1.6.13) to evaluate $\Phi(\bar{V}^{-1}\Sigma\bar{V}^{-1})$ and $\Phi(\Sigma^{-1})$ in (1.6.10) and (1.6.12).

It is important to bear in mind that *these simplifications are not valid if* $\gamma_{2,p} \neq 0$. If, however, x has an elliptical distribution, and if $\Sigma(\gamma)$ is invariant under a constant scaling factor, a simple modification of $\Phi(\Sigma^{-1})$ in (1.6.12) yields valid results when $\gamma_{2,p} \neq 0$ so that $\eta = \eta_{2,p} \neq 1$. Let

$$[\Phi_\eta(\Sigma^{-1})]_{k\ell} = \tfrac{1}{2}\eta^{-1}\{\mathrm{tr}[\Sigma^{-1} \frac{\partial\Sigma}{\partial\gamma_k} \Sigma^{-1} \frac{\partial\Sigma}{\partial\gamma_\ell}] - \frac{(\eta-1)}{(p+2)\eta-p} \mathrm{tr}[\Sigma^{-1}\frac{\partial\Sigma}{\partial\gamma_k}]\, \mathrm{tr}[\Sigma^{-1} \frac{\partial\Sigma}{\partial\gamma_\ell}]\}$$

so that $\Phi_\eta(\Sigma^{-1}) = \Phi(\Sigma^{-1})$ if $\eta = 1$. Then

$$\mathrm{LCov}(\hat{\delta}_\gamma,\hat{\delta}_\gamma^T) = \{\Phi_\eta(\Sigma^{-1})\}^{-1}$$

provided that plim $V = \Sigma$.

In general situations standard errors may be obtained from (1.6.4). If (1.5.15) holds, (1.6.4) may be written as

$$\mathrm{LCov}(\hat{\delta}_\gamma,\hat{\delta}_\gamma^T) = B\,\bar{T}\,B^T \qquad (1.6.14)$$

where

$$B = \{\Phi(\bar{V}^{-1})\}^{-1} B^*$$

and B^* is a q x $\tfrac{1}{2}p(p+1)$ matrix with typical element

$$[B^*]_{k,ij} = [\bar{V}^{-1} \frac{\partial\Sigma}{\partial\gamma_k} \bar{V}^{-1}]_{ij} \quad .$$

We now consider the case where conditions of the form (1.5.37) are employed. Let L be the r x q matrix of derivatives of elements of h with respect to elements of γ,

$$L = \frac{\partial h}{\partial\gamma^T} \quad , \qquad (1.6.15)$$

and let

$$\ddot{\Theta}(Z) = Q^{-1} - Q^{-1}L^{T}(LQ^{-1}L^{T})^{-1}LQ^{-1} \tag{1.6.16a}$$

where (cf. (1.6.5))

$$Q = Q(Z) = \Theta(Z) + L^{T}L \tag{1.6.16b}$$

in general, and (cf. (1.6.13))

$$Q = \Phi(T) + L^{T}L \tag{1.6.16c}$$

if (1.6.8) holds. Alternatively $\ddot{\Theta}(Z)$ is equal (Pringle & Rayner, 1971; Lee & Bentler, 1980) to the first q x q principal minor of

$$\begin{pmatrix} \Theta(Z) & L^{T} \\ L & 0 \end{pmatrix}^{-1} \tag{1.6.17a}$$

and also of

$$\begin{pmatrix} Q & L^{T} \\ L & 0 \end{pmatrix}^{-1} . \tag{1.6.17b}$$

The asymptotic covariance matrix of $\hat{\delta}_{\gamma}$ under (1.5.37) then (cf. (1.6.4) is

$$\mathrm{LCov}(\hat{\delta}_{\gamma},\hat{\delta}_{\gamma}^{T}) = \{\ddot{\Theta}(\bar{U}^{-1})\}\Theta(\bar{U}^{-1}\bar{T}\,\bar{U}^{-1})\{\ddot{\Theta}(\bar{U}^{-1})\} \tag{1.6.18}$$

which becomes (cf. (1.6.7))

$$\mathrm{LCov}(\hat{\delta}_{\gamma},\hat{\delta}_{\gamma}^{T}) = \{\ddot{\Theta}(\bar{T}^{-1})\} \tag{1.6.19}$$

when $\hat{\gamma}$ is a BGLS estimator. Alternative expressions for the case where $\gamma_{2,p}$ is zero may be obtained by replacing the $\Theta(\bar{U})$ and $\Theta(\bar{U}^{-1}\bar{T}\bar{U}^{-1})$ in (1.6.18) with (1.6.16) by $\Phi(\bar{V})$ and $\Phi(\bar{V}^{-1}\Sigma\bar{V}^{-1})$ and (cf. Lee & Bentler, 1980) replacing the $\Theta(\bar{T}^{-1})$ in (1.6.19) with (1.6.16) by $\Phi(\Sigma^{-1})$.

In practice an approximation to the covariance matrix of $\hat{\gamma}$ is of interest. In all cases this is obtained from

$$\mathrm{Cov}(\hat{\gamma},\hat{\gamma}^{T}) \approx n^{-1}\mathrm{LCov}(\hat{\delta}_{\gamma},\hat{\delta}_{\gamma}^{T}). \tag{1.6.20}$$

Estimates of the standard errors of the elements of $\hat{\gamma}$ are obtained from the square roots of diagonal elements of (1.6.20) with population para= meters replaced by estimates.

When a correlation structure is of interest, estimates of para= meters may be obtained by rescaling estimates of parameters in the cor= responding covariance structure (Krane & McDonald, 1978, p.227) provi= ded that it is invariant under changes of scale. For example, if the population covariance matrix Σ satisfies the unrestricted factor ana= lysis model

$$\Sigma = \Lambda\Lambda^T + D_\psi \tag{1.6.21}$$

then the population correlation matrix P satisfies the corresponding model

$$P = \Lambda^*\Lambda^{*T} + D^*_\psi \tag{1.6.22}$$

and estimates of Λ^* and D^*_ψ may be obtained from

$$\hat{\Lambda}^* = \text{Diag}^{-\frac{1}{2}}[\hat{\Sigma}]\hat{\Lambda} , \tag{1.6.23a}$$

$$\hat{D}^*_\psi = \text{Diag}^{-1}[\hat{\Sigma}]\hat{D}_\psi . \tag{1.6.23b}$$

If MWL estimates are employed it happens that $\text{Diag}[\hat{\Sigma}] = \text{Diag}[S]$ so that the scaling in (1.6.22) is carried out implicitly if the estimation procedure is applied to the sample correlation matrix R instead of to the sample covariance matrix S. Since the estimates are in fact scaled, and the scaling factors are stochastic, corrections to the standard errors are essential. Suitable formulae for MWL factor analysis are given by Lawley and Maxwell (1971, Sections 5.3 and 7.7) who also give examples illustrating the effect of the corrections. Here we shall give general formulae applicable to other estimators where $\text{Diag}[\hat{\Sigma}] \neq \text{Diag}[S]$ and to other models where scaling by $\hat{\sigma}_{kk}^{-\frac{1}{2}}$ or by $\hat{\sigma}_{kk}^{-1}$ occurs:

$$\text{Var}(\hat{\gamma}_i/\hat{\sigma}_{kk}^{\frac{1}{2}}) \approx \frac{1}{\sigma_{kk}}\left\{\text{Var}(\hat{\gamma}_i) - \frac{\gamma_i}{\sigma_{kk}}\text{Cov}(\hat{\sigma}_{kk},\hat{\gamma}_i) + \tfrac{1}{4}\frac{\gamma_i^2}{\sigma_{kk}^2}\text{Var}(\hat{\sigma}_{kk})\right\} \tag{1.6.24}$$

$$\text{Var}(\hat{\gamma}_i/\hat{\sigma}_{kk}) \approx \frac{1}{\sigma_{kk}^2}\left\{\text{Var}(\hat{\gamma}_i) - 2\frac{\gamma_i}{\sigma_{ii}}\text{Cov}(\sigma_{kk},\gamma_i) + \frac{\gamma_i^2}{\sigma_{kk}^2}\text{Var}(\hat{\sigma}_{kk})\right\} \tag{1.6.25}$$

where

$$\text{Cov}(\hat{\sigma}_{kk},\hat{\gamma}_i) \approx [\Delta\{\text{Cov}(\hat{\gamma},\hat{\gamma}^T)\}]_{ki} \tag{1.6.26}$$

$$\text{Var}(\hat{\sigma}_{kk}) \approx [\Delta\{\text{Cov}(\hat{\gamma},\hat{\gamma}^T)\}\Delta^T]_{kk} \tag{1.6.27}$$

and $[A]_{ki}$ represents the ki-th element of the matrix A.

The main purpose of the preceding results for variances of ele= ments of $\hat{\gamma}$ is to provide approximate confidence intervals for elements of γ. Let

$$\hat{\sigma}(\hat{\gamma}_i) = \sqrt{\text{Var}(\hat{\gamma}_i)}\,\Big|_{\gamma = \hat{\gamma}} \tag{1.6.28}$$

be an estimate of the standard error of $\hat{\gamma}_i$ obtained by replacing popu= lation parameters by estimates in the appropriate formula. Because of the asymptotic normal distribution of the estimators an approximate $(1-\alpha)\%$ confidence interval for γ_i is given by

$$\hat{\gamma}_i - c_\alpha \hat{\sigma}(\hat{\gamma}_i) < \gamma_i < \hat{\gamma}_i + c_\alpha \hat{\sigma}(\hat{\gamma}_i) \tag{1.6.29}$$

where c_α is the $(1-\frac{1}{2}\alpha)$ percentage point of the standard normal distri= bution for one-at-a-time confidence intervals, and c_α^2 is the $(1-\alpha)$ percentage point of the chi-squared distribution with q degrees of freedom for simultaneous confidence intervals on q identified parame= ters.

The confidence interval in (1.6.9) is suitable in situations when γ_i is not bounded. When γ_i is contained in some bounded interval, $\gamma_L < \gamma_i < \gamma_U$, the confidence interval (1.6.29) will not necessarily lie entirely in this interval even if $\hat{\gamma}_i$ does. A suitable confidence interval may then be obtained by using a transformation. Let $h(\gamma_i)$ be a monotonic increasing function of γ_i with unbounded range and let

$$\theta = h(\gamma_i) \quad , \qquad \gamma_L < \gamma_i < \gamma_U \, , \tag{1.6.30a}$$

$$\hat{\theta} = h(\hat{\gamma}_i) \quad . \qquad \gamma_L < \gamma_i < \gamma_U \, . \tag{1.6.30b}$$

An approximate $(1-\alpha)\%$ confidence interval for θ is now given by

$$\hat{\theta}_L < \theta < \hat{\theta}_U \tag{1.6.31a}$$

where

$$\hat{\theta}_L = \hat{\theta} - c_\alpha \hat{\sigma}(\hat{\theta}) \tag{1.6.31b}$$

$$\hat{\theta}_U = \hat{\theta} + c_\alpha \hat{\sigma}(\hat{\theta}) \tag{1.6.31c}$$

and

$$\hat{\sigma}(\hat{\theta}) = \left\{ \frac{\partial h}{\partial \gamma_i} \bigg|_{\gamma = \hat{\gamma}} \right\} \hat{\sigma}(\hat{\gamma}_i) \tag{1.6.32}$$

Since $\hat{\theta}$ is not bounded the distribution of $\hat{\theta}$ will generally be approximated more closely by a normal distribution than will the dis= tribution of $\hat{\gamma}_i$, particularly if γ_i is close to γ_U or γ_L. The confidence interval (1.6.31) for θ is usually not of primary interest, however. What is required is a confidence interval for γ_i. This may be obtained by applying the inverse transformation

$$\gamma_i = h^{-1}(\theta) \tag{1.6.33}$$

to (1.6.31). This provides the (1-α)% confidence interval for γ_i,

$$h^{-1}(\hat{\theta}_L) < \gamma_i < h^{-1}(\hat{\theta}_U). \qquad (1.6.34)$$

In general the confidence interval in (1.6.34) will not be symmetric about $\hat{\gamma}_i$ but unlike (1.6.9) cannot include inadmissible values of γ_i below γ_L or above γ_U.

Some particular choices for the transformation $h(\gamma_i)$ will now be considered. Suppose that γ_i represents a correlation coefficient so that $-1 < \gamma_i < 1$. This could occur if γ_i corresponds to an element of the factor correlation matrix Φ in (1.1.15) when $\varphi_{jj} = 1, j=1,\ldots,m$. Another situation where γ_i represents a correlation coefficient is when γ_i corresponds to a factor loading in a factor analysis model for a *correlation* matrix with uncorrelated factors, $\Phi = I$. Then one may use the Fisher z-transformation

$$\hat{\theta} = h(\hat{\gamma}_i) = \tfrac{1}{2}\log_e\left\{\frac{1+\hat{\gamma}_i}{1-\hat{\gamma}_i}\right\} \qquad (1.6.35)$$

with

$$\hat{\sigma}(\hat{\theta}) = \hat{\sigma}(\hat{\gamma}_i)/(1-\hat{\gamma}_i^2) \qquad (1.6.36)$$

and

$$h^{-1}(\hat{\theta}) = \{\exp(2\theta) - 1\}/\{\exp(2\theta) + 1\} \qquad (1.6.37)$$

If γ_i represents a variance (such as a diagonal element of D_ψ in (1.1.15) or a standard deviation so that $\gamma_i > 0$, a logarithmic trans= formation is suitable:

i.e. $$\hat{\theta} = h(\hat{\gamma}_i) = \log_e \hat{\gamma}_i \qquad (1.6.38)$$

with

$$\hat{\sigma}(\hat{\theta}) = \hat{\sigma}(\hat{\gamma}_i)/\hat{\gamma}_i \qquad (1.6.39)$$

and

$$h^{-1}(\hat{\theta}) = \exp(\hat{\theta}). \qquad (1.6.40)$$

Sometimes the interval $0 < \gamma_i < 1$ is appropriate. This occurs, for example, in the factor analysis of a correlation matrix when γ_i represents the ratio ψ_{ii}/σ_{ii} of a residual variance to an observed variable variance. One may then employ

$$\hat{\theta} = h(\hat{\gamma}_i) = -\log_e(\hat{\gamma}_i^{-1}-1) \qquad (1.6.41)$$

with

$$\hat{\sigma}(\hat{\theta}) = \hat{\sigma}(\hat{\gamma}_i)/\{\hat{\gamma}_i(1-\hat{\gamma}_i)\} \tag{1.6.42}$$

and

$$h^{-1}(\hat{\theta}) = 1/\{1 + \exp(-\hat{\theta}_i)\} \tag{1.6.43}$$

1.7 Tests of fit

We now consider the examination of the appropriateness of a co= variance structure model given a sample covariance matrix. The usual procedure is to test the null hypothesis

$$H_o : \Sigma = \Sigma(\gamma)$$

against the general alternative

$$H_1 : \Sigma \text{ is any } p \times p \text{ positive definite matrix.}$$

If H_o is true, if γ is identified and if

$$\text{plim } U = \bar{T} \tag{1.7.1}$$

then the asymptotic distribution of $nF(S;\hat{\Sigma}|U)$ (cf. (1.5.2)) is chi-squared with $d = p^* - q + r$ degrees of freedom,

$$\text{i.e.} \quad nF(S;\hat{\Sigma}|U) \sim \chi^2_{p^*-q+r} \tag{1.7.2}$$

If no conditions (1.5.37) are applied then $r = 0$.

If the distribution of x belongs to the class of elliptical distri= butions it is not necessary to make use of the large $p^* \times p^*$ matrix U. Let (cf. Mardia, 1974, p.116)

$$\hat{\eta} = \sum_i \sum_j \sum_k \sum_\ell w^{ij} w^{k\ell} w_{ijk\ell}/p(p+2) \tag{1.7.3a}$$

$$= \sum_{i=1}^{N} \{(x_i-\bar{x})^T W^{-1}(x_i-\bar{x})\}^2/Np(p+2) \tag{1.7.3b}$$

so that $\hat{\eta}$ is a consistent estimator of the multivariate coefficient of relative kurtosis η in (1.5.25). Then if

$$\text{plim } V = \Sigma \tag{1.7.4}$$

and if x has an elliptical distribution, the asymptotic distribution of $n\hat{\eta}^{-1}F(S;\hat{\Sigma}|V)$ (cf.1.5.17) under H_o is chi-squared,

$$\text{i.e.} \quad n\hat{\eta}^{-1}F(S;\hat{\Sigma}|V) \sim \chi^2_{p^*-q+r} , \tag{1.7.5}$$

provided that $\Sigma(\gamma)$ is invariant under a constant scaling factor. Since $nF_{MWL}(S;\hat{\Sigma})$ is asymptotically equivalent to $nF(S;\hat{\Sigma}|V = \hat{\Sigma}_{MWL})$ if (1.7.4) is true, it follows that asymptotically

$$n\hat{\eta}^{-1}F_{MWL}(S;\hat{\Sigma}) \sim \chi^2_{p^*-q+r} \quad (1.7.6)$$

under the same conditions. It is of interest to note that

$$\mathrm{tr}[S\hat{\Sigma}^{-1}] - p = 0 \quad (1.7.7)$$

if $\Sigma(\gamma)$ is invariant under a constant scaling factor and $\hat{\gamma}$ is a MWL estimate so that (1.5.20) simplifies to

$$F_{MWL}(S;\hat{\Sigma}) = \log|\hat{\Sigma}| - \log|S| \quad (1.7.8)$$

at the minimum.

The stipulation that $\Sigma(\gamma)$ be invariant under a constant scaling factor is no longer necessary if it is known that the distribution of x is multivariate normal (or has no multivariate kurtosis). Since the population coefficient of relative kurtosis is then unity we have asymptotically

$$nF(S;\hat{\Sigma}|V) \sim \chi^2_{p^*-q+r} \quad (1.7.9)$$

or

$$nF_{MWL}(S;\hat{\Sigma}) \sim \chi^2_{p^*-q+r} \quad (1.7.10)$$

This is the only situation which has received any real attention in the literature with attention concentrated on (1.7.10). For the spe= cific cases where Σ is known, where Σ is diagonal or where Σ satisfies the unrestricted factor analysis model (1.6.20) alternative multiply= ing factors n^* to n in (1.7.10) have been found (cf. Bartlett, 1954) which result in a substantial improvement of the approximation of the chi-squared distribution to the distribution of $n^*F_{MWL}(S;\hat{\Sigma})$ under H_o when n is not large. Since very large samples are seldom available there is a need for multipliers of this type for covariance structure models in general. Using heuristic arguments Swain (1975b, Section 3.4) proposed four alternatives for n which seem to result in an improve= ment of the approximation of the chi-squared distribution. The follow= ing general multiplying factor, which applies only to models which are invariant under a constant scaling factor, appeared slightly preferable to the others:

$$n^* = n - \{p(2p^2+3p-1) - y(2y^2+3y-1)\}/12(p^*-q+r) \quad (1.7.11a)$$

where

$$y = \tfrac{1}{2}[\{1 + 8(q-r)\}^{\frac{1}{2}}-1] \quad (1.7.11b)$$

and p^* is given by (1.4.6).

The test statistics we have considered up to this point apply only to estimates with minimum asymptotic variances and are simple functions of the discrepancy function minimised. At the expense of a fair amount of additional complication it is possible to obtain test statistics which can be applied in conjunction with GLS estimators in general. In particular these statistics can be applied in conjunction with MINRES estimators or with MWL estimators when the distribution of x is known to have substantial kurtosis.

We shall initially consider the situation where no conditions (1.5.37) are imposed. Let

$$\hat{\Delta}^* = \left.\frac{\partial\sigma^*}{\partial\gamma^T}\right|_{\gamma = \hat{\gamma}} \tag{1.7.12}$$

so that the $p^* \times q$ matrix $\hat{\Delta}^*$ is a consistent estimator of Δ^* and let

$$\hat{e}^* = s^* - \sigma^*(\hat{\gamma}) \tag{1.7.13}$$

If $\hat{T}$ is any consistent estimator of $\bar{T}$ (cf.(1.5.30),(1.5.31),(1.5.15)) then the limiting distribution of

$$n\ddot{F} = n\{\hat{e}^{*T}(\hat{T}^{-1} - \hat{T}^{-1}\hat{\Delta}^*(\hat{\Delta}^{*T}\hat{T}^{-1}\hat{\Delta})^{-1}\hat{\Delta}^*\hat{T}^{-1})\hat{e}^*\} \tag{1.7.14}$$

under H_o is chi-squared with $d = p^* - q$ degrees of freedom. In particu=lar, if

$$\hat{T} = U \tag{1.7.15}$$

then

$$\hat{\Delta}^{*T}\hat{T}^{-1}\hat{e}^* = \hat{\Delta}^T U^{-1}\hat{e}^* = 0 \tag{1.7.16}$$

so that the second term in (1.7.14) becomes zero and

$$n\ddot{F} = nF(S;\hat{\Sigma}|U). \tag{1.7.17}$$

Thus (1.7.14) reduces to the usual quadratic form statistic when (1.7.15) holds so that $\hat{\gamma}$ is a BGLS estimator.

If conditions of the form of (1.5.37) are applied let

$$\hat{L} = \left.\frac{\partial h}{\partial\gamma^T}\right|_{\gamma = \hat{\gamma}} \tag{1.7.18}$$

Then the asymptotic distribution of

$$n\ddot{F}_r = n\ddot{F} + n\hat{e}^{*T}\hat{T}^{-1}\hat{\Delta}^*(\hat{\Delta}^{*T}\hat{T}^{-1}\hat{\Delta}^*)^{-1}L\{L^T(\hat{\Delta}^{*T}\hat{T}^{-1}\hat{\Delta})^{-1}L\}^{-1}L^T(\hat{\Delta}^{*T}\hat{T}^{-1}\hat{\Delta}^*)^{-1}\hat{\Delta}^{*T}\hat{T}^{-1}\hat{e}^*, \tag{1.7.19}$$

where $n\ddot{F}$ is defined in (1.7.14), is chi-squared with $p - q + r$ degrees of freedom under H_o.

Suppose now that $\Sigma_1(\gamma_1)$ and $\Sigma_2(\gamma_2)$ are two covariance structures and that $\Sigma_1(\gamma_1)$ is a special case of $\Sigma_2(\gamma_2)$ possibly obtained by assigning specified values to some parameters. Suppose that nF_1 is a GLS test statistic distributed as chi-squared with d_1 degrees of freedom when $\Sigma = \Sigma_1(\gamma_1)$ and nF_2 is the corresponding GLS test statistic distributed as chi-squared with d_2 degrees of freedom when $\Sigma = \Sigma_2(\gamma_2)$. The null hypothesis

$$H_o : \Sigma = \Sigma_1(\gamma_1)$$

may be tested against the specific alternative

$$H_1 : \Sigma = \Sigma_2(\gamma_2)$$

by means of the test statistic $n(F_1 - F_2)$ which will have an asymptotic chi-squared distribution with $d_1 - d_2$ degrees of freedom when H_o is true.

When applying any of the tests described here it is important to bear in mind that results concerning asymptotic chi-squared distributions require that the parameter vector γ be identified and be an interior point of the parameter space G. In particular, if any elements of γ represent variances, none of the population variances should be zero. This point is discussed in the context of unrestricted factor analysis by Geweke and Singleton (1980) who also present the results of Monte Carlo experiments to illustrate how important it is that γ should be identified and should be an interior point of G.

In practical applications of tests of fit of covariance structure models, the tests which involve the assumption of no kurtosis have in the past been applied indiscriminately without examining estimates either of marginal coefficients of kurtosis or Mardia's coefficient of multivariate kurtosis. Unfortunately, if the distribution of x has substantial kurtosis, the chi-squared distribution may be a bad approximation for the distributions of these particular statistics under H_o so that no confidence can be placed on the results.

The sensitivity of tests concerning covariance matrices to non-normality has been pointed out by Layard (1972, 1974) and by Mardia (1974) who concentrated on tests for equality of covariance matrices. The sensitivity of tests concerning correlation coefficients to non-normality has been pointed out by Duncan & Layard (1973), Kraemer (1980) and Steiger and Hakstian (1980).

1.8 Selection of a model

There are substantial practical difficulties in the application of tests of fit. The main one is that a model can in practice only be regarded as an approximation to reality and that it is unreasonable to expect that the null hypothesis H_0 can be exactly true. Since the power of a test increases as sample size increases, tests of fit of covariance structure models invariably result in rejection of the null hypothesis when the sample size is large. This tempts practitioners either to avoid large samples or, if information on a large sample has been collected, to include additional "wastebasket parameters" purely to obtain a nonsignificant value of the test statistic. Subsequently no attempt is made to interpret these parameters and they are employed purely to hide lack of fit of the original model. Their values, in fact, tend to fluctuate widely from one sample to another. In factor analysis models of the form of (1.1.13) nondiagonal elements of Ψ are sometimes allowed to become nonzero if the original model with Ψ diagonal gives a value of the test statistic which exceeds the critical value employed. This is legitimate if a sound theoretical reason can be found why particular residual variates should be correlated. On the other hand there is always a temptation to employ this approach merely to legitimatize an analysis, with no interpretation being given of the correlated residual variates.

It is therefore clear that the usual hypothesis testing approach has some limitations in practice and that alternative approaches to the selection of a model will be of interest. One possibility is to adapt the cross validation approach, which has been so widely employed in regression, to the analysis of covariance structures, making use of the similarities pointed in Section 1.4. This involves a complete change of viewpoint. One no longer believes that the population covariance matrix Σ necessarily satisfies a model of the form $\Sigma = \Sigma(\gamma)$. Rather one considers which of a number of alternative structural forms

$$\hat{\Sigma}_1 = \Sigma_1(\hat{\gamma}_1) \quad , \quad \hat{\Sigma}_2 = \Sigma_2(\hat{\gamma}_2), \quad \hat{\Sigma}_3 = \Sigma_3(\hat{\gamma}_3) \ \ldots$$

will result in an approximation for the population covariance matrix Σ which is "best" in some sense. First an approximation $\hat{\Sigma}_i$, $i = 1,2,3,\ldots$, is chosen so as to minimise the discrepancy $F(S;\hat{\Sigma}_i)$ between the sample covariance matrix S and a reproduced covariance matrix, $\hat{\Sigma}_i$, of a specified structural form. One then attempts to select the reproduced covariance matrix $\hat{\Sigma}$ from the alternatives $\hat{\Sigma}_1,\hat{\Sigma}_2,\hat{\Sigma}_3\ldots$ which will most closely approximate Σ in the sense that the value of a discrepancy function $F(\Sigma;\hat{\Sigma})$ is a minimum. While there is no particular reason for using the same type of discrepancy function in the two stages, $F(S;\hat{\Sigma})$ and $F(\Sigma;\hat{\Sigma})$, there are no apparent disadvantages in doing so.

The main problem is that of estimating $F(\Sigma;\hat{\Sigma}_i)$, $i = 1,2,3,\ldots$. One promising approach is the classical cross validation approach frequently employed for validating a regression equation, in which two samples are employed. Let S_a and S_b represent the two sample covariance matrices. The fitted covariance matrix $\hat{\Sigma}_{ai}$, $i = 1,2,3,\ldots$, is first obtained by minimising $F(S_a;\hat{\Sigma}_{ai})$ in the calibration process. Then $F(S_b;\hat{\Sigma}_{ai})$ is employed as a biased but consistent estimate of $F(\Sigma;\hat{\Sigma}_{ai})$ in the validation process. The process may then be reversed with S_b used for calibration purposes and S_a for validation purposes. A new $\hat{\Sigma}_{bi}$ is obtained by minimising $F(S_b;\hat{\Sigma}_{bi})$ and $F(S_a;\hat{\Sigma}_{bi})$ is used as an estimate of $F(\Sigma;\hat{\Sigma}_{bi})$.

This approach has a disadvantage in that cases must be omitted from the calibration sample to be used for validation purposes. This disadvantage is minor if sufficient data are available for both samples to be large. An alternative worth consideration (D.M. Hawkins, personal communication) is to make use of Efron's (1979,Section 4) "bootstrap method".

One of the structural forms which should be included in any set to be compared is that of the general alternative hypothesis where Σ is any positive definite matrix and $q = \frac{1}{2}p(p+1)$. Then $\hat{\Sigma} = S$ and $F(S;\hat{\Sigma}) = 0$. Of course, this does not necessarily imply that $F(\Sigma;\hat{\Sigma}) = F(\Sigma;S)$ will be small, particularly if N is small. In factor analysis the set of structural models to be considered would also include several possible values for the number of factors.

This cross validation approach is not restricted to the comparison of nested models as in the hypothesis testing approach. Furthermore there is no attempt to find the "correct" model. One seeks the structural model which yields the best approximation $\hat{\Sigma}$ to Σ given a sample of a specified size. Just as in linear regression the number of predictors which will give optimal predictive precision tends to increase as the sample size increases it can be expected that structural models with more parameters will tend to be selected as N increases.

1.9 Computational considerations

A substantial amount of attention has been devoted to the computation of estimates of parameters in covariance structure models. In general an iterative algorithm is required although there are some simple models where MWL estimates can be expressed in closed form (cf. Mukherjee, 1970) while GLS estimates can be expressed in closed form for any linear covariance structure (e.g. Browne, 1974, Section 4).

Iterative algorithms from the class of quasi-Newton algorithms have been used predominantly to minimise discrepancy functions $F(S;\hat{\Sigma})$ or, equivalently, to minimise $\frac{1}{2}F(S;\hat{\Sigma})$, in order to avoid unnecessary multiplications by 2. These are of the form

$$\hat{\gamma}_{t+1} = \hat{\gamma}_t + \alpha_t H_t^{-1} g_t \qquad (1.9.1)$$

where

$\hat{\gamma}_t$ is the t-th successive approximation for the estimate $\hat{\gamma}$,

α_t is a step size parameter for the t-th step with $0 < \alpha_t \leq 1$ and with $\alpha_t = 1$ in the majority of cases,

H_t is an approximation to the Hessian $\frac{1}{2}\left.\frac{\partial^2 F}{\partial\gamma\partial\gamma^T}\right|_{\gamma = \hat{\gamma}_t}$ of $\frac{1}{2}F(S;\Sigma)$,

g_t is the negative gradient $-\frac{1}{2}\left.\frac{\partial F}{\partial\gamma}\right|_{\gamma = \hat{\gamma}_t}$ of $\frac{1}{2}F(S;\Sigma)$.

These methods differ mainly in the manner in which H_t is chosen. In the Newton method, in particular, H_t is defined as the Hessian of $\frac{1}{2}F$ and $\alpha_t = 1$ in the majority of situations although α_t is sometimes successively halved to ensure that $F(S;\Sigma(\hat{\gamma}_{t+1})) < F(S;\Sigma(\hat{\gamma}_t))$. This algorithm will converge to the minimum of F in the smallest number of iterations of all pseudo-Newton algorithms provided that the initial approximation $\hat{\gamma}_o$ is close to the minimum point $\hat{\gamma}$. Its main disadvantage is that the Hessian is often complicated (cf. (1.9.18), (1.9.7)) so that it can involve a really major programming task and its computation on each iteration can consume a substantial amount of computer time. An added disadvantage is that it can converge slowly or diverge if the initial approximation is poor so that it is generally preceded by an= other algorithm. Because of these disadvantages it has not been used extensively although it has been employed successfully for MWL estima= tes of parameters in linear covariance structures (Bock & Bargmann, 1966) and in algorithms for unrestricted factor analysis involving a "nesting" (Ross, 1970) procedure to reduce the number of parameters in= volved in the minimisation (Jennrich & Robinson, 1969; Clarke, 1970; Jöreskog & Goldberger, 1972; Swain, 1975b).

The class of conjugate gradient algorithms (e.g. Brodlie, 1977) avoids difficulties involved in the computation of the Hessian by build=

ing up an approximation to the inverse Hessian with an updating step carried out on each iteration. In theory this should converge to the inverse Hessian at the minimum but in practice the approximation af= ter termination of the algorithm is not close (Lee & Jennrich, 1979). A member of the class, the Davidon-Fletcher-Powell algorithm (Fletcher & Powell, 1963) was first used in the analysis of covariance struc= tures by Jöreskog (1967) and has been employed successfully by him in a variety of situations (Jöreskog, 1970a, 1971, 1977; Jöreskog & Sörbom, 1979). This algorithm requires a fairly complicated search on each iteration to choose α_t so as to minimise $F(S;\Sigma(\hat{\gamma}_{t+1}))$. Recently an alternative updating procedure for the inverse Hessian, known as the Broyden-Fletcher-Goldfarb-Shanno update, has appeared to be supe= rior in general applications (Brodlie, 1977). This finding was corro= borated by S.H.C. du Toit (personal communication) when developing computer programs for the analysis of covariance structures arising from nonlinear growth curves and from auto-regressive time series with moving average residuals (Du Toit, 1979). He found that when the Davi= don-Fletcher-Powell update leads to convergence the Broyden-Fletcher-Goldfarb-Shanno update leads to convergence *in essentially the same way*, but found cases where the Davidon-Fletcher-Powell update does not lead to convergence and the Broyden-Fletcher-Goldfarb-Shanno update does. A further advantage of the Broyden-Fletcher-Goldfarb-Shanno up= date is that a search to choose α_t is no longer always essential and it is often sufficient to let $\alpha_t = 1$.

The algorithm usually employed in nonlinear least squares pro= blems is the Gauss-Newton algorithm or a modification thereof (Dennis, 1977). This algorithm is readily adapted to generalised least squares problems such as those which arise in the analysis of covariance struc= tures (cf. (1.5.2)). Furthermore, a minor modification of the algorithm involving redefinition of the matrix U in (1.5.2) or, equivalently, V in (1.5.17), yields MWL estimates and is equivalent (Lee & Jennrich, 1979) to the well known Fisher scoring method for obtaining maximum likelihood estimates. In an interesting comparative study Lee and Jenn= rich (1979) found the Gauss-Newton or Fisher scoring algorithm to be superior to the Fletcher-Powell algorithm and essentially as effective as the more complicated Newton algorithm in a number of applications of restricted factor analysis models. The Gauss-Newton algorithm has also been used (Browne, 1974) for obtaining MWL estimates for linear covariance structures, and (Browne,1981) for obtaining generalised least squares estimates without assumptions of no kurtosis.

In the Gauss-Newton algorithm when the generalised least squares discrepancy function $F = F(S;\hat{\Sigma}|U)$ in (1.5.2) is being minimised we approximate the Hessian of $\frac{1}{2}F$ by (cf. (1.6.5))

$$H_t = \Theta(U^{-1})\Big|_{\gamma = \hat{\gamma}_t} \tag{1.9.2}$$

while the negative gradient of $\frac{1}{2}F$ is given (cf. (1.4.3); (1.6.1)) by

$$g_t = \Delta^{*T} U^{-1}(s^* - \sigma^*)\Big|_{\gamma = \hat{\gamma}_t} . \tag{1.9.3}$$

In the references cited earlier a modification of the Gauss-Newton algorithm due to Jennrich and Sampson (1968) was employed. The step size α_t in (1.9.1) is set equal to 1, but if an increase in the discrepancy function occurs, α_t is successively halved until

$$F(S;\Sigma(\hat{\gamma}_{t+1})|U) < F(S;\Sigma(\hat{\gamma}_t)|U). \tag{1.9.4}$$

It is unnecessary, and even undesirable (Lindqvist, 1980) to choose α_t so as to minimise $F(S;\Sigma(\hat{\gamma}_{t+1}))$. The Jennrich-Sampson algorithm makes use of a stepwise regression procedure to avoid a breakdown when H_t is near singular and includes a very useful facility for imposing upper and lower bounds on the elements of $\hat{\gamma}_t$.

Iteration of (1.9.1) is continued until some convergence crite= rion is met. A convergence criterion which was found to be particularly convenient because it is not affected by rescaling of the data was suggested by Dennis (1977, p.273). Let c_i represent the i-th "residual cosine" or cosine of the angle between the i-th column of Δ^* and the residual vector $(s^* - \sigma^*)$ in the metric of U^{-1}:

$$c_i = [g_t]_i / \{[H_t]_{ii} \; F(S;\Sigma(\hat{\gamma}_t)|U)\}^{\frac{1}{2}} \tag{1.9.5}$$

where $[g_t]_i$ is the i-th element of g_t and $[H_t]_{ii}$ is the i-th diagonal element of H_t. Iteration is terminated if either the discrepancy func= tion value falls below a certain tolerance limit ε_1 (e.g. $\varepsilon_1 = 10^{-6}$),

$$F(S;\Sigma(\hat{\gamma}_t)|U) < \varepsilon_1 \tag{1.9.6a}$$

or if all absolute residual cosines fall below another tolerance limit, ε_2 (e.g. $\varepsilon_2 = 10^{-3}$),

$$|c_i| < \varepsilon_2 \quad , \; i = 1,\ldots,p. \tag{1.9.6b}$$

It can be seen from (1.6.7) and (1.6.20) that estimates of the standard errors of the elements of $\hat{\gamma}$ can be obtained from the square roots of diagonal elements of $n^{-1}H_t^{-1}$ after convergence of the algo= rithm provided that it is known that U is a consistent estimator of $\bar{T}$. Singularity of H_t after convergence implies that $F(S;\Sigma(\gamma)|U)$ is flat and therefore has an infinity of local minima on a neighbourhood of $\hat{\gamma}$. This often, but not always, indicates that γ is not identified at its population value (see McDonald & Krane, 1979). In such a situa= tion it can be difficult to decide on the appropriate degrees of free= dom for the chi -squared test statistic.

It is still possible to make use of (1.9.2) and (1.9.3) when U has the special structure (1.5.15) under the assumption of no kurtosis, but this is wasteful of computer time and computer storage. In such situations typical elements of the approximate Hessian, H_t, and the gradient, g_t, may be calculated (Bargmann, 1967) from (cf. (1.6.13))

$$[H_t]_{k\ell} = \tfrac{1}{2}\mathrm{tr}\left[V^{-1} \frac{\partial\Sigma}{\partial\gamma_k} V^{-1} \frac{\partial\Sigma}{\partial\gamma_\ell} \right]\Bigg|_{\gamma = \hat{\gamma}_t} \tag{1.9.7}$$

and

$$[g_t]_k = \tfrac{1}{2}\mathrm{tr}\left[V^{-1} (S - \hat{\Sigma}) V^{-1} \frac{\partial\Sigma}{\partial\gamma_k} \right]\Bigg|_{\gamma = \hat{\gamma}_t} \tag{1.9.8}$$

noting that the nondiagonal elements of a matrix need not be computed if its trace only is required. The value of the discrepancy function is calculated using (1.5.17).

Sometimes in the literature on the analysis of covariance struc= tures, formulae for the approximate Hessian and the gradient are ex= pressed in a form which involves the Kronecker products $\hat{\Sigma}^{-1} \otimes \hat{\Sigma}^{-1}$ or $S^{-1} \otimes S^{-1}$. These formulae are variations of (1.9.2) and (1.9.3) using (1.6.9) with $T = \hat{\Sigma}^{-1}$ or $T = S^{-1}$. Consequently they are inefficient for computational purposes.

In the majority of structural models for covariance matrices which have been employed to date the derivatives of Σ may be expressed as symmetric matrices of rank 2 of the form

$$\frac{\partial\Sigma}{\partial\gamma_k} = x_k\, y_k^T + y_k\, x_k^T \tag{1.9.9}$$

The factor analysis model (1.1.15) may be considered as an example. If $[A]_{.i}$ represents the i-th column of a matrix A, we have for example

$$\frac{\partial\Sigma}{\partial\lambda_{ij}} : x = [I]_{.i} \qquad , \; y = [\Lambda\Phi]_{.j} \; , \tag{1.9.10a}$$

$$\frac{\partial\Sigma}{\partial\varphi_{ij}} : x = (1 + \delta_{ij})^{-1}[\Lambda]_{\cdot i} \quad , \quad y = [\Lambda]_{\cdot j} \tag{1.9.10b}$$

$$\frac{\partial\Sigma}{\partial\psi_{ii}} : x = \tfrac{1}{2}[I]_{\cdot i} \quad , \quad y = [I]_{\cdot i} \quad . \tag{1.9.10c}$$

When the derivatives have the form given in (1.9.9), formulae (1.9.7) and (1.9.8) may be simplified to:

$$[H_t]_{k\ell} = \{(x_k^T V^{-1} x_\ell)(y_k^T V^{-1} y_\ell) + (x_k^T V^{-1} y_\ell)(y_k^T V^{-1} x_\ell)\}\Big|_{\gamma = \hat{\gamma}_t} \tag{1.9.11}$$

$$[g_t]_k = x_k^T\{V^{-1}(S - \hat{\Sigma})V^{-1}\}y_k\Big|_{\gamma = \hat{\gamma}_t} \quad . \tag{1.9.12}$$

Note that (1.9.11) and (1.9.12) not only involve less computation than (1.9.7) and (1.9.8) but also involve less computer storage since the 2p elements of x_k and y_k replace the $\frac{1}{2}p(p+1)$ elements of $\partial\Sigma/\partial\gamma_k$ and many x_k and y_k vectors are common to several derivatives. Further gains in efficiency may be made by storing $V^{-\frac{1}{2}}x_k$, $V^{-\frac{1}{2}}y_k$ and $V^{-\frac{1}{2}}(S-\hat{\Sigma})V^{-\frac{1}{2}T}$, where $V^{-\frac{1}{2}}$ is the inverse of the Choleski square root of V. Use is then made of

$$x_k^T V^{-1} y_k = (V^{-\frac{1}{2}} x_k)^T (V^{-\frac{1}{2}} y_k) \tag{1.9.13}$$

and

$$x_k^T V^{-1}(S-\hat{\Sigma})V^{-1}y_k = (V^{-\frac{1}{2}}x_k)^T(V^{-\frac{1}{2}}(S-\hat{\Sigma})V^{-\frac{1}{2}T})(V^{-\frac{1}{2}}y_k). \tag{1.9.14}$$

Results (1.9.11) and (1.9.12) provide an alternative justification for a lemma due to Jöreskog (1977, Section 2.4).

The Fisher scoring algorithm for obtaining a MWL solution is ob= tained merely by redefining V in (1.9.11) and (1.9.12) or in (1.9.13) and (1.9.14) on each iteration as

$$V = \{\Sigma(\hat{\gamma}_t)\}^{-1} \tag{1.9.15}$$

replacing $F(S;\Sigma(\hat{\gamma}_t)|U)$ in (1.9.5) and (1.9.6) by $F(S;\Sigma(\hat{\gamma}_t)|V = \Sigma(\hat{\gamma}_t))$ and replacing $F(S;\Sigma(\hat{\gamma}_t)|U)$ in (1.9.4) by $F_{MWL}(S;\Sigma(\hat{\gamma}_t))$.

Similarly the biweight solution is obtained merely by redefining U^{-1} in (1.9.2), (1.9.3), (1.9.4), (1.9.5), (1.9.6) on each iteration to be the diagonal matrix with diagonal elements given by (1.5.36).

It is of interest to examine the terms to be added to the approxi= mate Hessian of $\frac{1}{2}F$ employed in the Gauss-Newton algorithm to yield the exact Hessian employed in the Newton method. The term to be added to the $k\ell$-th element of H_t in (1.9.2) is

$$[\delta H_t]_{k\ell} = -(s^* - \hat{\sigma}^*)^T U^{-1} \left.\frac{\partial^2 \sigma^*}{\partial\gamma_k \partial\gamma_\ell}\right|_{\gamma = \hat{\gamma}_t} \qquad (1.9.16)$$

which becomes

$$[\delta H_t]_{k\ell} = -\tfrac{1}{2}\mathrm{tr}\left[V^{-1}(S-\Sigma)V^{-1} \frac{\partial^2 \Sigma}{\partial\gamma_k \, \partial\gamma_\ell}\right]\Bigg|_{\gamma = \hat{\gamma}_t} \qquad (1.9.17)$$

if U has the structure (1.5.15) arising from the assumption of no kur= tosis and H_t can be expressed as in (1.9.7). Both (1.9.16) and (1.9.17) apply when U or V are fixed matrices. The appropriate correction to (1.9.7) to yield the Hessian of $\frac{1}{2}F_{MWL}(S;\Sigma)$ is

$$[\delta H_t]_{k\ell} = -\tfrac{1}{2}\mathrm{tr}\left[\Sigma^{-1}(S-\Sigma)\Sigma^{-1}\{ \frac{\partial^2 \Sigma}{\partial\gamma_k \, \partial\gamma_\ell} - 2\,\frac{\partial\Sigma}{\partial\gamma_k}\,\Sigma^{-1}\,\frac{\partial\Sigma}{\partial\gamma_\ell}\right]\Bigg|_{\gamma = \hat{\gamma}_t}. \qquad (1.9.18)$$

It can be seen from (1.9.16), (1.9.17) and (1.9.18) that the approximation to the Hessian employed in the Gauss-Newton and Fisher scoring methods will be close to the true Hessian at the minimum of the discrepancy function if the elements of $(S-\hat{\Sigma})$ are small and the second derivatives are not large. Cases where the elements of $(S-\hat{\Sigma})$ are large so that the fit of the model is poor are not of great interest but it is possible that the Gauss-Newton or Fisher scoring algorithms will not be effective if the covariance structure model is highly nonlinear and second derivatives are consequently large. The results given here may, however, be used to implement Newton's method at the expense of a more complicated computer program and more computation per iteration.

It is sometimes necessary to apply equality constraints of the form of (1.5.37). This can usually be done by means of a reparameteri= sation and subsequent unconstrained minimisation. It is, however, some= times convenient to impose constraints directly. One situation where this is so is in the analysis of a correlation structure $P(\gamma_\rho)$. A co= variance structure of the form of (1.2.5) is employed and the set of equality constraints

$$\mathrm{Diag}\{P(\gamma_\rho)\} = I$$

is imposed. While reparameterisations can also generally be employed in this situation, the use of (1.2.5) together with equality constraints avoids the need for using formulae of the type of (1.6.24) when obtain= ing estimates of standard errors of elements of $\hat{\gamma}_\rho$. Furthermore, re=

parameterisations sometimes result in a need for complicated inequality constraints.

Methods for constrained optimisation which employ equality constraints to reduce the degrees of freedom involved in the optimisation process analogously to what is done in reparameterisation are available. One of these is an effective adaptation of the Broyden-Fletcher-Goldfarb-Shanno procedure to constrained optimisation due to Powell (1978a, 1978b). Another is the following adaptation (Browne & Du Toit, 1979) of the Gauss-Newton algorithm.

Let

$$\begin{pmatrix} \delta_t \\ \lambda_t \end{pmatrix} = \begin{pmatrix} H_t & L_t^T \\ L_t & 0 \end{pmatrix}^{-1} \begin{pmatrix} g_t \\ -h_t \end{pmatrix} \qquad (1.9.19)$$

where (cf. (1.5.37), (1.6.15))

$$h_t = h(\hat{\gamma}_t) \qquad (1.9.20)$$

$$L_t = \left.\frac{\partial h}{\partial \gamma^T}\right|_{\gamma = \hat{\gamma}_t} . \qquad (1.9.21)$$

Then

$$\hat{\gamma}_{t+1} = \hat{\gamma}_t + \alpha_t \, \delta_t \qquad (1.9.22)$$

where (cf. Powell, 1978a, Section 4) α_t is chosen initially as 1 and is halved successively until

$$F(S;\Sigma(\hat{\gamma}_{t+1})) + 2 \sum_{i=1}^{r} \left| [\lambda_t]_i [h(\hat{\gamma}_{t+1})]_i \right|$$

$$< F(S;\Sigma(\hat{\gamma}_t)) + 2 \sum_{i=1}^{r} \left| [\lambda_t]_i [h(\hat{\gamma}_t)]_i \right| \qquad (1.9.23)$$

Note that the same Lagrange multipliers λ_t appear on both sides of the inequality. After convergence of the algorithm an estimate of the covariance matrix of the estimator $\hat{\gamma}$ may be obtained by multiplying the first q x q principal submatrix of the inverse matrix in (1.9.19) by n^{-1} (cf. (1.6.17), (1.6.19)).

This section has provided a number of general formulae for the gradient (1.9.3), (1.9.8), (1.9.12), for the approximate Hessian of the Gauss-Newton method (1.9.2), (1.9.7), (1.9.11) and corrections yielding the exact Hessian (1.9.16), (1.9.17), (1.9.18) which are easily specialised to specific covariance structures merely by obtaining deriva=

tives of the form $\frac{\partial \Sigma}{\partial \gamma_i}$ (and, for Newton's method only, $\frac{\partial^2 \Sigma}{\partial \gamma_i \partial \gamma_j}$). The mathematical aspects of providing an algorithm for obtaining estimates of parameters in a new covariance structure are therefore minimal (un= less further simplifications can be found for the specific covariance structure).

1.10 General Remarks

Before specific covariance structures are considered in Part II general points to be borne in mind will be discussed briefly.

One matter of importance is nonnormality of the data, and in particular kurtosis. In the majority of practical applications of mo= dels for covariance matrices reported to date no information concern= ing kurtosis of the data has been reported. While this is of less im= portance as far as the estimates are concerned since the estimates will generally be consistent in any case, it is crucial for the interpreta= tion of the test statistic and standard errors if the usual maximum likelihood (MWL) or generalised least squares (1.5.17) procedures are employed. The distribution of these particular test statistics under the null hypothesis will have the usually specified chi-squared dis= tribution and the usual estimates of standard errors will be consistent for the true standard errors only if the distribution of the data has no kurtosis. Consequently an estimate of a coefficient of multivariate kurtosis (Mardia, 1974) ought to be provided. If, as is generally the case, the model is invariant under a constant scaling factor, the cor= rection for kurtosis in (1.7.5) or (1.7.6) derived under the assump= tion of an elliptical distribution may provide a more robust test sta= tistic.

If the coefficient of kurtosis is appreciable and if p is not large an asymptotically distribution free procedure with U defined by (1.5.30) or (1.5.31) may be used. Examples are provided in Section 2.9.

An alternative is to employ (1.7.14) for the test statistic, and (1.6.4) or (1.6.14) with (1.6.20) for standard errors. These results may be employed in conjunction with MWL, MINRES or GLS estimates, and are asymptotically valid in the presence of kurtosis.

There are other possible approaches. One is to apply nonlinear transformations (e.g. Box & Cox, 1964) to individual variables so as to make estimated univariate coefficients of relative kurtosis (cf. Section 1.5) close to unity. This procedure may result in a value of the estimate of multivariate relative kurtosis (1.7.3) which is close to unity so that the standard MWL approach may be tried. There is no guarantee, however, that the estimate of multivariate relative kurto= sis will be close to unity. It should also be borne in mind that the same covariance structure will generally not hold before and after nonlinear transformations are applied. Furthermore, asymptotic dis= tributions for the resulting estimates and test statistic will not be known if the method employed for choosing a transformation can result in different transformations in different samples from the same popu= lation.

Another approach would be to make use of a robust estimate of the covariance matrix (cf. Devlin, Gnadadesikan & Kettenring, 1975; Campbell, 1980). Again, asymptotic standard errors for the resulting estimators of γ and a test statistic are not available.

It should be borne in mind that if a covariance structure holds for some continuous vector variate x and if x is converted into a discrete vector variate x^*, then x^* will not in general have the same covariance structure. This matter is discussed in detail by Olsson (1979) who provides examples of how the discretisation process affects the covariance structure.

Caution should be exercised when applying a computer program intended for covariance structures to a correlation matrix. One should verify that the model is invariant under changes of scale so that ana= lysing the correlation matrix is equivalent to analysing the covariance matrix. This is not always the case and the literature contains nume= rous examples where a correlation matrix is inappropriately treated as a covariance matrix. Even if the model is invariant under changes of scale, the estimates obtained from the correlation matrix should be rescaled so that the reproduced correlation matrix has unit diago= nals (Krane & McDonald, 1978). In some special cases, such as in MWL estimation for the unrestricted factor analysis model, this rescaling is not necessary. Finally it is important to remember that standard errors for estimators obtained from a correlation matrix differ from those obtained from a covariance matrix (cf. (1.6.24), (1.6.25)).

PART II

SPECIFIC APPLICATIONS

2.1 Introductory comments

Virtually all work on covariance structures has involved the assumption of multivariate normality and in the majority of instances the method of maximum likelihood has been employed. Because of the restrictiveness of this assumption and because Part I has provided general results for the analysis of covariance structures which are readily adapted to any specific situation, Part II will concentrate rather on possible models for covariance matrices and give less atten= tion to computational procedures.

2.2 Linear covariance structures

In linear covariance structures the elements of Σ are linear functions of certain unknown parameters. The generalised least squares estimates can always be expressed in closed form but MWL estimates in general require an iterative procedure (Anderson, 1969,1970,1973). There are exceptions, however, where MWL estimates for linear covariance structures can be expressed in closed form and these occur particular= ly in situations where Σ and Σ^{-1} have the same structural form. A com= prehensive review of linear covariance structures where MWL estimates can be expressed in closed form is given by Mukherjee (1970).

One class of linear covariance structures is the class of pat= terned covariance matrices (McDonald, 1974; Browne, 1977, Section 5). Some examples of patterned covariance matrices follow.

The structure of complete symmetry (Wilks, 1946)

This is the structure which arises when p parallel measurements are made on each subject. The diagonal elements of Σ are equal and the nondiagonal elements are equal so that Σ is of the form

$$\Sigma = \begin{pmatrix} \gamma_1 & \gamma_2 & \gamma_2 & \gamma_2 \\ \gamma_2 & \gamma_1 & \gamma_2 & \gamma_2 \\ \gamma_2 & \gamma_2 & \gamma_1 & \gamma_2 \\ \gamma_2 & \gamma_2 & \gamma_2 & \gamma_1 \end{pmatrix} \tag{2.2.1}$$

The MWL solution may be expressed in closed form.

The Toeplitz structure

This is the structure which arises when the rows of X represent independent replications of a *stationary* time series. The covariance matrix then is of the form

$$\Sigma = \begin{bmatrix} \gamma_1 & \gamma_2 & \gamma_3 & \gamma_4 \\ \gamma_2 & \gamma_1 & \gamma_2 & \gamma_3 \\ \gamma_3 & \gamma_2 & \gamma_1 & \gamma_2 \\ \gamma_4 & \gamma_3 & \gamma_2 & \gamma_1 \end{bmatrix} \tag{2.2.3}$$

The parallel batteries structure

When a battery of tests is applied on several different occa= sions or parallel batteries are employed, the covariance matrix Σ will have the structure

$$\Sigma = \begin{bmatrix} \Gamma & \dot{\Gamma} & \dot{\Gamma} & \dot{\Gamma} \\ \dot{\Gamma} & \Gamma & \dot{\Gamma} & \dot{\Gamma} \\ \dot{\Gamma} & \dot{\Gamma} & \Gamma & \dot{\Gamma} \\ \dot{\Gamma} & \dot{\Gamma} & \dot{\Gamma} & \Gamma \end{bmatrix} \tag{2.2.4}$$

where Γ is a symmetric matrix and $\dot{\Gamma}$ differs from Γ only in the diago= nal elements.

Another class of linear covariance structures is of the form of the factor analysis structure (1.1.15) in which Λ is known. Then Σ is of the form,

$$\Sigma = A \, \Phi \, A^T + D_\psi \tag{2.2.5}$$

where A is a known p x m matrix. Models of this type were introduced by Bock and Bargmann (1966) and also considered by Wiley, Schmidt and Bramble (1973) and Browne (1974, Section 4).

Consider an experiment in which a number of measurements under different treatment conditions are made on each of a number of subjects. If the usual assumptions are made for a mixed model analysis of vari= ance with subject effects random and all treatment effects fixed, the covariance structure is of the form of (2.2.5) with A a specified de= sign matrix with elements in the first column equal to one. In such a situation Φ is a diagonal matrix with subsets of equal diagonal ele= ments, D_ψ has equal diagonal elements and A need not be of full column rank. In practice, it is often not justifiable to assume homoscedasti= city of random treatment-subject interaction effects or of errors (Bock & Bargmann, 1966), or independence of interaction effects (Wiley, Schmidt & Bramble, 1973), and the less restricted model (2.2.5) is more plausible. In such a situation A is chosen as a basis of the design matrix of full column rank in order to ensure that the parameters are identified.

Another form of the linear covariance structure (2.2.5) is the quasi-Wiener simplex (Jöreskog, 1970b) in which Φ is diagonal and the known matrix A is lower triangular with all elements on and below the diagonal equal to unity. Additional information on simplex models is contained in Section 2.7.

2.3 Patterned correlation matrices

If the population correlation matrix P has a pattern of equal and/or null coefficients the corresponding covariance structure is

$$\Sigma = D_{\zeta} \; P \; D_{\zeta} \tag{2.3.1}$$

where the diagonal elements of D_{ζ} are standard deviations. McDonald (1975) has considered MWL estimates. Browne (1977) made use of the linearity of the correlation structure and the asymptotic distribution of the correlation coefficients to obtain generalised least squares estimates in closed form. Steiger (1980a, 1980b) considered the use of normalising z-transforms of correlation coefficients. Examination of the results of Steiger's (1980b) Monte Carlo experiments indicates that the z-transform does not seem to result in much improvement in the approximation of the chi-squared distribution to the distribution of the quadratic form statistic under the null hypothesis in general correlation patterns characterised by equality of correlation coeffi= cients or specified values of zero. The z-transformation did, however, result in a marked improvement when P was a known matrix with large elements.

Steiger and Hakstian (1980) provided formulae for asymptotic variances and covariances of correlation coefficients without assump= tions of normality. These may be employed for the analysis of patterned correlation matrices by generalised least squares.

2.4 Unrestricted factor analysis

The factor analysis model was discussed earlier in Section 1.1. This model is known as unrestricted when no equality constraints are imposed on the elements of Λ and D_{ψ}. It is then common practice to set $\Phi = I$, initially at least, to avoid lack of identification of the ele= ments of Λ and Φ. The covariance structure then is

$$\Sigma = \Lambda\Lambda^{T} + D_{\psi}. \tag{2.4.1}$$

Additional restrictions on Λ are also required to ensure identi= fication of Λ(cf. Section 1.3). One possibility which is mathematical= ly convenient for maximum likelihood estimation is that $\Lambda^T D_\psi^{-1} \Lambda$ should be a diagonal matrix. Both these restrictions and possibly those that $\Phi = I$ are relaxed at a later stage during the rotation process.

The estimation process will require an iterative algorithm. When the assumption of no kurtosis is made, however, it is possible to in= crease the efficiency of the algorithm by use of a nesting process (Jöreskog, 1967; Jennrich & Robinson, 1969; Clarke, 1970). The condi= tional minimum of the discrepancy function with respect to Λ given D_ψ can be expressed in closed form as a function of certain eigenvalues and eigenvectors. The minimisation is then carried out in terms of the p diagonal elements of D_ψ only. A clear description covering a variety of discrepancy functions is given by Swain (1975a). This sim= plification of the algorithm is no longer possible if the general forms for U in (1.5.30) or (1.5.31) are employed.

A fairly frequent problem in practical applications of factor analysis is the occurrence of estimates on the boundary of the para= meter space G with zero diagonal elements of $\hat{D}_\psi$. An interesting ap= proach in this situation (van Driel, Prins & Veltkamp, 1974; van Driel, 1978) is to no longer restrict the estimates to the interpretable parameter space and allow diagonal elements of $\hat{D}_\psi$ to be negative. Con= fidence intervals are employed subsequently to decide whether or not all diagonal elements of D_ψ are positive and consequently interpret= able.

Because of certain similarities in the computational procedures and in the results obtained, the method of principal components is quite frequently confused with factor analysis. There are, however, major differences in rationale. The method of principal components does not involve any covariance structure or any formal model. It is a descriptive data analytic procedure which entails the construction of a small set of linear combinations or principal components of a large set of observed variables in such a manner as to minimise the sum of the partial variances of the original variables given the components (cf. Okamoto, 1969, Section 4.2). In factor analysis on the other hand the factors are not functionally related to the observed variables and are chosen in such a way as to minimise partial correlations between the observed variables given the factors (cf. e.g. Lawley & Maxwell, 1971, Section 1.2). Thus the aim of principal components is to replace the original variables with a new smaller set of components so as to minimise loss of information in a specific sense, while the aim of factor analysis is to explain the correlation between observed vari= ables.

After an estimate, $\hat{\Lambda}$, has been obtained the numerically conve= nient identification conditions originally imposed (e.g. $\hat{\Lambda}\hat{D}_{\psi}^{-1}\hat{\Lambda}$ is a diagonal matrix) are discarded and are replaced by more interpretable conditions. This implies a transformation of the type

$$\hat{\Lambda}^{+} = \hat{\Lambda}T \qquad (2.4.2)$$

and an attempt is made to choose T in such a way that $\hat{\Lambda}^{+}$ has what was referred to by Thurstone (1935,1947) as simple structure. The concept of simple structure is not rigorously defined and there is no univer= sal agreement as to what exactly was intended by Thurstone (Yates,1979). Generally simple structure implies that there should be a fair number of zero or near zero elements of $\hat{\Lambda}^{+}$ thereby simplifying interpretation. The majority of methods of rotation in present use aim for what is called "perfect simple structure" in which each row of $\hat{\Lambda}^{+}$ has one substantial element and the remainder are near zero. Then $\hat{\Lambda}^{+}$ is of the form

$$\hat{\Lambda}^{+} = \begin{pmatrix} x & 0 & 0 \\ x & 0 & 0 \\ x & 0 & 0 \\ 0 & x & 0 \\ 0 & x & 0 \\ 0 & x & 0 \\ 0 & 0 & x \\ 0 & 0 & x \\ 0 & 0 & x \end{pmatrix} \qquad (2.4.3)$$

where x represents a substantial loading and 0 represents a near zero loading. This view of simple structure, however, is not held by Yates (1979) who requires at least one near zero loading per row.

Generally the rotation is accomplished by optimising some func= tion $\varphi(\hat{\Lambda}^{+})$ of the elements of $\hat{\Lambda}^{+}$. Two classes of rotation procedure may be distinguished. In orthogonal rotation the assumption that the factors are uncorrelated is retained and T is chosen to be orthogonal, i.e.

$$TT^{T} = I.$$

In oblique rotation the assumption that the factors are uncorrelated is discarded but the assumption that they have unit variances is retained in order to facilitate the comparison of loadings within rows of $\hat{\Lambda}^+$. Then T is chosen so that the diagonal elements of

$$\hat{\Phi} = (T^T T)^{-1} \tag{2.4.4}$$

are all equal to one.

The most popular method of orthogonal rotation is the Varimax method (Kaiser, 1958, 1959) in which the sum of within column varian= ces of squared elements of $\hat{\Lambda}^*$ is maximised. An appealing modification involving a differential weighting system has been proposed by Cureton and Mulaik (1975).

Although oblique rotation is more appealing than orthogonal ro= tation, the lack of a really satisfactory automatic procedure preclu= ded its general use for a number of years. The main shortcoming of the earlier methods was that the condition that the diagonal elements of $\hat{\Phi}$ be equal to one was not imposed during the course of the itera= tive process. A method for doing this was discovered by Jennrich and Sampson (1966) who concentrated primarily on the minimisation of the quartimin criterion

$$\varphi(\hat{\Lambda}^+) = \sum_{i=1}^{p} \sum_{j=1}^{m} \sum_{k \neq j} \hat{\lambda}_{ij}^{+2} \hat{\lambda}_{ik}^{+2}. \tag{2.4.5}$$

While modifications of this criterion have been proposed, careful exa= mination of the results of artificial experiments carried out by Haks= tian and Abell (1974) and by Crawford (1975) will not reveal any con= sistent advantage of the other criteria.

Bargmann (1956) has suggested a test for simple structure and corresponding tables are given by Cattell (1978, Appendix A6). This test involves the null hypothesis that points representing measure= ments are scattered at random on a hypersphere. It does not take into account random sampling fluctuations in estimates of factor loadings resulting in imprecise location of measurement points on the hyper= sphere.

Methods for orthogonal (Browne, 1972a) and oblique (Browne,1972b) rotation of a factor matrix to a partially specified target are avail= able. The sum of squares of factor loadings in specified positions of the factor matrix is minimised. These procedures may be employed either to improve on a previously obtained simple structure or, as an alter= native to restricted factor analysis (Section 2.6) when knowledge of the measurements enables one to specify the positions of near zero factor loadings.

When MWL estimates are employed, approximate standard errors of rotated factor loadings may be obtained using methods given by Archer and Jennrich (1973) and Jennrich (1974) in orthogonal rotation and by Jennrich (1973a) in oblique rotation, provided that the assumption of no kurtosis is justified. A computationally less expensive method in= volving an application of (1.6.14) is given by Jennrich and Clarkson (1980) and applies to MWL estimates whether or not the assumption of no kurtosis is justified. Jennrich (1973b) gives an interesting dis= cussion of the effect of rotation on standard errors.

2.5 Multiple battery factor analysis

The multiple battery factor analysis model is employed when re= lationships between k batteries of measurements are of primary interest and relationships within each battery are unimportant. It is of the form

$$\Sigma = \Lambda\Lambda^T + \Psi \qquad (2.5.1)$$

where Ψ is a symmetric block diagonal matrix of the form

$$\Psi = \begin{pmatrix} \Psi_{11} & 0 & \dots & 0 \\ 0 & \Psi_{22} & \dots & 0 \\ & & & \\ \cdot & \cdot & & \cdot \\ 0 & 0 & \dots & \Psi_{kk} \end{pmatrix} \qquad (2.5.2)$$

This model is invariant (cf. Section 1.2) under block diagonal trans= formations and is consequently also invariant under changes of scale.

Maximum likelihood estimation is considered by Browne (1980). In particular when the number of batteries is equal to two the model is known as Tucker's (1958) inter-battery factor analysis model. Maxi= mum likelihood estimates may be obtained (Browne, 1979) by merely re= scaling canonical loadings obtained from a canonical correlation ana= lysis.

Similar rotation procedures to those of standard unrestricted factor analysis may be employed.

2.6 Nested models

Jöreskog (1969) considered a restricted factor analysis model of the form

$$\Sigma = \Lambda\Phi\Lambda^T + \Psi \qquad (2.6.1)$$

where any of the elements of Λ or Φ and the diagonal elements of the diagonal matrix Ψ can have specified values (usually zero) or be equal to other elements. Subsequently (Jöreskog, 1970a, 1974) he considered a more general second order factor analysis model of the form

$$\Sigma = \Lambda_2(\Lambda_1\Phi_1\Lambda_1^T + \Psi_1)\Lambda_2^T + \Psi_2 \qquad (2.6.2)$$

where again elements of the parameter matrices Λ_2, Λ_1, Φ_1 and the dia= gonal parameter matrices Ψ_2, Ψ_1 may have elements with specified values or be equal to other elements. Clearly (2.6.1) is a special case of (2.6.2) with $\Lambda_2 = I$ and $\Psi_2 = 0$. Many examples of the application of (2.6.1) and (2.6.2) are contained in Part I of Jöreskog & Sörbom (1979).

A structural equation model involving latent variables was pro= posed by Jöreskog (1977). Suppose that the factors can be separated into two sets z_1 and z_2 which satisfy the linear structural relations

$$z_1 = B\,z_1 + \Gamma\,z_2 + u \qquad (2.6.3)$$

where the elements of u represent uncorrelated errors and

$$\begin{aligned} \mathrm{Cov}(u,u^T) &= \Psi \\ \mathrm{Cov}(z_1,u^T) &= 0 \\ \mathrm{Cov}(z_2,u^T) &= 0 \\ \mathrm{Cov}(z_2,z_2^T) &= \Phi. \end{aligned}$$

Thus z_1 represents latent dependent variates and z_2 represents latent independent variates.

Suppose now that the observed variates x_1 are influenced by la= tent dependent variates z_1 and the observed variates x_2 are influenced by the latent independent variates z_2:

$$x_1 = \Lambda_1\,z_1 + e_1 \qquad (2.6.4a)$$

$$x_2 = \Lambda_2\,z_2 + e_2 \qquad (2.6.4b)$$

where $\mathrm{Cov}(e_1,e_1^T) = \Theta_1$

$\mathrm{Cov}(e_2,e_2^T) = \Theta_2$

and e_1 and e_2 are uncorrelated with all other latent variates.

It then follows that the covariance matrix satisfies Jöreskog's LISREL (Linear Structural Relations) model:

$$\Sigma = \mathrm{Cov}(x,x^T) = \begin{pmatrix} \Sigma_{11} & \Sigma_{12} \\ \Sigma_{21} & \Sigma_{22} \end{pmatrix}$$

$$= \begin{bmatrix} \Lambda_1(I-B)^{-1}(\Gamma\Phi\Gamma^T + \Psi)(I-B^T)^{-1}\Lambda_1^T + \Theta_1 & \Lambda_1(I-B)^{-1}\Gamma\Phi\Lambda_2^T \\ \Lambda_2\Phi\,\Gamma^T(I-B^T)^{-1}\Lambda_1^T & \Lambda_2\Phi\,\Lambda_2^T + \Theta_2 \end{bmatrix} \quad (2.6.5)$$

It can be seen that (2.6.2) may be regarded as a special case of (2.6.5) by omitting x_2 and letting $B = 0$. Numerous applications of the LISREL model are contained in Part II of Jöreskog and Sörbom (1979).

Apparently more general models were suggested by Bentler (1976) and by McDonald (1978,1980). It has recently been found, however, that any covariance structure arising from linearly related latent and ob= served variables, including these two models, may be expressed as a special case of any one of several mathematically equivalent simpli= fied forms of (2.6.5)(McArdle, 1979; Bentler & Weeks, 1979,1980). One of these, proposed by Bentler and Weeks (1980, equation (2.5)), is obtained from (2.6.5) by omitting Ψ, Θ_1 and Θ_2 and replacing Λ_1 and Λ_2 by known selection matrices formed from some of the rows of an iden= tity matrix. An alternative mathematically equivalent general model with fewer, but larger and sparser, parameter matrices is advocated by McArdle (1979, equation (14)). These models, although they are special cases of (2.6.5), also contain (2.6.5) as a special case.

2.7 Time series structures

Guttman (1954) proposed his "simplex" as a model for correlation matrices obtained from measurements of increasing complexity. A per= fect simplex is a correlation matrix P whose elements satisfy the mo= del

$$\rho_{ij} = \tau_i/\tau_j \quad (2.7.1)$$

where τ_i is a complexity parameter associated with the i-th measure= ment with

$$\tau_i > 0 \quad , \quad i = 1...p \quad (2.7.2)$$

and

$$\tau_1 \le \tau_2 \le \ldots \tau_p \,. \quad (2.7.3)$$

Equation (2.7.1) involves a knowledge of the complexity order (2.7.3) which may not be available a priori in practice. Schönemann (1970,equation 15) expressed (2.7.1) in the equivalent form

$$\rho_{ij} = \min(\tau_i,\tau_j)/\max(\tau_i,\tau_j) \quad (2.7.4)$$

to make an arbitrary ordering of the variables possible.

Anderson (1960) showed that P, with elements defined by (2.7.1), is the correlation matrix of a Markov process. Alternatively P is the correlation matrix of a first order autoregressive time series with nonhomogeneous weights.

It was pointed out by Jöreskog (1970b) that, if (2.7.3) holds, the covariance matrix corresponding to P may be reparameterised as

$$\Sigma = D_\alpha\ A\ D_\varphi\ A^T\ D_\alpha \tag{2.7.5}$$

where D_α and D_φ are diagonal matrices with D_φ nonnegative definite and A is a lower triangular matrix with all elements on and below the diagonal equal to unity. He referred to the covariance structure (2.7.5) as the Markov simplex and to the special case with $D_\alpha = I$ as the Wiener simplex.

It is usually the case in practice that one has to allow for errors of measurement in observed variables. If the correlation ma= trix of true scores is a simplex the covariance matrix of observed variables will have the general structure

$$\Sigma = D_\alpha(A\ D_\varphi\ A^T + D_\psi)D_\alpha \tag{2.7.6}$$

where D_ψ is diagonal. This is known as a quasi-Markov simplex and the special case with $D_\alpha = I$ as a quasi-Wiener simplex (Jöreskog,1970b). The quasi-Markov simplex has three indeterminacies (cf.Jöreskog,1970b) requiring identification conditions.

The parameterisation of the quasi-Markov simplex in (2.7.6) is a special case of (2.6.2). When applied to psychological tests, rather than measurements repeated over time, it has the disadvantage that prior knowledge of the ordering (2.7.3) is required. This may be avoi= ded by making use of Schönemann's formulation and expressing the quasi-Markov simplex as

$$\Sigma = D_\zeta(P + D_\nu)D_\zeta \tag{2.7.7}$$

where D_ζ and D_ν are diagonal, D_ν is nonnegative definite and the ele= ments of P are functions of $\tau_1 \ldots \tau_p$ given by (2.7.4). Note that P here represents a true score correlation matrix and not an observed variable correlation matrix.

There are some indeterminacies in the model (Anderson, 1960,pp. 211-212; Schönemann, 1970,p.9) which should be borne in mind when es= timates of the complexity parameters are interpreted. First of all, the direction of complexity is unknown since (2.7.4) will hold if τ_i

is replaced by τ_i^{-1}, $i = 1,2,\ldots,p$. Then the scale employed is arbi= trary since all the τ_i may be multiplied by any positive constant. The scale may be fixed by imposing the identification condition that the geometric mean of the τ_i be equal to unity. Finally the smallest and the largest of the τ_i are indeterminate and their values will depend on the identification conditions employed. Possible identifi= cation conditions are to require that the diagonal elements, ν_{ii}, of D_ν corresponding to the two smallest of the τ_i be equal, and also those corresponding to the two largest of the τ_i.

When the same measurement process is repeated over time, and the time ordering of variables is consequently known, an autoregres= sive time series with moving average residuals may be an appropriate model. Under certain assumptions on the initial state of the process (Anderson, 1975) the covariance structure will be of the form (cf. (2.6.5))

$$\Sigma = (I-B)^{-1}\Gamma D_\varphi \Gamma^T (I-B^T)^{-1} \tag{2.7.8}$$

where (I-B) is a lower triangular matrix of autoregressive weights, Γ is a lower triangular matrix of moving average weights, and D_φ is a diagonal matrix of residual variances. The number of non null diago= nals below the principal diagonals of (I-B) and Γ respectively give the autoregressive and moving average orders of the process. Anderson (1975,1977) considered MWL estimation for stationary processes in which (I-B) and Γ have homogeneous diagonals and the elements in the principal diagonal of D_φ are equal. These assumptions may be relaxed to give time series with nonhomogeneous weights and nonhomogeneous variances (Du Toit, 1979). The Markov simplex is the covariance struc= ture of a first-order autoregressive time series with nonhomogeneous parameters (Jöreskog, 1970b) while the quasi-Markov simplex is the covariance structure of a first-order autoregressive time series with first-order moving average residuals and nonhomogeneous parameters (Du Toit,1979). Alternatively the quasi-Markov simplex may be regarded as the covariance structure of a first-order autoregressive time se= ries with nonhomogeneous parameters contaminated by error of measure= ment (Jöreskog, 1970b).

2.8 Direct product structures

In some situations measurements are made under all combinations of two sets of conditions. An example of this is Campbell & Fiske's (1959) multitrait-multimethod approach where each of several psycholo=

gical traits is measured by each of a number of methods. While various restricted factor analysis models may be applied to data of this type (Jöreskog, 1974; Browne, 1980) alternative models involving direct products have been suggested.

Swain (1975b) proposed a simple model in which the covariance matrix is expressed as the direct product of two symmetric parameter matrices (cf. (1.2.3))

$$\Sigma = \Sigma_m \otimes \Sigma_t \tag{2.8.1}$$

where, in the case of multitrait-multimethod data, Σ_m is interpreted as a method covariance matrix and Σ_t as a trait covariance matrix. He provided an easily implemented computational procedure for obtaining MWL estimates and gave practical examples where the model proved useful for the interpretation of data.

Tucker (1966) provided a generalisation of principal components analysis to three modes which was expressed as a factor analysis model by Bentler and Lee (1979) who also considered the estimation of parameters. This model is:

$$\Sigma = (\Lambda_m \otimes \Lambda_t)\Gamma\Phi\Gamma^T(\Lambda_m^T \otimes \Lambda_t^T) + D_\psi \tag{2.8.2}$$

where, in the case of multitrait-multimethod data, Λ_m is a factor matrix for methods, Λ_t is a factor matrix for traits, Γ is a matrix representation of Tucker's (1966) core box, Φ is a symmetric factor covariance matrix and D_ψ is a diagonal residual covariance matrix. The large number of parameters in this model can lead to difficulty in interpretation.

A simpler version of (2.8.2) in which $\Gamma\Phi\Gamma^T = I$ was investigated by Bentler and Lee (1978) and may be expressed in the form (cf.(2.8.1))

$$\Sigma = (\Lambda_m\Lambda_m^T) \otimes (\Lambda_t\Lambda_t^T) + D_\psi \tag{2.8.3}$$

A modification of this model which involves discarding certain columns of $(\Lambda_m \otimes \Lambda_t)$ and replacing D_ψ by a matrix with some non-null nondiagonal elements was employed by McDonald (1980, Section 4).

A common feature of all models considered in this section is that they are not invariant under changes of scale (cf. Section 1.2). Consequently the use of sample correlation matrices is not appropriate when testing their fit.

2.9 Some numerical examples

This section will provide some examples of covariance struc= tures fitted to data. These have been chosen mainly to illustrate cer= tain points made in Part I.

We first consider some artificial data generated to illus= trate the sensitivity of tests based on the assumption of multivariate normality to kurtosis. An 8 x 8 population covariance matrix with all diagonal elements equal,

$$\sigma_{ii} = 1 \qquad , i = 1 \dots 8$$

and all nondiagonal elements equal

$$\sigma_{ij} = .5 \qquad , \quad i \neq j$$

was chosen. Two random samples of size 500 from distributions with this covariance matrix were generated. The first was from an appropria= tely rescaled multivariate chi-squared distribution (Johnson & Kotz, 1972, Section 40.3; Krishnaiah & Rao, 1961) with 2 degrees of freedom and all marginal relative kurtosis coefficients equal to 3. Table 1 shows the sample covariance matrix, S, obtained.

Table 1: *S : Rescaled multivariate chi-squared distribution*

1.10							
.57	1.06						
.62	.63	1.14					
.65	.61	.64	1.07				
.61	.53	.69	.62	1.06			
.51	.51	.58	.56	.50	.92		
.65	.69	.69	.72	.68	.59	1.31	
.65	.61	.71	.66	.64	.56	.69	1.12

The second random sample was generated from a multivariate nor= mal distribution. Table 2 shows the sample covariance matrix obtained. Examination of these tables will show that the sample covariances tend to be closer to .5 and the sample variances tend to be closer to 1 in the case of the multivariate normal distribution than in that of the rescaled multivariate chi-squared distribution. The estimate, S, of Σ obtained from a sample of size 500 from a multivariate normal distribution was more precise than the estimate of Σ obtained from a sample of the same size from the leptokurtic rescaled multivariate chi-squared distribution. This is to be expected since a leptokurtic

Table 2: *S :Multivariate normal distribution*

.97							
.51	1.02						
.48	.50	1.05					
.43	.52	.47	.88				
.50	.50	.48	.46	.88			
.48	.50	.50	.45	.45	.97		
.46	.48	.47	.42	.42	.46	.90	
.51	.50	.48	.47	.45	.50	.42	.92

distribution results in higher variances of the estimators, s_{ij}, of elements of Σ(cf. (1.5.10), (1.5.11)). The estimate $\hat{\eta}$ (cf. (1.7.3)) of the multivariate coefficient of relative kurtosis was 1.94 for the multivariate chi-squared distribution and 1.00 for the multivariate normal distribution.

The population covariance matrix Σ which was chosen satisfies the model (cf. (1.1.16))

$$\Sigma = 1\,\varphi 1^T + \psi\, I \qquad (2.9.1)$$

with $\varphi = .5$ and $\psi = .5$. Table 3 shows the estimates $\hat{\varphi}$ and $\hat{\psi}$, the es= timates $\hat{\sigma}(\hat{\varphi})$and $\hat{\sigma}(\hat{\psi})$ of standard errors of the estimators of φ and ψ, the value of the test statistic $nF(S;\hat{\Sigma})$, and the corresponding upper tail probability for a chi-squared distribution with 34 degrees of freedom, obtained by each of three methods applied to each of the two samples. The three methods were the method of maximum likelihood under normality assumptions (MWL) in which (1.5.20) is minimised, a method of generalised least squares under normality assumptions (GLS) in which (1.5.17) with V defined by (1.5.18) is minimised, and an asymptotically distribution free (ADF) generalised least squares method in which (1.5.2) with U defined by (1.5.31) is minimised. The test statistic in each case was obtained by multiplying the minimised dis= crepancy function by n = 499. Also shown in the case of the first sample are corrections for kurtosis (MWL*) to the likelihood ratio test statistic (1.7.5) and standard errors of MWL estimates (Section 1.6) obtained under the assumption of an elliptical distribution. This assumption is not strictly valid since the multivariate chi-squared distribution is not elliptical although it does have equal univariate coefficients of kurtosis for its marginal distributions. The model has the necessary property of being invariant under a constant scaling factor. Corresponding results for the sample from a normal distribution are not shown since $\hat{\eta}$ was very close to unity and the corrections did not result in appreciable changes.

Table 3: *Estimates and test statistics for two samples with N=500*

Distribution	Method	$\hat{\varphi}$	$\hat{\sigma}(\hat{\varphi})$	$\hat{\psi}$	$\hat{\sigma}(\hat{\psi})$	$nF(S;\hat{\Sigma})$	Prob.
Rescaled	MWL	.62	.043	.48	.011	65.9	.001
Multivariate	GLS	.56	.041	.45	.011	65.0	.001
Chi-squared	ADF	.41	.052	.42	.019	26.9	.80
	MWL*		.066		.026	34.0	.47
	MWL	.47	.034	.47	.011	34.4	.45
Multivariate	GLS	.46	.033	.46	.011	36.3	.36
Normal	ADF	.48	.031	.46	.011	39.9	.22

It can be seen that the incorrect assumption of no kurtosis was accompanied by a definite *rejection* of a *true* null hypothesis when the MWL and GLS procedures were applied to the sample from the rescaled multivariate chi-squared distribution. While an example does not prove anything beyond doubt, it does lend support to the earlier plea for caution (Section 1.10) when routinely applying methods involving an assumption of no kurtosis. It is of interest to note that the ADF test statistic and the likelihood ratio test statistic corrected for kurtosis (MWL*) did not even approach an incorrect rejection of the null hypothesis in this sample. The estimated standard errors for the MWL and GLS estimators are less than those for the ADF estimators. This reflects the fact that use of (1.6.12) when obtaining the MWL and GLS standard errors involves an incorrect assumption of no kurtosis and can be expected to underestimate the true standard errors for these estimators. The ADF estimates of φ and ψ are both below the true population values. This point is noted here because it has been observed in other examples that ADF estimators of parameters in covariance structures tend to be noticeably biased below the true values. The bias is not so noticeable when parameters in correlation structures are estimated and estimates are scaled so as to yield a reproduced correlation matrix with unit diagonals. MWL estimators of φ and ψ are linear functions of the elements of S (Wilks, 1946) and are unbiased.

The three methods yielded very similar estimates and estimated standard errors in the sample from a multivariate normal distribution. Because S was a precise estimate of Σ, the model fitted S closely and different methods consequently gave similar results. The test statis=

tics differed a little more but all led to the same correct conclu= sion that the model could not be rejected.

An application to real data will now be considered. Huba, Win= gard & Bentler (1980) applied a questionnaire concerning frequency of usage of thirteen intoxicating substances to a sample of N=1634 Los Angeles high school students. The correlation matrix is reported by Huba, Wingard & Bentler (1980,Table 1). Table 4 presents estimates of the univariate and multivariate coefficients of relative kurtosis.

Table 4 : *Coefficients of relative kurtosis: Huba,Wingard & Bentler data*

Cigarettes	.76
Beer	.73
Wine	.70
Liquor	.87
Cocaine	15.20
Tranquilizers	12.07
Medication	4.49
Heroin	43.37
Marijuana	1.40
Hashish	6.26
Inhalants	5.29
Hallucinogenics	22.94
Amphetamines	8.35
Multivariate relative kurtosis, $\hat{\eta}$	4.19

It can be seen that some of the marginal distributions are high= ly leptokurtic. This applies mostly to those substances with low fre= quencies of use. The multivariate coefficient of relative kurtosis also is substantial.

The factor analysis model with three uncorrelated factors was employed as a model for the *correlation* matrix. The corresponding co= variance structure then is (cf. (1.6.22),(1.2.5))

$$\Sigma = D_\zeta(\Lambda^*\Lambda^{*T} + D_\psi^*)D_\zeta \tag{2.9.2a}$$

where

$$\mathrm{Diag}(\Lambda^*\Lambda^{*T} + D_\psi^*) = I \tag{2.9.2b}$$

so that the diagonal elements of D_ζ represent standard deviations of the observed variables. In order to identify Λ^* three elements were restricted to be zero. ADF and MWL estimates of parameters were obtained by first assuming the covariance structure (1.6.21) and, in the case of ADF estimates, rescaling according to (1.6.23). This occurred im= plicitly to MWL estimates since the estimation process was applied to the sample correlation matrix.

The resulting estimates of elements of Λ^* appear in Table 5. The three factor loadings restricted to be zero are shown. These were

Table 5 : *Factor loadings: Huba, Wingard & Bentler data*

	I		II		III		$\hat{\xi}_i/s_i$
	MWL	ADF	MWL	ADF	MWL	ADF	ADF
Cigarettes	.53	.52	.10	.09	.32	.34	.98
Beer	.79	.78	.01	.00	.10	.12	.99
Wine	.79	.80	.00*	.00*	.00*	.00*	.99
Liquor	.74	.74	.13	.13	.14	.14	.98
Cocaine	.07	.07	.47	.46	.06	.02	.78
Tranquilizers	.18	.17	.63	.66	.11	.10	.82
Medication	.13	.13	.35	.28	.01	.01	.90
Heroin	.07	.08	.50	.41	.00*	.00*	.71
Marijuana	.46	.44	.23	.23	.77	.74	.95
Hashish	.33	.30	.46	.52	.36	.32	.87
Inhalants	.24	.27	.48	.47	.10	.15	.89
Hallucinogenics	.09	.08	.64	.66	.02	-.05	.79
Amphetamines	.23	.22	.70	.72	.17	.14	.84

*Factor loadings restricted to be zero.

chosen in such a manner as to allow only one factor to load on wine and to prevent one factor from loading on heroin. Certain groupings of variables are apparent in this matrix. The first factor has high loadings on beer, wine and liquor and also loads on cigarettes, mari= juana and hashish. Tranquilizers, cocaine, heroin, hashish, inhalants, hallucinogenics and amphetamines load on the second factor. The third factor has a high loading on marijuana with smaller but non-negligible loadings on cigarettes and hashish. Huba, Wingard & Bentler (1980) re= port restricted MWL factor analyses with correlated factors for this data.

It can be seen from Table 5 that the MWL and ADF factor loadings are too similar for conclusions derived from the two analyses to differ. Inspection of the last column of Table 5 will show that the ADF esti=

mates of standard deviations $\hat{\zeta}_i = \sqrt{\hat{\sigma}_{ii}}$ assuming the factor analysis model were smaller than the usual estimates of standard deviations $s_i = \sqrt{s_{ii}}$ when Σ is assumed only to be positive definite. The MWL so=lution on the other hand has the property that $\hat{\zeta}_i = s_i$. Comparison of Tables 4 and 5 will show that the ratio of the ADF $\hat{\zeta}_i$ to s_i tends to be smallest for those variables which are most highly leptokurtic. Since s_i is only a slightly biased estimate of $\zeta_i = \sqrt{\sigma_{ii}}$ these results seem symptomatic of the substantial bias, mentioned earlier, of ADF estimates of parameters related to variances and covariances. In this example the bias is unimportant since the elements of $\hat{\Lambda}^*$ are of primary interest and since s_i can be used to estimate a ζ_i of interest.

Table 6 shows estimates of standard errors of the MWL and ADF estimators of factor loadings. Formula (1.6.24) has been used in con=

Table 6: *Standard errors: Huba, Wingard & Bentler data*

	I		II		III	
	MWL	ADF	MWL	ADF	MWL	ADF
Cigarettes	.023	.025	.032	.037	.037	.047
Beer	.015	.019	.026	.027	.029	.034
Wine	.016	.019	-	-	-	-
Liquor	.016	.016	.026	.030	.030	.037
Cocaine	.029	.027	.023	.071	.035	.046
Tranquilizers	.029	.029	.020	.059	.040	.057
Medication	.028	.031	.025	.050	.032	.039
Heroin	.029	.029	.023	.078	-	-
Marijuana	.027	.028	.045	.059	.055	.070
Hashish	.028	.029	.029	.060	.043	.062
Inhalants	.028	.031	.023	.055	.036	.057
Hallucinogenics	.030	.035	.019	.066	.040	.064
Amphetamines	.029	.033	.020	.048	.042	.066

junction with (1.6.12) with Σ replaced by $\hat{\Sigma}$ for MWL estimators and (1.6.7) with $\bar{T}$ replaced by U for ADF estimators. Since the multivariate coefficient of relative kurtosis, $\hat{\eta}$, is substantially greater than 1 little reliance can be placed on standard errors for MWL factor load=ings obtained using the inverse information matrix as is done in (1.6.12). It can be seen that the standard errors reported for MWL factor load=ings of leptokurtic variates on the second and third factors are sub=stantially smaller than those for corresponding ADF factor loadings. This results from the erroneous application of the inverse information matrix since ADF estimators are BGLS estimators and consequently other estimators cannot have smaller asymptotic variances.

Table 7 shows values of the test statistics. The likelihood ratio test statistic $nF_{MWL}(S;\hat{\Sigma})$ has a value of 230.5 which with 42 degrees of freedom indicates a definite rejection of the null hypothesis. It is, however, not correct to make use of this test statistic in the

Table 7 : *Test statistics : Huba, Wingard & Bentler data*

	Test statistic	Probability
MWL : $nF_{MWL}(S;\hat{\Sigma})$	230.5	< .00001
ADF : $nF(S;\hat{\Sigma}\|U)$	62.3	.02
MWL*: $n\hat{\eta}^{-1}F_{MWL}(S;\hat{\Sigma})$	55.0	.09

present situation because of the large value of $\hat{\eta}$. The ADF test statistic $nF(S;\hat{\Sigma}|U)$ is very much smaller in magnitude and is significant at the 5% level but not the 1% level. It is doubtful that the assumption of an elliptical distribution would be reasonable here because of the disparity of the coefficients of relative kurtosis in Table 4. The likelihood ratio test statistic corrected for kurtosis $n\hat{\eta}^{-1}F_{MWL}(S;\hat{\Sigma})$ does, however, happen to be closer in magnitude to the ADF test statistic than is the uncorrected likelihood ratio test statistic.

Finally we shall consider an example which illustrates the danger of treating a covariance structure which is not invariant under changes of scale as a correlation structure. An alternative model for the same data will also be examined.

McDonald (1980) found that a factor analysis model with two factors where loadings on the second factor are linearly related to loadings on the first often gives a good fit to correlation matrices with an approximate simplex structure. The correlation structure may be expressed as a covariance structure of the form of (2.9.2) with

$$\lambda^*_{i2} = \alpha^* + \beta^*\lambda^*_{i1} \quad , \quad i = 1,\ldots,p. \qquad (2.9.3)$$

It is important to bear in mind that this is not equivalent to assuming a similar factor analysis model (1.6.21) for the covariance matrix with

$$\lambda_{i2} = \alpha + \beta\lambda_{i1} \quad , \quad i = 1,\ldots,p \qquad (2.9.4)$$

One can easily verify that the model given by (1.6.21) with (2.9.4) is not invariant under changes of scale because of the constant term α in (2.9.4). Consequently, unlike the previous example, it is incorrect to disregard D_ζ in (2.6.2) and treat the correlation struc= ture as a covariance structure.

A linear relationship of the form of (2.9.3) remains under ortho= gonal rotation of the axes or, equivalently, after postmultiplication of Λ^* by an orthogonal matrix. Consequently an identification condi= tion must be imposed. This was done by specifying that

$$\beta^* = -1 \qquad (2.9.5)$$

so that the straight line through points representing variables forms an isosceles triangle with the two axes representing factors.

McDonald (1980) reported a 6 x 6 correlation matrix with N = 326 which he attributed to "Kinzer and Kinzer". MWL estimates of para= meters in the model given by (2.9.2) with (2.9.3) and (2.9.5) were ob= tained. The algorithm of (1.9.19) was employed in order to impose the constraints (2.9.2b) and correctly treat the parameters, Λ^*, D_ψ^* and α^* as parameters in a correlation structure. Results are shown in Table

Table 8 : *Kinzer & Kinzer data. Linearly related factor loadings*

	Correlation structure: $\hat{\alpha}^* = .974$				Covariance structure: $\hat{\alpha} = .978$		
i	$\hat{\lambda}^*_{i1}$	$\hat{\lambda}^*_{i2}$	$\hat{\psi}^*_{ii}$	$\hat{\zeta}_i/s_i$	$\hat{\lambda}_{i1}$	$\hat{\lambda}_{i2}$	$\hat{\psi}_{ii}$
1	.34	.63	.48	1.01	.36	.62	.53
2	.32	.65	.47	1.00	.32	.66	.45
3	.49	.48	.53	.99	.49	.49	.49
4	.58	.40	.51	.99	.57	.40	.47
5	.78	.20	.36	1.00	.80	.18	.31
6	.67	.30	.46	1.02	.67	.30	.54
Statistic	14.63				10.34		
Degrees of freedom	8				8		
Probability	.07				.24		

8. Also shown in Table 8 are results obtained by incorrectly applying the covariance structure of (1.6.21) with (2.9.4) and $\beta = -1$ to the sample correlation matrix. In this case the covariance structure does not yield MWL estimates with the property $\mathrm{Diag}[\hat{\Sigma}] = \mathrm{Diag}[S]$ so that the reproduced correlation matrix does not have unit diagonals. Re= scaling estimates would destroy the linear relationship between load= ings on the two factors.

It can be seen that differences between the two sets of estima= tes are negligible. Although incorrectly applying the method of maxi= mum likelihood to the correlation matrix does not produce maximum likelihood estimates, the estimators still are consistent for parame= ters in the correlation structure. The difference between the correct value of 14.63 of the likelihood ratio statistic and the incorrect value of 10.34 is however, not negligible. Since the value of the mul= tivariate coefficient of relative kurtosis is not known, it is not possible to judge whether or not use of the chi-squared distribution with (cf. (1.7.10))

$$21 - (6 + 6 + 6 + 2) + (6 + 1) = 8$$

degrees of freedom was appropriate for obtaining the first upper tail probability reported in Table 8. The second probability of .24 is in any case inappropriate but is given merely for comparative purposes.

It is interesting to fit the quasi-Markov simplex to the Kinzer and Kinzer correlation matrix. Schönemann's order free formulation (2.7.4) was used together with (2.7.7) and the identification condi= tions described in Section 2.7. MWL estimates were obtained using a starting point with all $\hat{\tau}_i$ equal to 1, all $\hat{\nu}_{ii}$ equal to .5, and all $\hat{\zeta}_i/s_i$ equal to 1 in a Fisher scoring algorithm. This was done to avoid any possibility of predetermining the order of the $\hat{\tau}_i$. (Since $\partial\rho_{ij}/\partial\tau_i$ does not exist if $\tau_i = \tau_j$ the partial derivative on the right was em= ployed at the starting point).

Results are shown in Table 9. The fit of the quasi-Markov sim=

Table 9 : *Kinzer & Kinzer data:Quasi-Markov simplex*

i	$\hat{\tau}_i$	$\hat{\nu}_{ii}$	$\hat{\zeta}_i/s_i$
1	.78	.84	.74
2	.84	.84	.74
3	.94	.70	.77
4	1.03	.64	.78
5	1.17	.54	.81
6	1.35	.54	.81
Statistic		7.13	
D.F.		6	
Probability		.31	

plex to the Kinzer and Kinzer data appears a little better than that of the factor analysis model with linearly related loadings but an objective judgment is difficult. Conclusions concerning the ordering of the variables derived from the two models would differ since the order according the magnitude of the $\hat{\tau}_i$ in Table 9 differs from that of the $\hat{\lambda}^*_{i1}$ in Table 8.

All models considered in this section have been invariant under a constant scaling factor so that Swain's (1.7.11) correction for the likelihood ratio test statistic would be applicable. The effect on the upper tail probabilities would be small because all sample sizes are substantial. For example application of the correction to the likeli= hood ratio test for the fit of the quasi-Markov simplex to the Kinzer and Kinzer data yields a value of 7.06 instead of 7.13. The upper tail probability is increased by .006 (if the assumption of no kurtosis is valid).

ACKNOWLEDGEMENTS

Part of the research reported here was carried out at the Center for the Study of Adolescent Drug Abuse Etiologies of the University of California, Los Angeles under support of grant number DA01070 from the National Institute on Drug Abuse.

I am indebted to P.M. Bentler, D.M. Hawkins, G.J. Huba and J.A. Woodward for thought provoking discussions. I am also grateful to G.J. Huba for making the data used in the second example of Section 2.9 available to me and to S.H.C. du Toit for help with the third example.

REFERENCES

ANDERSON, T.W. (1960): Some stochastic process models for intelligence test scores. In: K.J.Arrow et al. (eds.) *Mathematical Methods in the Social Sciences*. Stanford: Stanford University Press, 205-220.

ANDERSON, T.W. (1969): Statistical inference for covariance matrices with linear structure. In: P.R. Krishnaiah (ed.) *Multivariate Analysis, Volume 2*. New York: Academic Press, 55-56.

ANDERSON, T.W. (1970): Estimation of covariance matrices which are linear combinations or whose inverses are linear combinations of given matrices. In: R.C. Bose et al. (eds.). *Essays in Proba= bility and Statistics*. Chapel Hill: University of North Caro= lina Press, 1-24.

ANDERSON, T.W. (1973): Asymptotically efficient estimation of covariance matrices with linear structure. *Annals of Statistics*, 1, 135-141.

ANDERSON, T.W. (1975): Maximum likelihood estimation of parameters of autoregressive processes with moving average residuals and other covariance matrices with linear structure. *Annals of Statistics*, 3, 1283-1304.

ANDERSON, T.W. (1977): Estimation for autoregressive moving average models in the time and frequency domains. *Annals of Statistics*, 5, 842-865.

ANDERSON, T.W. & RUBIN, H. (1956): Statistical inference in factor analysis. In: J. Neyman (ed.), *Proceedings of the Third Berkeley Symposium on Mathematical Statistics and Probability, Vol.V.* Berkeley: University of California Press, 111-150.

ARCHER, C.O. & JENNRICH, R.I. (1973): Standard errors for rotated factor loadings. *Psychometrika*, 38, 123-140.

BARGMANN, R.E. (1956): Une epreuve statistique de la stabilité de la structure simple. In: *L'Analyse Factorielle et ses Applications.* Paris: Centre National de la Recherche Scientifique. 143-156.

BARGMANN, R.E. (1967): Matrices and determinants. In: S.M. Selby (ed.) Chemical Rubber Company Handbook of Tables for Mathematics. Cleveland: Chemical Rubber Company: 144-166.

BARTHOLOMEW, D.J. (1980): Factor analysis for categorical data. *Journal of the Royal Statistical Society*, Series B, 144, 293-321.

BARTHOLOMEW, D.J. (1981): Posterior analysis of the factor model. *British Journal of Mathematical and Statistical Psychology*, 34, 93-99.

BARTLETT, M.S. (1954): A note on the multiplying factors for various χ^2 approximations. *Journal of the Royal Statistical Society, Series* B, 16, 296-298.

BENTLER, P.M. (1976): Multistructure statistical model applied to factor analysis. *Multivariate Behavioural Research*, 11, 3-25.

BENTLER, P.M. (1980): Multivariate analysis with latent variables: Causal modelling. *Annual Review of Psychology*, 31, 419-456.

BENTLER, P.M. & LEE, S.Y. (1978): Statistical aspects of a three-mode factor analysis model. *Psychometrika*, 43, 343-352.

BENTLER, P.M. & LEE, S.Y. (1979): A statistical development of three-mode factor analysis. *British Journal of Mathematical and Statistical Psychology*, 32, 87-104.

BENTLER, P.M. & WEEKS, D.G. (1979): Interrelations among models for the analysis of moment structures. *Multivariate Behavioural Research*, 14, 169-185.

BENTLER, P.M. & WEEKS, D.G. (1980): Linear structural equations with latent variables. *Psychometrika*, 45, 289-308.

BOCK, R.D. & BARGMANN, R.E. (1966): Analysis of covariance structures. *Psychometrika*, 31, 507-534.

BOX, G.E.P. & COX, D.R. (1964): An analysis of transformations. *Journal of the Royal Statistical Society, Series B*, 26, 211-252.

BRODLIE, K.W. (1977): Unconstrained minimization. In: D. Jacobs (ed.) *The State of the Art in Numerical Analysis*. London: Academic Press, 229-268.

BROWNE, M.W. (1972a): Orthogonal rotation to a partially specified target. *British Journal of Mathematical and Statistical Psychology*, 25, 115-120.

BROWNE, M.W. (1972b): Oblique rotation to a partially specified target. *British Journal of Mathematical and Statistical Psychology*, 25, 207-212.

BROWNE, M.W. (1974): Generalized least squares estimators in the ana= lysis of covariance structures. *South African Statistical Journal*, 8, 1-24 (Reprinted in Aigner, D.J. & Goldberger, A.S. (eds.). *Latent Variables in Socio-economic Models*. Amsterdam. North Holland, 1977).

BROWNE, M.W. (1977): The analysis of patterned correlation matrices by generalized least squares. *British Journal of Mathematical and Statistical Psychology*, 30, 113-124.

BROWNE, M.W. (1979): The maximum likelihood solution in inter-battery factor analysis. *British Journal of Mathematical and Statistical Psychology*, 32, 75-86.

BROWNE, M.W. (1980): Factor analysis of multiple batteries by maximum likelihood. *British Journal of Mathematical and Statistical Psychology*, 33, 184-199.

BROWNE, M.W. (1981): Asymptotically distribution free methods for the analysis of covariance structures. Manuscript in Preparation.

BROWNE, M.W. & Du Toit, S.H.C. (1979): Simplified analysis of nonstan= dard structures for mean vectors and covariance matrices. Paper presented at the 1979 annual conference of the South African Statistical Association.

CAMPBELL, D.T. & FISKE, D.W. (1959): Convergent and discriminant vali= dation by the multitrait-multimethod matrix. *Psychological Bulle= tin*, 56, 81-105.

CAMPBELL, N.A. (1980): Robust procedures in multivariate analysis.I: robust covariance estimation. *Applied Statistics*, 29, 231-237.

CATTELL, R.B. (1978): *The Scientific Use of Factor Analysis in Beha= vioural and Life Sciences*. New York: Plenum Press.

CLARKE, M.R.B. (1970): A rapidly convergent method for maximum likeli= hood factor analysis. *British Journal of Mathematical and Statis= tical Psychology*, 23, 1970, 43-52.

CRAWFORD, C. (1975): A comparison of direct oblimin and primary parsi= mony methods of oblique rotation. *British Journal of Mathematical and Statistical Psychology*, 28, 201-213.

CURETON, E.E. & MULAIK, S.A. (1975): The weighted varimax rotation and the promax rotation. *Psychometrika*, 40, 183-196.

DENNIS, J.E. (1977): Non-linear least squares and equations. In: D.Jacobs (ed.). *The State of the Art in Numerical Analysis*. London: Academic Press, 269-312.

DEVLIN, S.J., GNADADESIKAN, A. & KETTENRING, J.R. (1975): Robust estimation and outlier detection with correlation coefficients. *Biometrika*, 62, 531-545.

DEVLIN, S.J., GNADADESIKAN, R. & KETTENRING, J.R. (1976): Some multivariate applications of elliptical distributions. In: S.Ideka (ed.). *Essays in Probability and Statistics*. Tokyo: Shinko Tsusho, 365-395.

DU TOIT, S.H.C. (1979): *The Analysis of Growth Curves*. Unpublished Ph.D. thesis. University of South Africa.

DUNCAN, G.T. & LAYARD, M.W.J. (1973: A Monte Carlo study of asymptotically robust tests for correlation coefficients. *Biometrika*, 60, 551-558.

EFRON, B. (1979): Bootstrap methods: another look at the jackknife. *The Annals of Statistics*, 7, 1-26.

FLETCHER, R. & POWELL, M.J.D. (1963): A rapidly convergent descent method for minimization. *Computer Journal*, 2, 163-168.

GALLANT, A.R. (1975): Nonlinear regression. *The American Statistician*, 29, 73-81.

GEWEKE, J.F. & SINGLETON, K.J. (1980): Interpreting the likelihood ratio statistic in factor models when sample size is small. *Journal of the American Statistical Association*, 75, 133-137.

GUTTMANN, L. (1954): A new approach to factor analysis: the radex. In: Lazarsfeld, P. (ed.). *Mathematical Thinking in the Social Sciences*. Glencoe, Ill.: Free Press, 258-348.

HAKSTIAN, A.R. & ABELL, R.A. (1974): A further comparison of oblique factor transformation methods. *Psychometrika*, 39, 429-444.

HARMAN, H.H. (1976): *Modern Factor Analysis, (3rd edition)*. Chicago: University of Chicago Press.

HARMAN, H.H. & JONES, W.H. (1966): Factor analysis by minimising residuals (minres). *Psychometrika*, 31, 351-368.

HUBA, G.J., WINGARD, J.A. & BENTLER, P.M. (1980): A comparison of two latent variable causal models for adolescent drug use. *Journal of Personality and Social Psychology*. In Press.

JENNRICH, R.I. (1973a): Standard errors for obliquely rotated factor loadings. *Psychometrika*, 38, 593-604.

JENNRICH, R.I. (1973b): On the stability of rotated factor loadings: the Wexler phenomenon. *British Journal of Mathematical and Statistical Psychology*, 26, 167-176.

JENNRICH, R.I. (1974): Simplified formulae for standard errors in maximum likelihood factor analysis. *British Journal of Mathematical and Statistical Psychology*, 27, 122-131.

JENNRICH, R.I. & CLARKSON, D.B. (1980): A feasible method for standard errors of estimate in maximum likelihood factor analysis. *Psychometrika*, 45, 237-247.

JENNRICH, R.I. & ROBINSON, S.M. (1969): A Newton-Raphson algorithm for maximum likelihood factor analysis. *Psychometrika*, 34, 111-123.

JENNRICH, R.I. & SAMPSON, P.F. (1966): Rotation for simple loadings. *Psychometrika*, 31, 313-323.

JENNRICH, R.I. & SAMPSON, P.F. (1968): Application of stepwise regression to non-linear estimation. *Technometrics*, 10, 63-72.

JOHNSON, N.L. & KOTZ, S. (1972): *Distributions in Statistics: Continuous Multivariate Distributions*. New York: Wiley.

JÖRESKOG, K.G. (1967): Some contributions to maximum likelihood factor analysis. *Psychometrika*, 32, 443-482.

JÖRESKOG, K.G. (1969): A general approach to confirmatory maximum likelihood factor analysis. *Psychometrika*, 34, 183-202.

JÖRESKOG, K.G. (1970a): A general method for analysis of covariance structures. *Biometrika*, 57, 239-251. (Reprinted in Aigner, D.J. & Goldberger, A.S. (eds.). *Latent Variables in Socioeconomic Models*. Amsterdam: North Holland, 1977).

JÖRESKOG, K.G. (1970b): Estimation and testing of simplex models. *British Journal of Mathematical and Statistical Psychology*, 23, 121-145.

JÖRESKOG, K.G. (1971): Simultaneous factor analysis in several populations. *Psychometrika*, 36, 409-426.

JÖRESKOG, K.G. (1974): Analyzing psychological data by structural analysis of covariance matrices. In: Atkinson, R.C. et al.(eds.). *Contemporary Developments in Mathematical Psychology*. San Francisco: Freeman.

JÖRESKOG, K.G. (1977): Structural equation models in the social sciences: specification, estimation and testing. In: Krishnaiah,P.R. (ed.). *Applications of Statistics*. Amsterdam: North Holland, 265-287.

JÖRESKOG, K.G. & GOLDBERGER, A.S. (1972): Factor analysis by generalized least squares. *Psychometrika*, 37, 243-260.

JÖRESKOG, K.G. & SÖRBOM, D. (1979): *Advances in Factor Analysis and Structural Equation Models*. Cambridge, Mass.: Abt Books.

KAISER, H.F. (1958): The varimax criterion for analytic rotation in factor analysis. *Psychometrika*, 23, 187-200.

KAISER, H.F. (1959): Computer program for varimax rotation in factor analysis. *Educational and Psychological Measurement*, 19, 413-420.

KELKER, D. (1970): Distribution theory of spherical distributions and a location-scale parameter generalisation. *Sankhya A*, 32, 419-430.

KENDALL, M.G. & BUCKLAND, W.R. (1971): A Dictionary of Statistical Terms (3rd Edition). London: Longman.

KENDALL, M.G. & STUART, A. (1969): *The Advanced Theory of Statistics, Vol.1*, (3rd Edition). London: Griffin.

KRAEMER, H.C. (1980): Robustness of the distribution theory of the product moment correlation coefficient. *Journal of Educational Statistics*, 5, 115-128.

KRANE, W.R. & McDONALD, R.P. (1978): Scale invariance and the factor analysis of correlation matrices. *British Journal of Mathematical and Statistical Psychology*, 31, 218-228.

KRISHNAIAH, P.R. & RAO, M.M. (1961): Remarks on a multivariate gamma distribution. *American Mathematical Monthly*, 68, 342-346.

LAWLEY, D.N. & MAXWELL, A.E. (1971): *Factor Analysis as a Statistical Method (2nd Edition)*. London: Butterworth.

LAYARD, M.W.J. (1972): Large sample tests for the equality of two co= variance matrices. *Annals of Mathematical Statistics*, 43, 123-141.

LAYARD, M.W.J. (1974): A Monte Carlo comparison of tests for equality of covariance matrices. *Biometrika*, 61, 461-465.

LEE, S.Y. & BENTLER, P.M. (1980): Some asymptotic properties of con= strained generalized least squares estimation in covariance structure models. *South African Statistical Journal*, 14, 121-136.

LEE, S.Y. & JENNRICH, R.I. (1979): A study of algorithms for covariance structure analysis with specific comparisons using factor analy= sis. *Psychometrika*, 44, 99-113.

LINDQVIST, S. (1980): A new step-strategy for Marquardt's iterative nonlinear least squares algorithm. In: M.M. Barrit & Wishart,D. (eds.). *Compstat 1980*. Wuerzburg: Physica-Verlag, 375-381.

MALINVAUD, E. (1970): *Statistical Methods of Econometrics* (2nd Edition). London: North Holland.

MARDIA, K.V. (1970): Measures of multivariate skewness and kurtosis with applications. *Biometrika*, 57, 519-530.

MARDIA, K.V. (1974): Applications of some measures of multivariate skewness to testing normality and to robustness studies. *Sankhya*, Series B, 36, 115-128.

McARDLE, J.J. (1979):The development of general multivariate soft= ware. In: J.J. Hirschbuhl (ed.). *Proceedings of the Association for the development of computer-based instructional systems.* Akron, Ohio: University of Akron. 824-862.

McDONALD, R.P. (1974): Testing pattern hypotheses for covariance ma= trices. *Psychometrika*, 39, 189-201.

McDONALD, R.P. (1975): Testing pattern hypotheses for correlation ma= trices. *Psychometrika*, 40, 253-255.

McDONALD, R.P. (1978): A simple comprehensive model for the analysis of covariance structures. *British Journal of Mathematical and Statistical Psychology*, 31, 59-72.

McDONALD, R.P. (1980): A simple comprehensive model for the analysis of covariance structures: Some remarks on applications. *British Journal of Mathematical and Statistical Psychology*, 33, 161-183.

McDONALD, R.P. & KRANE, W.R. (1979): A Monte-Carlo study of local identifiability and degrees of freedom in the asymptotic likeli= hood ratio test. *British Journal of Mathematical and Statistical Psychology*, 32, 121-132.

MOSTELLER, R. & TUKEY, J.W. (1977): Data Analysis and Regression. Read= ing, Mass.: Addison-Wesley.

MUIRHEAD, R.J. & WATERNAUX, C.M. (1980): Asymptotic distributions in canonical correlation analysis and other multivariate procedures for nonnormal populations. *Biometrika*, 67, 31-43.

MUKHERJEE, B.N. (1970): Likelihood ratio test of statistical hypotheses associated with patterned covariance matrices in psychology.*British Journal of Mathematical and Statistical Psychology*, 23, 89-120.

MULAIK, S.A. (1972): *The Foundations of Factor Analysis*.New York:McGraw-Hill.

MUTHÉN, B. (1978): Contributions to factor analysis of dichotomous var= ables. *Psychometrika*, 43, 551-560.

NEL, D.G. (1980): On matrix differentiation in statistics. *South African Statistical Journal*, 14, 137-193.

OKAMOTO, M. (1969): Optimality of principal components. In: P.R.Krishnaiah (ed.). *Multivariate Analysis II*, New York, Academic Press.

OLSSON, O. (1979): On the robustness of factor analysis against crude classification of the observations. *Multivariate Behavioural Research*, 14, 485-500.

POWELL, M.J.D. (1978a): A fast algorithm for nonlinearly constrained op= timization calculations. In: Watson, G.A.(ed.). *Numerical Analysis: Proceedings of the Biennial Conference held at Dundee,1977*.Lecture Notes in Mathematics No 630. Berlin: Springer-Verlag,144-157.

POWELL, M.J.D. (1978b): Subroutine VFO2AD. In: Hopper, M.J. *Harwell Subroutine Library: A Catalogue of Subroutines*, (1978).

PRINGLE, R.M. & RAYNER, A.A. (1971): *Generalized Inverse Matrices with Applications to Statistics*. Griffin's Statistical Monographs and Courses No.28. London: Griffin.

ROSS, G.J.S. (1970): The efficient use of function minimisation in nonlinear maximum likelihood estimation. *Applied Statistics*, 19, 205-221.

SCHÖNEMANN, P.H. (1970): Fitting a simplex symmetrically. *Psychometrika*, 35, 1-21.

STEIGER, J.H. (1979): Factor indeterminacy in the 1930's and the 1970's: some interesting parallels. *Psychometrika*, 44, 157-167.

STEIGER, J.H. (1980a): Tests for comparing elements of a correlation matrix. *Psychological Bulletin*, 87, 245-251.

STEIGER, J.H. (1980b): Testing pattern hypotheses on correlation ma= trices: Alternative statistics and some empirical results. *Multivariate Behavioural Research*, 15, 335-352.

STEIGER, J.H. & HAKSTIAN, A.R. (1980): The asymptotic distribution of elements of a correlation matrix. Technical Report No.80-3, In= stitute of Applied Mathematics and Statistics, University of British Columbia, Vancouver.

SWAIN, A.J. (1975a): A class of factor analytic estimation procedures with common asymptotic sampling properties. *Psychometrika*, 40, 315-335.

SWAIN, A.J. (1975b): *Analysis of Parametric Structures for Variance Matrices*. Unpublished Ph.D.thesis, University of Adelaide.

THURSTONE, L.L. (1935): *The Vectors of Mind*. Chicago: University of Chicago Press.

THURSTONE, L.L. (1947): *Multiple Factor Analysis*. Chicago: University of Chicago Press.

TUCKER, L.R. (1958): An inter-battery method of factor analysis. *Psychometrika*, 23, 111-136.

TUCKER, L.R. (1966): Some mathematical notes on three mode factor ana= lysis. *Psychometrika*, 32, 443-482.

VAN DRIEL, O.P. (1978): On various causes of improper solutions in maximum likelihood factor analysis. *Psychometrika*, 43, 225-243.

VAN DRIEL, O.P., PRINS, H.J. & VELTKAMP, G.W. (1974): Estimating the parameters of the factor analysis model without the usual con= straints of positive definiteness. *Proceedings of the Symposium on Computational Statistics (COMPSTAT)*, Vienna. 225-265.

WILEY, D.E., SCHMIDT, W.H., BRAMBLE, W.J. (1973): Studies of a class of covariance structure models. *Journal of the American Statistical Association*, 68, 317-323.

WILKS, S.S. (1946): Sample criteria for testing equality of means, equality of variances and equality of covariances in a normal multivariate distribution. *Annals of Mathematical Statistics*, 17, 257-281.

YATES, A.T. (1979): *Multivariate Exploratory Data Analysis: True Exploratory Factor Analysis*. In Press.

THE LOG-LINEAR MODEL AND ITS APPLICATION TO MULTI-WAY CONTINGENCY TABLES

THEUNIS J. V.W. KOTZE, *INSTITUTE FOR BIOSTATISTICS*

1. INTRODUCTION

Categorical variables can be divided into two main sub-classes:- nominal and ordinal. *Ordinal* variables, such as social status, num= ber of children, categorized continuous variables, etc. have some sort of inherent grading which allows one to say, for example, that social class II is on the same "side" of social class I as is social class III, but is "nearer" to it than is social class III. *Nominal* variables, on the other hand do not have this inherent grading. They are typified by eyecolour, the type of disorder diagnosed in a psy= chiatric patient, and the city in which one resides.

The binomial, multinomial and Poisson distributions are com= monly used for modelling the frequency of occurrence of a particular value of both nominal and ordinal variables.

All three distributions belong to the broader class of the exponential family. This family has the unifying property that the logarithms of the frequencies of occurrence of a particular value of the variable may be expressed as a linear function of the parameters of the distribution. Such models for log frequencies are termed log-linear models. They may be used for the analysis of two-way or multi-way contingency tables and data on rates and proportions (both of which arise naturally with nominal variables), and may also be extended to logit analysis and log-odds ratios which are the natu= ral tools for dealing with ordinal variables.

The aim of this review is to enable the reader to understand and apply log-linear models to the analysis of data; to highlight the various pitfalls in the application of the models; and to men= tion further references useful for a more thorough study of the sub= ject. No attempt will be made to provide an in-depth knowledge of the theory of log-linear models, but we shall try to give the reader an understanding of the basic theory and the wide field of application of these models.

While they are also useful for estimation and prediction, log linear models are used most commonly to test hypotheses about the nature of the relationships between two or more categorical vari= ables.

The resulting test statistics generally have asymptotic chi-squared distributions as the sample size tends to infinity. While these asymptotic chi-squared distributions may easily be derived theoretically, their applicability to samples of finite size is by no means clear, and so this problem is discussed at some length in this review, because it is fundamental in the application of log-linear models.

Over the years a very large number of papers have been publi= shed on log-linear models and/or their applications to the analysis of contingency tables, and it would be impossible to mention all of them. Important handbooks in this field are, for example, Kullback (1959), Bishop, Fienberg and Holland (1975), Gokhale and Kullback (1978), Goodman (1978). An important bibliography on the subject of contingency table literature from 1900 to 1974 is given by Killion and Zahn (1976).

2. NOTATION

Count or frequency data usually originate when categorical (nominal or ordinal) data are summarized in a contingency table. The one-way contingency table is the simplest table possible, but it can easily be extended to higher dimensional tables, for example, the absence or presence of two characteristics A and B in males and fe= males, and with the cells listed within the Table 1 below.

Table 1 - *Three-way contingency table*

	Males		Females	
	A absent	A present	A absent	A present
B absent	(1,1,1)	(1,2,1)	(2,1,1)	(2,2,1)
B present	(1,1,2)	(1,2,2)	(2,1,2)	(2,2,2)

Table 1 contains eight cells in which various counts appear. These counts are denoted by x(ijk) with the following notation:

Characteristic	Index	1	2
Sex	i	Male	Female
A	j	Absent	Present
B	k	Absent	Present

The one-way and two-way marginals are denoted by:

$$\sum_{j,k} x(ijk) = x(i..) \qquad i = 1,2$$

and

$$\sum_{j} x(ijk) = x(i.k) \qquad i = 1,2; \;\; k = 1,2.$$

The separate cell probabilities will be denoted by p(ijk), and the marginal totals of the probabilities in a similar fashion.

We shall also use notations such as p_{ij}^{AB}. This symbol denotes the probability of cell ij in the cross classification by factors A and B. If factors A and B are merely two of a larger set, then p_{ij}^{AB} refers to the two-way marginal table obtained by summing over all other factors.

The symbol ω will be used to denote cells (i,j) in a two-way table, (i,j,k,ℓ) in a four-way table and so on. For example, in the 2x2x2 table displayed previously the symbol $x(\omega)$ will replace x(ijk), being one of the 2x2x2 = 8 cells. The general order in which the cells of the contingency table are listed is the lexicographic order in which the last index varies the fastest, followed by the index next to it, and so on. The complete set of cells of a contingency table is denoted by Ω. Ω is also used to denote the number of cells in the table. It will be clear from context which meaning should be attached to Ω.

3. INTERACTION IN CONTINGENCY TABLES

A contingency table contains counts in the different cells that are described by the simultaneous occurrence of a single category of each of several factors. The influence exerted by these factors on each other is usually described in terms of their probability distribution. For a simple two-way contingency table the hypothesis of no interaction between the factors I and J is defined by

$$p_{ij}^{IJ} = p_i^I \, p_j^J$$

for all possible indices i and j. The definition has a *multiplicative* character. It may be shown, however, to be equivalent to the following *additive* definition of lack of interaction:

$$\frac{p_{ij}^{IJ}}{p_i^I p_j^J} = \alpha_i^I + \beta_j^J$$

for all i,j and some $\{\alpha_i^I\}$ and $\{\beta_j^J\}$.

Turning to three-way tables, we have the following multiplicative definition of lack of three-way interaction (Darroch, 1974):-

$$p_{ijk}^{IJK} = \theta_{jk}^{JK} \quad \phi_{ik}^{IK} \quad \psi_{ij}^{IJ}$$

for all i,j and k and for some $\{\theta_{jk}^{JK}\}$, $\{\phi_{ik}^{IK}\}$ and $\{\psi_{ij}^{IJ}\}$.
This definition is equivalent to

$$\frac{p_{ijk}^{IJK}\, p_{ist}^{IJK}\, p_{rjt}^{IJK}\, p_{rsk}^{IJK}}{p_{rst}^{IJK}\, p_{rjk}^{IJK}\, p_{isk}^{IJK}\, p_{ijt}^{IJK}} = 1.$$

Darroch (1974) further provided a definition of no additive interaction. It is formulated as follows in a three-way contingency table, for all i,j and k

$$\frac{p_{ijk}^{IJK}}{p_i^I p_j^J p_k^K} = \alpha_{jk}^{JK} + \beta_{ik}^{IK} + \delta_{ij}^{IJ} \quad .$$

An expansion of the theory of additive interaction in the case when one factor is selected as a response variable, together with its application, is presented in Chapter 7 of Gokhale and Kullback (1978).

For three or more factors, the multiplicative and additive definitions of lack of interaction are no longer equivalent, and it is necessary to elect to use one or the other.

In view of the many desirable properties peculiar to the multiplicative definition such as having conditional independence as a special case, sub-table invariance and the absence of constraints on marginal probabilities, we concur with Darroch (1974) in preferring the multiplicative definition, which will be used in the sequel.

In order to explain the terms in a log-linear model associated with multiplicative interactions between the factors, a four-way contingency table with factors A, B, C and D can be used as an example. The saturated log-linear model can be written as

$$\begin{aligned} \ell n\, x_{i\,j\,k\,\ell}^{A\,B\,C\,D} = 1 &+ \lambda_i^A + \lambda_j^B + \lambda_k^C + \lambda_\ell^D \\ &+ \lambda_{ij}^{AB} + \lambda_{ik}^{AC} + \ldots + \lambda_{k\ell}^{CD} \\ &+ \lambda_{ijk}^{ABC} + \lambda_{ij\ell}^{ABD} + \ldots + \lambda_{jk\ell}^{BCD} \\ &+ \lambda_{ijk\ell}^{ABCD} \end{aligned}$$

with certain restrictions on the λ's, for example that they sum to zero, being required for identifiability of the parameters.

In the non-saturated model, some of the λ terms will be omitted. We shall depict the form of non-saturated model fitted to data with a shorthand notation perhaps best illustrated by example. The nota= tion (ABC,D) will refer to a model containing the λ^{ABC} interaction and all interactions implied by ABC, namely λ^{AB}, λ^{AC}, λ^{BC}, λ^{A}, λ^{B} and λ^{C}, together with λ^{D}. Similarly, (AB,CD) would denote the model containing the λ^{AB}, λ^{A}, λ^{B}, λ^{CD}, λ^{C} and λ^{D} terms.

Note the hierarchy implied by this notation. If, say, the ACEF interaction is included in the model, then our notation implies that all one-, two- and three-factor λ terms with superscripts selec= ted from A, C, E, F are also included in the model.

4. TWO TYPES OF INVESTIGATIONS

As in other branches of multivariate analysis, it is profitable to distinguish two types of investigations.

In the first, which we term a "correlation" type of investiga= tion, the different variables are regarded as equivalent in importance. A likely starting point for the analysis of such a study would be a test for the mutual independence of all the factors.

In a second type of investigation, which we may term a "regres= sion"-type investigation, one of the variables is of particular im= portance, and we wish to study its dependence on the others. For example if, of four factors A, B, C, D we regard A as dependent, then the real concern is with the prediction of A from B, C and D, and not the details of the interrelationships between B, C and D.

Thus in this situation, we might start by testing the null hypothesis of independence of A on B, C and D by fitting the model (A,BCD). If this null hypothesis is rejected, we might then fit suc= cessive hierarchic models containing interaction terms between A and one or more of B, C and D.

Finally if B, C or D were ordinal and not just nominal, we might test whether its effects could be explained by low order polynomials (linear or quadratic) in its levels.

5. THE MINIMUM DISCRIMINATION INFORMATION (MDI) APPROACH

There are two broad approaches to the fitting of log-linear models to contingency tables, and we shall now sketch out the first of them - the Minimum Discrimination Information approach pioneered by Kullback (1959).

5.1 Formulation of the MDI statistic

Suppose we have two discrete frequency distributions $p(\omega)$ and $\pi(\omega)$ defined over the set of cells Ω, so that

$$\sum_{\Omega} p(\omega) = \sum_{\Omega} \pi(\omega) = 1 .$$

The discrimination information is defined by

$$I(p : \pi) = \sum_{\Omega} p(\omega) \,\ell n \frac{p(\omega)}{\pi(\omega)} .$$

It is largely immaterial whether p and π are both frequency distributions or relative frequency distributions, as I for the for= mer case is identically N times I for the latter case. $I(p : \pi)$ is greater than or equal to zero for all possible distributions $p(\omega)$, $\omega \in \Omega$. It can also be seen as a measure of divergence between $p(\omega)$ and $\pi(\omega)$ and increases as the agreement deteriorates.

The distribution $\pi(\omega)$ arises from the problem of interest, where it might be the observed relative frequency distribution or some other distribution fitted to it. The distribution $p(\omega)$ is a member of a family p of distributions. We will be concerned with finding that member of p which minimizes $I(p : \pi)$ subject to linearly independent constraints of the form

$$C\, p = \theta$$

where C is a known $(r+1) \times \Omega$ matrix, θ a known vector of r+1 compo= nents, $r+1 \leq \Omega$ and p has been written as an $\Omega \times 1$ vector in lexico= graphic order.

The determination of the desired $p(\omega)$ may be made using Lagrange multipliers, and some simple algebra shows the $p(\omega)$ minimizing $I(p : \pi)$ to be of the form

$$p^*(\omega) = \exp\,(\tau_o\, C_o\,(\omega) + \tau_1\, C_1\,(\omega) + \ldots + \tau_r C_r(\omega))\;\pi(\omega)$$

or equivalently (a form bringing out the log-linearity of the solution) as

$$\ell n\left(\frac{p^*(\omega)}{\pi\;(\omega)}\right) = \tau_o\, C_o(\omega) + \ldots + \tau_r\, C_r(\omega), \qquad \omega = 1,2,\ldots,\Omega$$

where $C_o(\omega)$, $C_1(\omega),\ldots,C_r(\omega)$ are the elements of the ωth column of C.

Now writing $\ell n(p^*/\pi)$ as an $\Omega \times 1$ vector, we may rephrase this result in matrix form as

$$\ell n\left(\frac{p^*}{\pi}\right) = C'\tau$$

The (r+1) as yet undetermined Lagrange multipliers $\tau_o \ldots \tau_r$ must satisfy the constraints $Cp^* = \theta$, which then yield the necessary r+1 equations to determine them.

From the distributional point of view, and anticipating the results of a later section, the τ_i are the natural parameters of the distribution p^*.

The constraint matrix C and the right-hand side θ are implied by the models to be fitted to the data which in turn depend on the hypothesis to be tested.

Suppose for example that one wishes to test the null hypothesis of marginal homogeneity in a 3x3 contingency table:- i.e. whether

$$\sum_{j=1}^{3} p_{ij} = \sum_{j=1}^{3} p_{ji} \quad , \quad i = 1,2,3.$$

This specification, together with the usual requirement that $\Sigma_{ij}\, p_{ij}=1$, may be rephrased into the constraints

$$\overset{C}{\begin{bmatrix} 1 & 1 & 1 & 1 & 1 & 1 & 1 & 1 & 1 \\ 0 & 1 & 1 & -1 & 0 & 0 & -1 & 0 & 0 \\ 0 & -1 & 0 & 1 & 0 & 1 & 0 & -1 & 0 \end{bmatrix}} \qquad \overset{\theta}{\begin{bmatrix} 1 \\ 0 \\ 0 \end{bmatrix}}$$

(Note that r=2, the constraint for i=3 being implied by those for i=1 and 2).

The first constraint on p is always that

$$\sum_{\Omega} p(\omega) = 1$$

This can also be written

$$C_1 p = 1.$$

The constraints matrix can then be partitioned as

$$C = \begin{bmatrix} C_1 \\ C_2 \end{bmatrix}$$

where C_1 is $1 \times \Omega$, C_2 is $r \times \Omega$.

The constraints vector can be partitioned correspondingly as

$$\theta = \begin{Bmatrix} 1 \\ \theta^* \end{Bmatrix}$$

where θ^* is $r \times 1$.

The constraints are now $C_1 p = 1$ and $C_2 p = \theta^*$.

Now let $\pi(\omega) = x(\omega)/n$ with $n = \Sigma_\Omega\, x(\omega)$ be the observed relative frequency distribution, and let $x^*(\omega) = np^*(\omega)$ be the fitted frequency distribution. Then the MDI statistic (Gokhale and Kullback 1978) is defined by

$$2I(x^* : x) = 2nI(p^* : \pi) = 2 \sum_{\Omega} x^*(\omega) \, \ell n \, \frac{x^*(\omega)}{x(\omega)}$$

which is asymptotically distributed as a central χ^2 with the appro= priate degrees of freedom if the observed table $x(\omega)$ is consistent with the null hypothesis (constraints) or the estimate $x^*(\omega)$. The degrees of freedom of the test statistic is equal to r, the number of the para= meters constrained by C. This is then a method of determining whether the observed table agrees with the estimates $x^*(\omega)$ which satisfy the null hypothesis.

In some cases, the x^* minimizing $I(x^* : x)$ and satisfying the constraints $C\, x^* = n\theta$ may be written as an explicit product of the marginal totals. This situation quite often arises with internal con= straints problems (see below). In other cases, however, x^* can not be found explicitly, and then it is necessary to obtain it iteratively. This is usually done by finding an initial guess of x^* and then using constrained optimization procedures to refine this initial distribution iteratively.

An important distinction may be drawn between problems where the constraints solely fix certain marginals (Internal Constraints Problems) and others where outside restrictions are imposed on the theoretical distribution (External Constraints Problems). The problems where cer= tain marginals are restricted to be equal to the corresponding observed marginals and which are associated with generalized independence hypo= theses are classified as Internal Constraints Problems (ICP).

A property of ICP's is that the observed frequency table satis= fies the constraints, i.e.

$$C\, x = n\theta$$

and furthermore trivially minimizes $I(x^*: x)$. Thus in an ICP, if an iterative minimization of $I(x^* : x)$ is needed, the initial guess $x^* = x$ is unsatisfactory, and so the initial guess of a uniform distribution is used for x^*.

In the more general class of External Constraints Problems (ECP's) the observed distribution x does not necessarily satisfy the constraints (apart of course from the constraint $\Sigma\, x^*(\omega) = n$), and so may be used as the initial guess in an iterative solution.

ECP's arise in such problems as testing for the equality of means of two internal integer-valued random variables; testing for homogeneity of marginals in square contingency tables; and (in gene=

tics) testing whether the marginals equal some prespecified values.

5.2 Partitioning of the test-statistic

The question now arises as to the improvement that can be achieved in model fitting by describing less stringent constraints. Suppose that two hypotheses H_1 and H_2 need to be compared and the question is asked whether the improvement is statistically significant. Let the constraints describing H_1 be summarized by

$$C_1 p = \theta_1$$

and the constraints describing H_2 by

$$C_2 p = \theta_2 .$$

Suppose further that H_2 is stronger than H_1, and that therefore $C_2 p = \theta_2$ implies $C_1 p = \theta_1$, so that if there are m_1 and m_2 constraints (apart from $\Sigma_\Omega\, p = 1$) contained in C_1 and C_2 respectively, then $m_2 > m_1$. Let $2I(x_1^* : x)$ and $2I(x_2^* : x)$ be the MDI statistics corresponding to H_1 and H_2. They are respectively asymptotically distributed as chi-squared with m_1 and m_2 degrees of freedom. It may then be demonstrated that

$$2I(x_2^* : x) = 2I(x_2^* : x_1^*) + 2I(x_1^* : x).$$

This equation can be illustrated diagrammatically as follows:

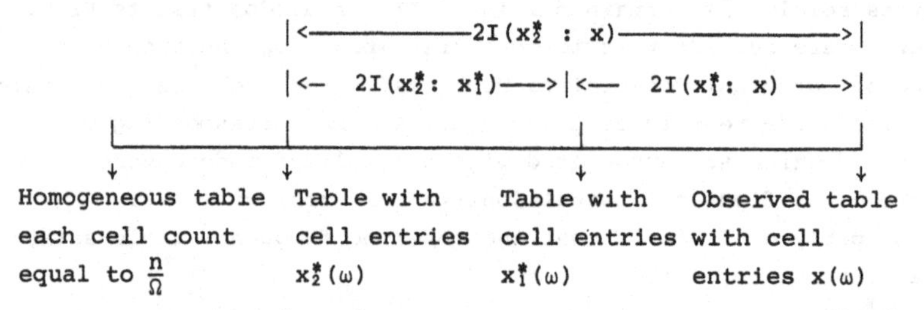

The statistic $2I(x_2^* : x_1^*)$ measures the contribution to $2I(x_2^* : x)$ of the additional constraints in C_2 but not in C_1, and is asymptotically distributed as chi-squared with $m_2 - m_1$ degrees of freedom.

This partitioning of the test statistic can be illustrated by a 2x2x2x2 contingency table. We consider an Internal Constraints problem in which the constraints fix certain marginals of the estimated cell counts to be equal to the corresponding observed marginals. The uniform distribution is used as the initial distribution of the iterative solution and its properties are conserved except for those implied by the constraints. The C_1 matrix and θ_1 vector that correspond to the hypothesis of mutual independence of the four factors,

$$H_1 : p(ijk\ell) = p(i...)p(.j..)p(..k.)p(...1),$$

are

$$\overset{C_1}{\begin{bmatrix} 1&1&1&1&1&1&1&1&1&1&1&1&1&1&1&1 \\ 1&1&1&1&1&1&1&1&0&0&0&0&0&0&0&0 \\ 1&1&1&1&0&0&0&0&1&1&1&1&0&0&0&0 \\ 1&1&0&0&1&1&0&0&1&1&0&0&1&1&0&0 \\ 1&0&1&0&1&0&1&0&1&0&1&0&1&0&1&0 \end{bmatrix}} \quad \overset{\theta_1}{\begin{bmatrix} 1 \\ x(1...)/N \\ x(.1..)/N \\ x(..1.)/N \\ x(...1)/N \end{bmatrix}}$$

Observe that the dimensions of C_1 are 5 x 16, where 16 is the number of cells in the four-way table and the five constraints consisting of the common constraint $\Sigma_\Omega\, p(\omega) = 1$, and four other constraints which relate to the hypothesis H_1.

A stronger hypothesis is that of pairwise independence of the first two factors and last two. It is formulated as

$$H_2 : p(ijk\ell) = p(ij..)\; p(..k\ell).$$

The additional constraints necessary for this hypothesis can be summarized in

$$\overset{C_3}{\begin{bmatrix} 1&1&1&1&0&0&0&0&0&0&0&0&0&0&0&0 \\ 1&0&0&0&1&0&0&0&1&0&0&0&1&0&0&0 \end{bmatrix}} \quad \overset{\theta_3}{\begin{bmatrix} x(11..)/N \\ x(..11)/N \end{bmatrix}}$$

This hypothesis is stronger with regard to the uniform distribution, and not to the observed distribution.

The constraints describing H_2 are

$$\begin{bmatrix} C_1 \\ C_3 \end{bmatrix} p = \begin{bmatrix} \theta_1 \\ \theta_3 \end{bmatrix}$$

The method of partitioning enables one to evaluate the improvement achieved in the test of fit by specifying the stronger hypothesis H_2. In ordinary multiway contingency tables it is not difficult to partition the test statistic when stronger independence hypotheses are investigated. The constraints C_2 fix the marginals np(ij..) and np(..kℓ) and this implies that the marginals np(i...), np(.j..), np(..k.) and np(...ℓ) are also fixed to be equal to the respective observed marginals.

6. A GENERAL APPROACH VIA EXPONENTIAL DISTRIBUTIONS

The second broad approach leading to log-linear models is through the general exponential family of distributions (Nelder and Wedderburn, 1972). This family includes several continuous and discrete distributions:- among the latter are the binomial, multinomial and Poisson distributions. First, we shall review some properties of the family as a whole.

6.1 Some properties of the exponential family of distributions

We shall consider probability functions within the general exponential family of distributions defined by

$$p(z\ ;\ \theta,\phi) = \exp[\alpha(\phi)\ \{z\theta - g(\theta) + h(z)\} + \beta(\phi,z)],\ \alpha(\phi) > 0$$

for fixed ϕ(Nelder and Wedderburn 1972)
The parameter ϕ could represent a nuisance or scale parameter such as the variance of a normal distribution.

The natural logarithm of the likelihood function of a single observation z is

$$L = \ell n\ p(z\ ;\theta,\phi) = \alpha(\phi)\{z\theta - g(\theta) + h(z)\} + \beta(\phi,z).$$

Then for all exponential distributions of this type it is possible to show that

$$\mu = E(z) = g'(\theta) \quad , \text{ and}$$

$$\frac{d\mu}{d\theta} = g''(\theta) = V \quad , \text{ say.}$$

6.2 Fitting the generalized linear model

To apply this distribution to the problems at hand, let us suppose that we have available m independent or predictor variables w_1, $w_2,\ldots,w_m$ on which a variable y may be regressed linearly,

$$y = \sum_{i=1}^{m} \beta_i\ w_i$$

and that the parameter θ is determined by y

$$\theta = f(y)$$

f being a linking function.

In a multi-way contingency table, we will have associated with each cell ω a parameter value of θ_ω, with the linking function

$$\theta_\omega = \exp\ (y_\omega)$$

$$\text{or}\quad y_\omega = \ell n\ (\theta_\omega)$$

which is to be regressed on the predictors w_i.

Arranging the cells in lexicographic order, we may then write the model as

$$y = W\beta.$$

The joint distribution of the cell counts in a contingency table is multinomial. However it is equivalent (and mathematically more convenient) to describe the $x(\omega)$ as Poisson random variables with mean $\theta(\omega)$, and subject to the constraint $\Sigma_{\Omega}\, x(\omega) = n$ (Nelder 1974). The latter constraint however is removed by incorporating a "general mean" term into the design matrix.

Rather analogously with the constraints matrix C of the MDI approach, the design matrix W is fixed by the hypothesis or model under consideration.

For example, in a 2x2 contingency table under the model of independence, a suitable design matrix is

$$W = \begin{bmatrix} 1 & 1 & 1 \\ 1 & 1 & 0 \\ 1 & 0 & 1 \\ 1 & 0 & 0 \end{bmatrix}$$

and the corresponding β_i constitute the trivial general mean, a column effect and a row effect for the table.

More generally, the predictor variables may include not only the binary indicator variables of this example, but also interval predictors. Thus the general linear model approach has a power and applicability extending far beyond contingency table analysis.

Using the Poisson distribution, section 6.1, and general theory of maximum likelihood, we find that the log likelihood of the contingency table and its first and expected second derivatives are:

$$L = \text{Constant} + \sum_{\omega=1}^{\Omega} \{x(\omega)y_{\omega} - \exp(y_{\omega})\}$$

$$\frac{\partial L}{\partial \beta_i} = \sum_{\omega=1}^{\Omega} w_{j\omega}\, \{x(\omega) - \exp(y_{\omega})\} = C_i \text{ say}$$

$$E\{\frac{\partial^2 L}{\partial \beta_i \partial \beta_j}\} = \sum_{\omega=1}^{\Omega} - w_{i\omega}\, w_{j\omega} \exp(y_{\omega}) = A_{ij} \text{ say.}$$

The Newton-Raphson algorithm using the expected second derivative then proceeds from some initial estimates β^0 using

$$\beta^{r+1} = \beta^r - A^{-1}\, C$$

A suitable starting value β^0 may be obtained by regressing the vector $\ell n\, x$ on W.

The adequacy of a particular model is obtained by maximizing L in this way to get L_{max}, and, using the full saturated model, to get L_{cmax}. The *deviance* is defined by $-2(L_{max}-L_{cmax})$ and follows an asymptotic chi-squared distribution with degrees of freedom equal to $\Omega - 1 - m$. This distribution is central under the null hypothesis that the contingency table is described by the current model.

It may be shown (Nelder 1974) that the deviance is simply

$$2\ \Sigma_{\Omega}\ x(\omega)\ \ell n\ \frac{x(\omega)}{x^*(\omega)} = 2I(x^*: x\ )$$

where x^* is the fitted frequency table.

Asymptotically, the maximum likelihood estimate $\hat{\beta}$ is distributed normally with covariance matrix $(W'VW)^{-1}$, where

$$V = \text{diag}\{\exp - (y_{\omega})\}$$

In accordance with the subject of this paper, the above discussion has been specialized to the contingency table application of the general linear model. The reader wishing to see a broader exposition should refer to Nelder and Wedderburn (1972) and Nelder (1974).

Despite the apparent dissimilarity between the MDI and the generalized linear model outlined above, there are in fact strong similarities, and not only in that both give rise to asymptotically chi-squared distributed test statistics.

To be more specific, if the model specifies the probability of any cell in terms of linear functions of observed cell frequencies, then the MDI estimates are identical to the maximum likelihood estimates, and the MDI test statistic is identical to the deviance.

This is the case in the important class of problems in which the parameters of the log-linear model depend only on marginal totals of the observed cell frequencies, which is true for multiplicative interaction models (Gokhale and Kullback 1978 p.40). With more general constraints, however, it is not necessarily the case. Example of the latter situation are some of the tests associated with square contingency tables (Ireland, Ku and Kullback, 1969); the analysis of mixed discrete and continuous data (Afifi and Elashoff, 1969) some logit models (Berkson 1972), and tests for marginal homogeneity.

Despite the greater generality of the general linear model (Nelder 1974), the MDI approach continues to be the more commonly applied approach, and our subsequent discussion will be largely confined to that approach.

7. STRATEGIES FOR SELECTING A FINAL LOG-LINEAR MODEL

The number of potential models increases dramatically as the dimension of the table increases. If as is normally desirable one considers only hierarchical models then there are 17 possible models for a three-way table, 166 models for a four-way table and several thousand possible hierarchical models for a five-way table. It is therefore necessary to have strategies to select a final model.

Whether one follows the MDI approach or the general linear model, there are very explicit asymptotically chi-squared distributed test statistics for testing both the adequacy of fit of a proposed model, and the improvement in fit on adding a particular interaction to it. Nevertheless, it is no trivial matter to decide on which of the multi= tude of possible models to fit. Some relevant concepts, sketched below, are discussed more fully in Brown (1976), Benedetti and Brown (1978) and Gokhale and Kullback (1978).

The concepts of marginal and partial association are due to Brown (1976). Suppose for example that there is a four-way contingency table indexed by the factors A,B,C and D. The *marginal* association be= tween A and B is then tested by the difference between the tests-of-fit of the models (AB) and (A,B). The test for marginal association is also identical to the difference between the tests-of-fit of the models (AB,C,D) and (A,B,C,D). This is an unconditional test of interaction and ignores the effect of the other variables. The *partial* association between A and B for example is tested by the difference between the tests-of-fit of the models (AB,AC,AD,BC,BD,CD) and (AC,AD,BC,BD,CD). This is a conditional test of the AB - interaction adjusted for all other effects of the same order.

The marginal and partial effects give a good indication of the relative importance of each possible interaction in a multiway table, and so provide a very useful filter in the search for a good model. By concentrating on the fitting of models containing terms whose marginal or partial effects are large, one may obtain good fits with a minimum of trial and error.

In selecting the final model, formal significance testing of the individual terms is a useful tool, but must not be the only one used. In very large data sets, for example, one finds that most interactions are statistically significant even if many of them contribute little to the quality of fit of the model. In these circumstances, to rely solely on formal significance tests leads to cumbersome, pedantic models, and an alternative approach is desirable.

One such is the method of percentage of information explained. This method was first suggested by Kullback (1973) and later again dis= cussed by Benedetti and Brown (1978). In this method it is necessary to select an initial baseline model. In the 'correlation' type of in= vestigation a good choice of the baseline model may be the model that contains only the main effects (all factors mutually independent), where=

as in the 'regression' type of investigation an appropriate baseline model is one where the factor to be predicted is independent of all the remaining ones.

Let x^*_B be the estimate under the hypothesis that describes the baseline model. The quantity $2I(x^*_B : x)$ is a test of fit of the base= line estimate to the observed table. Let x^*_1 be estimates associated with a hypothesis H_1 which is described by less stringent constraints than the baseline model. The quantity

$$2I(x^*_B : x) - 2I(x^*_1 : x) = 2I(x^*_B : x^*_1)$$

measures the improvement obtained on the baseline model by the model de= scribed by the hypothesis H_1. The following percentage gives a compara= tive measure of improvement :

$$\frac{(2I(x^*_B : x) - 2I(x^*_1 : x)) .100}{2I(x^*_B : x)}$$

$$= \frac{I(x^*_B : x^*_1).100}{I(x^*_B : x)}$$

The above measure can help the practical data analyst, when faced with large data sets, to decide whether a higher interaction term contri= butes enough in explaining the relationships present. For example, if a model already explains 95.0% of $I(x_B : x)$ and if the inclusion of a further significant interaction will push this percentage up to 95.4%, then there is certainly no justification for including that particular interaction, unless it simplifies the fitted model.

With smaller samples, strategies based on formal significance testing are reasonable. There are two broad approaches to these tests, with some subsidiary issues as well.

The broad classification rests on the fact that the MDI statistic and its partitionings enable one to test for the adequacy of fit of a model, and also for the significance of the change in fit on adding or deleting terms in the model.

The *principle of parsimony* is based on fitting the model contain= ing fewest terms which does not yield a statistically significant lack of fit. Despite the name, it is possible for this principle to lead to a model containing (or indeed made up entirely of) terms which are not individually statistically significant. It is also possible, and more common, for it to lead to models which exclude terms which are statisti= cally significant.

The other broad approach is based on the significance of the con= tribution of each of the terms to the model. This approach may be implemented stepwise using forward stepping, backward stepping, or a combination of the two. In forward stepping, one examines all hier= archical models that consist of adding a single term to the current model. The "best" of these terms is then added to the model if its contribution is sufficiently large. The "best" terms to add may be defined as the most significant; that leading to the largest decrease in the overall test of fit; or that giving the largest decrease per degree of freedom added to the model by the term.

Backward stepping is done in essentially the same way. Hierarchic models differing from the current model only by the omission of a single term are fitted, and one deletes the "worst" such term.

It is a defect of these stepwise procedures that the significance of a term depends on which other terms are already in the model. Thus the final model selected is sensitive to the initial model used to begin the stepping, and to minor fluctuations in the data which might alter the relative importance of two terms, and hence the subsequent course of the stepping. This defect may be avoided by the use of Brown's statistics for marginal and partial association, which essen= tially use a standardized order of inclusion for testing the signifi= cance of any term. Thus the strategy of selecting the model on the ba= sis of the significance of marginal and partial association leads to a final model which is better understood, and which is not subject to minor vagaries of the data.

A final model having been selected, the practical issue of inter= pretability then comes to the fore. Suppose, for example, that one of the formal procedures outlined above led to the model (AB,BC,AC,DE) in a five-way contingency table. If one were to augment this model with the ABC interaction (whose inclusion would not be justified on purely statistical grounds), the model (ABC,DE) would result. The latter model has the great virtue of having a simpler interpretation than the former, and so would often be preferred on practical grounds. This addition of non-significant terms to a model to attain better interpretability is a common and sensible procedure, but involves judgement rather than formalistic reasoning.

Since the model selection procedures all involve multiple tests on the data, they are fraught with inferential problems about the over= all significance level attained. To circumvent these difficulties, Aitken (1979) proposed a simultaneous test procedure for multi-way contingency table models. To apply this test procedure it is first necessary to set up a hierarchy of models, starting from a model that contains only the main effects, then a model that contains main effects as well as two-way interactions ..., and lastly the saturated model. An analysis of deviance table is set up in which the test statistics are partitioned. Type I error rates are defined for each set of terms e.g. γ_2 for two-way interactions, γ_3 for three-way interactions, etc.

Reduction of the model proceeds by testing the highest possible interactions, at level γ_m(m-way contingency table,say). If this inter= action is not significant, the group of (m-1)-way interactions is tes= ted at the appropriate level. When a group of interactions (i-way) is found to be significant, permutations of terms in the group are evaluated. The subset of models which are not significant at level γ_i is called adequate at level γ_i. A further subset out of adequate models are those models which yield a significant test at level γ_i but have no proper adequate subset at that level, and are called the mini= mal adequate models. Further reduction of the minimal adequate models may be possible.

The Type I error rates are set at

$$\gamma_i = 1 - (1 - \alpha)^{\binom{r}{i}}$$

so that the Type I error rate for the global test that all the inter= actions are null is

$$\gamma = 1 - (1 - \alpha)^{2^r - r - 1}$$

Aitken (1979) recommends a choice of α which results in a γ between 0.25 and 0.5. This method usually leads to a very parsimonious model.

To summarize, then, the following criteria can be followed in selecting a final log-linear model:

i) Parsimony refers to the choice of a model that contains the lowest number of terms and yields a model that fits the data.

ii) Another strategy that leads to a parsimonious model, is the method of simultaneous testing as developed by Aitken (1979).

iii) A model with a simple interpretation is less complicated to inter= pret, but may have more terms than a parsimonious model.

iv) A hierarchical model that contains all significant effects may be more complicated than a model with a simple interpretation. For this model the change in the test-of-fit due to the addition of any single term is non-significant, while the change due to the deletion of any single effect is significant. This criterion may be satisfied by several different models.

v) The criterion of information explained can be used in the case of extremely large data sets, where almost every interaction term is significant.

8. PROBLEMS OF APPLYING LOG-LINEAR MODELS TO MULTI-WAY CONTINGENCY TABLES

In the preceding section it was assumed that there exists a suit= able multi-way table to analyse, but in practice this is often not the case as there are initially too many variables available in a survey. Freeman and Jekel (1980) provides a simple algorithm to select factors to include in a multi-way table. Firstly all pairwise associations among the available factors are investigated. From these the signifi= cant pairs are selected and compounded variables constructed. The associations of all the single factors and the compounded variables are then investigated. The process is repeated until the expected cell count becomes too small for a reasonable analysis. Out of the signifi= cant interactions a set of factors are selected. On this subset the ordinary log-linear analysis can be performed. The selection process has one major disadvantage, namely the marginality of the tests used.

As in other fields of statistics, problems occur when the tech= niques developed theoretically are applied to the analysis of real life data. Two major test statistics for the various hypotheses are avail= able, namely the Pearson chi-square and the log-likelihood ratio, the latter sometimes also called the information statistic. Both test statistics follow the chi-square distribution asymptotically. The question that arises is which of the two distributions associated with these test statistics is best approximated by the chi-square distribu= tion. Another major distributional problem is created by low cell fre= quencies. It may partly be overcome by collapsing in two ways, firstly over categories of a particular factor or factors, secondly by omitting certain factors so as to reduce the dimension of the table.

The problem areas of distributional difficulties, empty cells or sparse tables and the collapsing of tables, will be treated separately. Sparse tables give rise, however, to distributional difficulties and it is difficult to decide whether or not it is permissible to collapse such a table.

8.1 Aspects of sparse tables

In the practical analysis of multi-way tables it is usually diffi= cult to decide how many variables, and which, should be included in the table. Even after such a decision has been made, the problem of sparse tables can still arise.

The problem of empty cells is a real one even in one- and two-way tables. If the factors of a two-way contingency table are categorized continuous variables or ordinal variables which are strongly related, this may give rise to areas in the contingency table with zero or very low counts. Collapsing over categories to eliminate this problem will lead to a loss of information. For example, Kendall (1977) gives a sparse table resulting from the World Fertility Survey in Fiji of 6779 women, and displays the distribution of their age by the number of their children. There are two reasons for the large sparse areas in the table, one being that there was a small proportion of women in the up= per age groups, and the other that young women had not yet been able to produce large families.

It is valuable to study a table with as many as possible of its categories and factors, because it enables one to spot particular con= centrations and structures in the table that would not be recognized in collapsed versions of the original table.

8.2 Outliers in Contingency Tables

As with other areas of statistics the distribution determines whether or not a certain cell count is an outlier. In the case of con= tingency tables outliers may lead one to reject a model which fits to the remaining observations. Gokhale and Kullback (1978) provide a procedure to test for an outlier on a single cell count at a time. Let x_a^* be the MDI estimates under the hypothesis H_a (i.e. x_a^* subjected to certain marginal constraints). Define now a new set of estimates x_b^* derived from the same marginal constraints as x_a^* except that $x(\omega_1)$, say, is not included, so that $x_b^*(\omega_1) = x(\omega_1)$. From the additive pro= perty of the MDI statistics it follows that

$$2I(x : x_a^*) - 2I(x : x_b^*) = 2I(x_b^* : x_a^*).$$

The above quantity has an asymptotic chi-square distribution with 1 degree of freedom and provides an outlier test for the count in cell ω_1. By making use of the convexity property of information (Kullback, 1959) the following inequality can be derived:

$$2I(x_b^* : x_a^*) \geq 2\left\{x(\omega_1)\ \ln \frac{x(\omega_1)}{x_a^*(\omega_1)} + (n-x(\omega_1))\ \ln \frac{n-x(\omega_1)}{n-x_a^*(\omega_1)}\right\} .$$

The last value is much easier to calculate than $2I(x_b^* : x_a^*)$ and is helpful in identifying the outlier cells. If the few cells that are identified as outliers are omitted, then a simpler model with fewer higher order interaction terms may be fitted. If however a consider=able number of cells can be classified as outliers, higher order models need to be investigated.

The description of the contingency table as a simple model together with a few identified outlying cells is analogous to Brown's (1975) procedure in analysis of variance. It is not necessarily implied that the outlying cells are the result of blunders in the data collection - this model merely provides a convenient summary of the table.

8.3 Collapsing of tables

In most practical applications of the analysis of multi-way con=tingency tables the question of collapsing tables arises. In research projects a number of nominal and ordinal variables may be recorded, together with interval variables. The sample sizes are usually so li=mited that attention in the statistical analysis is necessarily restric=ted to smaller subsets of variables. A problem that occurs frequently is that the resulting contingency table has too few observations in the cells and/or too many empty cells. The evident solution of this pro=blem is to apply some collapsing to the table.

There are two types of collapsing procedures:- collapsing over categories, and reducing the dimension of the original table. In both collapsing procedures there is the danger that information about rela=tionships among the factors still present in the table may be lost. The problem has been addressed by a number of researchers, inter alia: Simpson (1951), Darroch (1962), Plackett (1969), Bishop (1971) and Whittemore (1978). If the dimension is kept as high as possible dis=tributional problems can arise; on the other hand, valuable informa=tion can be lost by collapsing, and thus a move in either direction results in difficulties in the case of sparse tables.

No exposition of the pitfalls of collapsing can be complete without quoting Simpson's paradox (Simpson 1951) which is illustrated in Table II.

Table II - *Sex of patient with treatment for a disease and cure. All the counts in thousands*

	Males			Females			Males and Females together		
	Treatment			Treatment			Treatment		
Cure	Yes	No		Yes	No		Yes	No	
Yes	10	100	110	100	50	150	110	150	260
No	100	730	830	50	20	70	150	750	900
	110	830	940	150	70	220	260	900	1160

In both the male and the female groups, association between TREAT=MENT and CURE is negative, but if the table is collapsed over sex, this negative association changes into a positive one.

A further artificial example can be constructed to illustrate both ways of collapsing and some of the associated problems. Let factor A be an ordinal variable with four levels, and factors B and C two binary variables.

Table III

		FACTOR A			
Factor B	Factor C	1	2	3	4
1	1	80	20	20	80
	2	20	80	80	20
2	1	20	80	80	20
	2	80	20	20	80

The combination of categories 1 and 2, and 3 and 4 of factor A results in a table without any interactions at all. If the table is collapsed over factor B, the interaction between factors A and C, pre=sent in the original table, vanishes. This also happens for factors A and B if the table is collapsed over factor C.

It is thus important to know whether a contingency table is col=lapsible, i.e. whether one factor can be eliminated without affecting interactions among the remaining factors. Whittemore (1978) provides necessary and sufficient conditions for the collapsibility of an n-way table. These conditions are less restrictive than those given by Bishop, Fienberg and Holland (1975). It remains, however, a problem how to decide whether these conditions listed by Whittemore (1978) hold in sparse tables or not.

Practical advice on collapsing will be not to collapse except where sparseness makes it absolutely necessary. A table will become too sparse when more than 20% of the cells have an expected cell count less than one. The order of collapsing should be as follows: combine the categories; and if that does not alleviate the problem it will be necessary to reduce the number of factors.

8.4 Distributional aspects

As previously seen, the distributions of the test statistics used are closely related to the sparseness of the contingency table. A fur= ther related aspect is to determine which of the different test statis= tics follow the chi-square distribution most closely. There are various ways of evaluating discrete test statistics, and since this is such an important topic it will be covered in considerable depth.

The best evaluations can be obtained in the cases where it is possible to have an exact test of the null hypothesis to which the asymptotic tests ought to be equivalent. Where it is not possible to determine the exact distribution under the null hypothesis it is more difficult to evaluate the test statistic objectively.

A number of guidelines have been proposed in the literature for the minimum expected value in any cell of an n x k contingency table for use of the Pearson χ^2 test. Opinions differ greatly, however: some writers consider that the minimum expected should be 20 or more, while others reduce this minimum to 5. The main reason for this difference of opinion is the arbitrary manner in which acceptable differences be= tween the exact and asymptotic distributions have been defined.

One of the main reasons for comparative studies between the Pear= son chi-square and the log-likelihood ratio (or information statistic) is the fact that they have the same asymptotic distribution but are mathematically defined in totally different ways, namely

$$\chi^2 = \Sigma \frac{(x_F - x_E)^2}{x_E}$$

and

$$2I(x_F, x_E) = 2\Sigma\ x_F\ \ell n \frac{x_F}{x_E}$$

where x_F is the observed, and x_E the expected cell count, under the null hypothesis. If $x_F - x_E < x_E$ for all the cells in the contingency table one can get the following approximation:

$$2\Sigma\, x_F\, \ell n\, \frac{x_F}{x_E} = 2\Sigma\, \{x_E + (x_F - x_E)\}\, \ell n\, \{1 + \frac{(x_F - x_E)}{x_E}\}$$

$$= \Sigma \left[\frac{(x_F - x_E)^2}{x_E} + o\, \{\frac{(x_F - x_E)^3}{x_E^2}\} \right] .$$

If $x_F - x_E << x_E$, the last term can be neglected, but otherwise it cannot, and then there may be a marked difference between the two test statistics.

In a few cases it is possible to find the exact probability of an outcome under the null hypothesis by generating all possible out= comes and determining the exact probability and the value of the test statistic(s) under the null hypothesis for each outcome. For example if all the outcomes in a multinomial experiment of sample size 20 and four classes need to be listed then the sequence may be as follows (20,0,0,0); (19,1,0,0); (18,1,1,0) ... (5,5,5,5). (If the null hypo= thesis of equal cell probabilities is tested only ordered generations need to be considered.) The exact cumulative distribution $F_E(x)$ of the outcomes can then be determined by ranking the outcomes in ascending order according to their exact probabilities and adding the probabili= ties of the less probable events to that of the event under considera= tion.

Katti (1973) compared the asymptotic cumulative distribution $F_T(x)$, $(=\alpha$, say) with the exact cumulative distribution of the test statistic, (the exact probability of having a test statistic greater or equal to x). The absolute value of the difference between these two distributions becomes smaller as the sample size increases, but is con= siderable in smaller sizes, when testing for equal cell probabilities in a multinomial distribution.

Katti (1973) further used this type of exact distribution to show that in the case of ordinary multinomial data where the null hypothesis of equal cell probabilities is being tested, any continuity correction that only reduces the Pearson chi-square, will make the test unnecessa= rily conservative. Another way in which the exact probability can be used is to obtain a probability ordering of the outcomes and to com= pare that with the ordering obtained from the test statistics. The ordering obtained from the exact probability is executed in an ascending manner and will be used as the base for correct ranks. Two test statis= tics compared by Kotze and Gokhale (1980) were the Pearson chi-square and log-likelihood ratio. The outcomes according to these test statis= tics have to be placed in descending order of the test statistic, so that this sequence may be comparable to the base ordering. If the

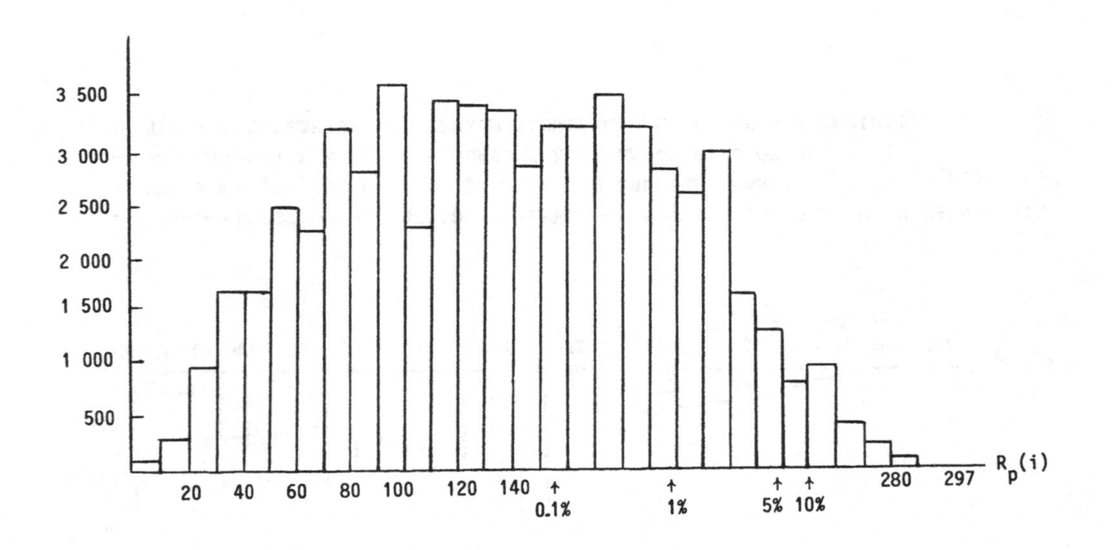

FIG. IA : A GRAPH RELATING THE ABSOLUTE DIFFERENCE BETWEEN THE RANKING OBTAINED FROM THE PEARSON CHI-SQUARE, $R_p(i)$, AND THE BASE RANKING, i, TO $R_p(i)$ FOR A MULTINOMIAL WITH FOUR CLASSES AND SAMPLE SIZE OF 30.
RANKS NEAREST TO CERTAIN SIGNIFICANCE LEVELS ARE INDICATED.

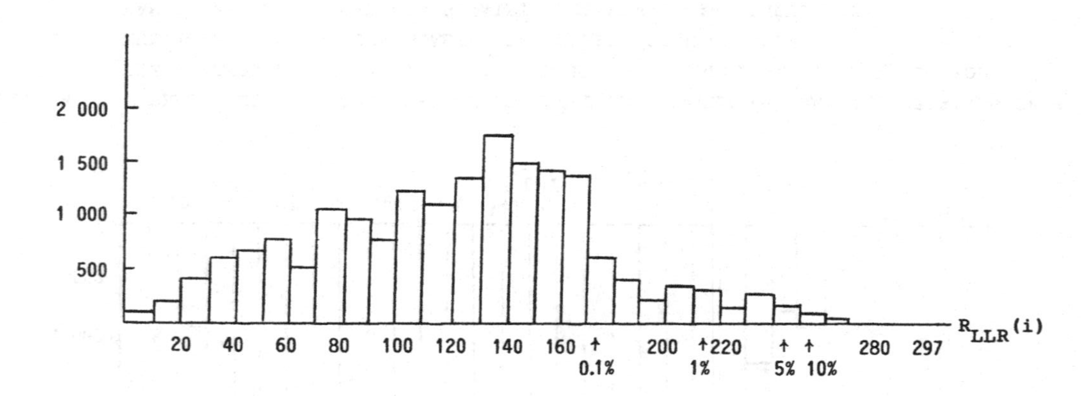

FIG. IB : GRAPH RELATING THE ABSOLUTE DIFFERENCE BETWEEN THE RANKING OBTAINED FROM THE LOG-LIKELIHOOD RATIO $R_{LLR}(i)$ AND THE BASE RANKING, i, TO $R_{LLR}(i)$ FOR A MULTINOMIAL WITH FOUR CLASSES AND SAMPLE SIZE OF 30.
THE RANKS NEAREST TO CERTAIN SIGNIFICANCE LEVELS ARE INDICATED.

asymptotic test statistics are completely equivalent to the exact tests, then the different ways of ordering will be the same.

The discrepancies between the base ordering and the ordering obtai= ned by the test statistic can be summarized in a histogramlike figure. As an example, Fig. I shows this for the Pearson chi-square and log-likelihood statistics in the case of a multinomial with four classes and a sample size of 30. This example shows how the asymptotic test statistics relate to the exact test and enables one to compare the two tests. For other sample sizes and numbers of categories the log-likeli= hood ratios have performed consistently better than the Pearson chi-squares.

By pairwise graphical comparison of the Pearson chi-square and the log-likelihood ratio, it was found that the Pearson chi-square is slightly more conservative than the log-likelihood ratio. In overall assessment the log-likelihood ratio would be preferable to the Pearson chi-square statistic. A bibliography on small samples is provided by Hutchinson (1979).

9. COMPUTER PROGRAMS AVAILABLE

Since 1970, a number of programs have been developed under the guidance of Prof S Kullback. The program KULLITR is the most general program in this set and allows the user to set up his own design matrix. Another important program is P3F in the BMDP program library, which has many options for the analysis of multi-way contingency tables. The FUNCAT procedure of the Statistical Analysis System (SAS) utilizes the method of Grizzle, Starmer and Koch (1969) and models functions of nominal and ordinal responses as a linear model. The most general pro= gram available presently is the GLIM system whuch may be used to ana= lyse ANOVA models and contingency tables and for probit analysis.

10. THE APPLICATION OF LOG-LINEAR MODELS TO THE ANALYSIS OF RATES AND PROPORTIONS

10.1 Introduction

In biological sciences rates and proportions play an important role, and are sometimes linked to a quantitative variable, e.g. time or age. Concepts such as log-odds, relative risks and the odds-ratio arise and can easily be linked to log-linear models. For example, in a two-way table with k age categories showing coronary heart deaths in a sample over a fixed period of time, the common hypothesis that the log-odds change linearly with age can be written as

$$\ln \frac{p_i}{1 - p_i} = \alpha + \beta x_i \qquad i = 1,\ldots,k$$

with p_i denoting the probability of a coronary heart death at age x_i. The x_i's are usually equally spaced.

If this survey is carried out on males as well as females, the two log-odds can be written as

$$\ln \frac{p_{i1}}{1 - p_{i1}} = \Phi_{i1} \quad , \qquad \text{for males}$$

$$\ln \frac{p_{i2}}{1 - p_{i2}} = \Phi_{i2} \quad , \qquad \text{for females}$$

where p_{ij} is the probability of death from heart disease for a person of sex j and age i, $\quad i = 1,\ldots,k$.

The log-odds ratio can now be defined by

$$\psi_i = \Phi_{i1} - \Phi_{i2} \qquad i = 1,\ldots,k$$

$$= \ln \{(\frac{p_{i_1}}{1 - p_{i_1}})/(\frac{p_{i_2}}{1 - p_{i_2}})\} \quad .$$

A common null hypothesis considered is that the log-odds ratio remains constant over age, i.e.

$$\psi_i = \psi \qquad i = 1,\ldots,k$$

If the contents of the above table are stored in a three-way con= tingency table with the factors AGE (A), SEX (S) and DEATH (D), then the saturated log-linear representation of a cell entry x^{ASD}_{ijk} is

$$\ln x^{ASD}_{ijk} = \theta + \lambda^A_i + \lambda^S_j + \lambda^D_k + \lambda^{AS}_{ij} + \lambda^{AD}_{ik} + \lambda^{SD}_{jk} + \lambda^{ASD}_{ijk}$$

For the log-odds it is easy to find

$$\Phi_{ij} = \ln \frac{p_{ij}}{1 - p_{ij}} = \lambda^D_1 - \lambda^D_2 + \lambda^{AD}_{i1} - \lambda^{AD}_{i2} + \lambda^{SD}_{j1} - \lambda^{SD}_{j2} + \lambda^{ASD}_{ij1} - \lambda^{ASD}_{ij2}$$

$$i=1,\ldots,k, \quad j=1,2$$

and for the log-odds ratio

$$\psi_i = \Phi_{i1} - \Phi_{i2} = \lambda^{SD}_{11} - \lambda^{SD}_{12} - \lambda^{SD}_{21} + \lambda^{SD}_{22} + \lambda^{ASD}_{i11} - \lambda^{ASD}_{i12} - \lambda^{ASD}_{i21} - \lambda^{ASD}_{i22}$$

The first four terms of the expression for ψ_i stay constant as i changes, but the remainder do not. With the usual restrictions on the λ's it can be shown that the null hypotheses of a constant log-odds ratio and no three-way interaction are equivalent. However, hypotheses concerning proportions associated with ordinal factor(s) can more easily be analyzed by logistic regression.

10.2 Logistic regression

Armitage (1955) proposed a test to determine whether proportions change linearly with respect to an ordinal or continuous predictor variable y :- i.e.

$$p_i = \alpha + \beta y_i$$

This differs from the previously mentioned hypothesis on the log-odds. One point in common in the relationship between these two hypotheses is that there exists a relationship between the binary response and the y-variable, but the relationship is restricted to be a linear one. Fewer parameters are necessary to express this relationship than in the case of ordinary interaction in a 2 x k contingency table.

Due to the unified approach offered by log-linear models for the analysis of multi-way tables, a different way of expressing the rela= tionship between the binary response and the y_i's has been formulated. To repeat the hypothesis given in the introduction.

$$\ell n \frac{p_i}{1 - p_i} = \alpha + \beta y_i \qquad i = 1,2,\ldots,k \quad .$$

The k can have a wider meaning than the number of classes of an ordinal variable, it may represent the cells of a contingency table in which the binary responses were observed. The factors can be binary, multinomial or ordinal in nature. It will therefore be necessary to adapt the number of model parameters according to the hypothesis under consideration. The model can more generally be expressed as

$$\ell ogit(p) = \ell n \frac{p}{1 - p} = C\beta$$

where $\ell n \frac{p}{1 - p}$ is a g x 1 vector

C is a g x n matrix

and β is a n x 1 vector.

When the binary response is part of a contingency table $g = \Omega/2$ where Ω is the number of cells.

While β may be estimated using maximum likelihood applied to the binomial distribution with probability p_i of the outcomes in row i, it is more common to transform the x_{ji} and use standard regression proce= dures. For example, the logistic transform may be applied, giving

$$\ell n \frac{x_{1i}}{x_{\cdot i} - x_{1i}} = \ell n \frac{x_{1i}}{x_{2i}}$$

Asymptotically, the logistic transform has a variance which may be estimated by $x_{.i}/(x_{1i}\ x_{2i})$ and mean $\ell n\{p_i/(1-p_i)\}$.

Thus the logit transformed x_{ji} may be used directly in an ordinary weighted regression to estimate, test and set confidence intervals on β. This approach is quick, and gives acceptable results unless x_{1i} or x_{2i} is very small.

There are several minor adaptations of the logit transform which use $\ell n\ (x_{1i} + a)/(x_{2i} + b)$. In addition, there are other linearizing transforms of the p_i such as $\Phi^{-1}(p_i)$, (Φ denoting the cumulative normal distribution) which gives probit analysis, and the standard variance stabilizing transform $\sin^{-1}(p_i^{\frac{1}{2}})$. However the logit transform remains the most common linearizing transform.

To illustrate the potential parsimony attainable using these mo= dels, consider the analysis of a 2x4 contingency table using first the saturated log-linear model, and then logistic regression. For the for= mer model, we fit

$$\ell n(x^*/\pi) = C\tau$$

where C is defined as

$$\begin{array}{cccccccc} \tau_0 & \tau_1 & \tau_2 & \tau_3 & \tau_4 & \tau_5 & \tau_6 & \tau_7 \end{array}$$
$$\begin{bmatrix} 1 & 1 & 1 & 0 & 0 & 1 & 0 & 0 \\ 1 & 1 & 0 & 1 & 0 & 0 & 1 & 0 \\ 1 & 1 & 0 & 0 & 1 & 0 & 0 & 1 \\ 1 & 1 & 0 & 0 & 0 & 0 & 0 & 0 \\ 1 & 0 & 1 & 0 & 0 & 0 & 0 & 0 \\ 1 & 0 & 0 & 1 & 0 & 0 & 0 & 0 \\ 1 & 0 & 0 & 0 & 0 & 0 & 0 & 0 \\ 1 & 0 & 0 & 0 & 0 & 0 & 0 & 0 \end{bmatrix}$$

Therefore the estimated log-odds are given by

$$\ell n \frac{x^*_{11}}{x^*_{.1} - x^*_{11}} = \ell n \frac{x^*_{11}}{x^*_{21}}$$

$$= \tau_1 + \tau_5$$

$$\ell n \frac{x^*_{12}}{x^*_{22}} = \tau_1 + \tau_6$$

$$\ell n \frac{x^*_{13}}{x^*_{23}} = \tau_1 + \tau_7$$

$$\ell n \frac{x^*_{14}}{x^*_{24}} = \tau_1$$

To fit a logistic regression, the last three columns of C are replaced by $[0\ 1\ 2\ 3\ 0\ 0\ 0\ 0]^T$ and $\tau_5\ \tau_6\ \tau_7$ replaced by τ_8 to get

$$\ell n \frac{x_{11}^*}{x_{21}^*} = \tau_1$$

$$\ell n \frac{x_{12}^*}{x_{22}^*} = \tau_1 + \tau_8$$

$$\ell n \frac{x_{13}^*}{x_{23}^*} = \tau_1 + 2\tau_8$$

$$\ell n \frac{x_{14}^*}{x_{24}^*} = \tau_1 + 3\tau_8$$

Five parameters are necessary to describe the independence model and eight parameters for the full saturated model. The model associated with the hypothesis that there exists a linear relationship between the logits needs only 6 parameters, one more than the independence mo= del. The use of functional relationships of ordinal variables in pre= dicting the expected cell counts, may thus lead to a considerable saving in the number of parameters in the fitted model. One of the disadvan= tages of logistic regression is that if several binary factors are pre= sent only one can be selected to be the response variable. Most of these hypotheses can be tested by using the more general programs such as GLIM, KULLITR and FUNCAT.

A comprehensive introduction to the analysis of rates and propor= tions is provided by Fleiss (1973). A more general approach to binary data and the logistic function is described by Cox (1970). Breslow (1976) shows that an 'exact' analysis, based on the conditional likeli= hood obtained by fixing all the marginals, is asymptotically equivalent to the use of an unconditional log-linear model. A thorough review of regression models for ordinal data is given by McCullach (1980). Holford (1980) describes how to use log-linear models in the analysis of rates and survivorship. The key to most of these applications of log-linear models is to rewrite the hypothesis to be tested in terms of a linear model of the logarithm of the expected frequencies. To obtain estimates of the parameters the generalized iterative scaling procedures of Darroch and Ratcliff (1972) can be used.

11. THE GRAPHICAL REPRESENTATION OF THE RELATIONSHIPS PRESENT IN A MULTI-WAY TABLE

11.1 Determination of associations present in the appropriate model

We assume that an appropriate model has been chosen by using the strategies discussed in section 7. For this model new terms similar to marginal and partial association of Brown (1976), called 'model marginal' and 'model partial' association, will be defined. The models used in the method of Brown (1976) to obtain tests for the marginal and partial associations may contain redundant terms, i.e. non-significant interactions, and to eliminate this defect, we have attempted to re= late the newly defined interactions as closely as possible to the chosen model. Illustrations of the definitions will be provided by means of an example : a five-way table with the model (ABC,BDE,AE).

(1) Start with the highest interaction or interactions (say k factor interactions). For example, in the model (ABC,BDE,AE) the model marginal and model partial associations of ABC and BDE must be deter= mined first.

(2) After all the tests for terms that contain k factors have been determined, the process is repeated for the (k-1) factor interac= tions, starting with the selected model with all k factors and higher interactions removed. The last tests to be calculated are for the two-way effects present in this model. In the model that is used for the example, the two-way interactions are AB, AC, BC, BE, DE, AE.

There are various necessary and sufficient conditions for collap= sibility of multi-way contingency tables as set out by Whittemore (1978), and to avoid any problems in collapsing which may occur in the defini= tion of the marginal association by Brown (1976), the procedure descri= bed below will be used for determining the model marginal association. If it is necessary to determine the model marginal association of the term ABC... which contains k factors then all the other terms present in the current model must contain (k-1) factors or fewer. The test-of-fit of the model in which the k factor term has been eliminated by breaking it up into various (k-1) factor subsets is determined. All the other terms present in the model under consideration are retained, but redundant terms are removed. The difference between tests-of-fit of the model containing the k factor term and the reduced model is defined to be the test for the *model marginal association* of ABC... for the model selected.

For example the model marginal association of ABC in the model (ABC,BDE,AE), is the difference between the tests-of-fit of the models (ABC,BD,BE,DE,AE) and (AB,AC,BC,BD,BE,DE,AE).

The test for model partial association for the model selected is defined in a way similar to Brown's (1976). If the test for model par= tial association of ABC... (k factor) is to be determined, the current model may not contain terms with more than k factors. The difference between the test-of-fit of the current model, and that of the model which differs only in that the ABC... effect has been eliminated, de=

fines the model partial association of ABC...

For example, the model partial association of ABC in the model (ABC,BDE,AE), is the difference between the tests-of-fit of the models (ABC,BDE,AE) and (AB,AC,BC,BDE,AE).

11.2 Graphical representation

To visualize the final model, we have devised a method for dis= playing the interactions graphically. The mean of the tests for the model marginal and model partial associations is calculated for each effect (two-way and higher). The chi-square tail probability of each mean is then determined to remove the effect of the degrees of freedom. This method of combining these two tests is arbitrary and other methods can be devised. These probabilities are used to set up a figure in the following fashion: first, all the factors present in the multi-way contingency table are circled. The factors for which two-way in= teractions are present in the final model are connected with lines whose thickness is proportional to the absolute value of the logarithm of the probability associated with the two-way interaction. The fac= tors related by three-way or higher interactions are enclosed by a line whose thickness is determined in the same way as before.

11.3 Practical example

The data of this example are taken from a survey carried out by Dr I L Carstens of the University of Stellenbosch. In a sample of 249 schoolchildren, the dental status of a particular molar was measured. The relatively few children in whom that molar was missing were exclu= ded. It was noted whether or not the tooth was decayed (S) and plaque (P) and calculus (C) indices were determined. This information toget= her with ethnic group (R) was summarized in a four-way contingency table (Table IV).

The marginal and partial effects according to the definition by Brown (1976) are given in Table V. From this it was decided to consider the models (CPR,S) with LR statistic = 19.27,11 degrees of freedom (tail probability = 0.056); and (CPR,RS) with LR statistic = 9.30,9 degrees of freedom (tail probability = 0.410). These two models were singled out be= cause (CPR,S) contained most of the significant effects ($p < 0.05$), and gave a non-significant test-of-fit, and (CPR,RS) had the additional sig= nificant interaction RS.

The analysis will be considered to be of the 'correlation' type because one is not only interested in the relationship of S with the other variables R, P and C, but also in the interrelationships between R, P and C. The model (CPR,RS) was selected because it gave a much better fit to the data. For this model, tests for the proposed model partial and marginal associations were determined. The model partial and marginal associations are the same for CPR, and are given by the difference between the tests-of-fit of the models (CPR,RS) and (CP,CR, PR,RS). The model partial effect CP is determined by the difference between the tests-of-fit of the models (CP,CR,PR,RS) and (CR,PR,RS). Similarly the model marginal effect CP is determined by the difference between the tests-of-fit of the models (CP,R,S) and (C,P,R,S). From Table VI it can be seen that the differences between the tests for the model partial and marginal associations are possible. This is further evidence of the possible large confounding role played by the order of fitting in the determination of the separate interactions using any stepwise procedure.

Since both the tests for the model partial and marginal associa= tions are equally valid in determining the interaction, the means of the LR tests have been calculated and tabulated in Table VII. As pre= viously described, the absolute values of the natural logarithms of the associated probabilities are used to set up Fig.II which graphi= cally represents the model (CPR,RS) and the interactions within it.

In Fig.II the factors CALCULUS (C) and PLAQUE (P) are placed in rectangular blocks because these substances form on the tooth. CARIES (S) was enclosed in a different shape because it describes permanent damage to the tooth. Another shape was used for RACE (R) as it descri= bes a personal characteristic. From Fig.II it can be seen that there is a strong two-way interaction between PLAQUE and CALCULUS, indica= ting that PLAQUE and CALCULUS are closely related. The same is true for PLAQUE and RACE. There are minor interactions between CALCULUS and RACE, and RACE and CARIES. There is also a three-way interaction between CALCULUS, PLAQUE and RACE, suggesting that the relationship between CALCULUS and PLAQUE varies from RACE to RACE. An interesting fact is that there is no direct link between CARIES and PLAQUE or CARIES and CALCULUS.

TABLE V : THE DISTRIBUTION OF 218 CHILDREN ACCORDING TO THE FACTORS RACE, CARIES, PLAQUE AND CALCULUS

RACE (R)	CARIES (S)	PLAQUE (P)	CALCULUS (C)	
			Low	High
Black	Sound	Low	4	7
		High	10	6
	Not sound	Low	4	10
		High	5	14
White	Sound	Low	13	1
		High	1	2
	Not sound	Low	43	13
		High	2	8
Coloured	Sound	Low	5	3
		High	4	9
	Not sound	Low	18	4
		High	6	26

TABLE V : MARGINAL AND PARTIAL ASSOCIATIONS ACCORDING TO THE DEFINITION BY BROWN (1976)

		Test of partial association		Test of marginal association	
Effect	DF	LR CHISQ	PROB.	LR CHISQ	PROB.
C	1	0.66	0.42		
P	1	4.71	0.03		
R	2	3.83	0.15		
S	1	36.56	0.00		
CP	1	21.08	0.00	34.21	0.00
CR	2	6.65	0.04	18.95	0.00
CS	1	3.60	0.06	0.65	0.42
PR	2	26.96	0.00	42.99	0.00
PS	1	0.84	0.36	1.63	0.20
RS	2	10.01	0.01	9.97	0.05
CPR	2	14.99	0.00	18.74	0.00
CPS	1	1.91	0.17	4.49	0.03
CRS	2	2.74	0.25	2.39	0.30
PRS	2	0.60	0.74	0.02	0.99
CPRS	2	1.22	0.54		

TABLE VI : TESTS FOR MODEL PARTIAL AND MARGINAL ASSOCIATIONS

Effect	DF	Test for model partial association	Test for model marginal association
CPR	2	18.73	18.73
CP	1	20.35	34.21
CR	2	5.09	18.95
PR	2	29.14	43.00
RS	2	9.98	9.97

TABLE VII : MEANS OF THE LR TESTS ASSOCIATED WITH THE MODEL PARTIAL AND MARGINAL EFFECTS

Effect	DF	LR	PROB	$\lvert \ell n(\text{PROB}) \rvert$
CPR	2	18.73	$8.57\text{x}10^{-5}$	9.36
CP	1	27.28	$1.76\text{x}10^{-7}$	15.55
CR	2	12.02	$2.45\text{x}10^{-3}$	6.01
PR	2	36.07	$1.47\text{x}10^{-8}$	18.04
RS	2	9.975	$6.82\text{x}10^{-3}$	4.99

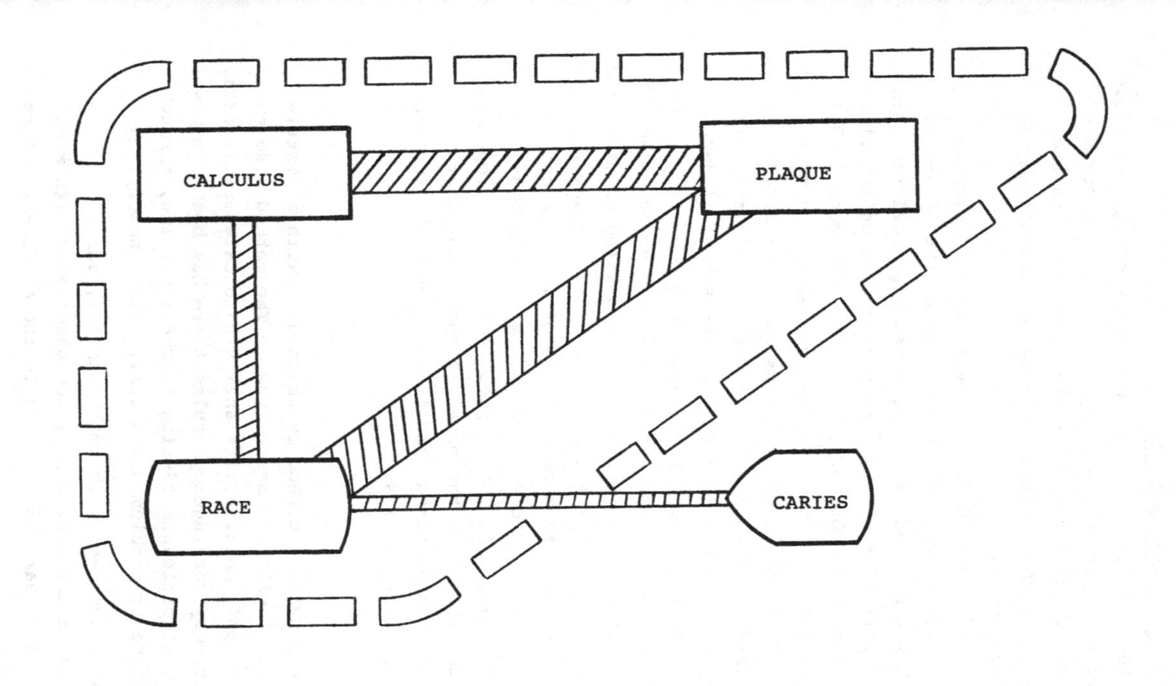

FIG. II : GRAPHICAL REPRESENTATION OF RELATIONSHIPS AMONGST CATEGORICAL FACTORS

11.4 Discussion of graphical methods

The method proposed tackles the problem of decomposing the selec= ted model into the various interactions present. It also suggests a technique to overcome the problem caused by the order in which the models are fitted, which can change the magnitude of the test statistic associated with each interaction.

Davis (1971) provided a method for a 'causal system' diagram to be used in the analysis of surveys. In these diagrams the causal di= rection is shown, which was omitted in our proposed graphical represen= tation because it is so difficult in some cases to decide on the direc= tion of the causal relationship. Darroch, Lauritzen and Speed (1980) give a method of constructing a simple, visual representation of any decomposable model. Goodman (1970) calls the decomposable subset 'models with an elementary hypothesis', and provides an algorithm for deciding whether a hierarchical model is part of the subset of decom= posable models. For elementary models the iterative procedure for ob= taining estimates converges at the end of the first cycle and these estimates can be written as explicit expressions consisting of marginals.

The premise in developing this graphical method is, however, that various strategies in model selection exist, some of which do not de= pend on significance testing for deciding on an appropriate model (see Benedetti and Brown 1978). In the proposed method an attempt was made to present the interactions in the model in such a way that the implications of the model can be summarized in one figure showing the strengths of the various relationships.

12. CONCLUSION

Log-linear models provide a method by which data with a distri= bution in the exponential family can be analyzed. The method also en= ables the researcher to apply multivariate analysis to categorical data. In the analysis of multi-way contingency tables there has been a move= ment away from ordinary significance testing to model fitting, through log-linear models. Two major problem areas still exist, namely:

i) distributional inaccuracies in small samples, and
ii) the differences in the test statistics associated with a para= meter in the log-linear model, owing to the different orders of model fitting.

Despite problems with the application of log-linear models, they provide new methods of analyzing data.

REFERENCES

AFIFI, A.A. and ELASHOFF, R.M. (1969). Multivariate two sample tests with dichotomous and continuous variables. I. The location model. *Ann. Math. Statist.*,40, 290-298.

ARMITAGE, P. (1955): Tests for linear trends in proportions and frequencies. *Biometrics*, 11, 375-385.

AITKEN, M. (1979). A simultaneous test for contingency table models. *Appl. Statist.*, 28, No.3, 233-242.

BENEDETTI, J.K. and BROWN, M.B. (1978). Strategies for the selection of log-linear models. *Biometrics*, 34, 680-686.

BERKSON, J. (1972). Minimum discrimination information: The "no-inter= action" problem and the logistic function. *Biometrics*, 28, 433-468.

BIRCH, J. (1972). Maximum likelihood in three-way contingency tables. *J.Roy.Statist.Soc.*,B, 25, 220-233.

BISHOP, Y.M.M. (1971). Effects of collapsing multidimensional contin= gency tables. *Biometrics*, 27, 545-562.

BISHOP, Y.M.M., FIENBERG, S.E. and HOLLAND, P.W. (1975). *Discrete Multivariate Analysis: Theory and Practice.*Cambridge, Massachusetts, MIT Press.

BRESLOW, N. (1976). Regression analysis of the log-odds ratio: a method for retrospective studies. *Biometrics*, 32, 409-416.

BROWN, M.B. (1975). Exploring interaction effects in the analysis of variance. *Appl. Statist.*, 24, 288-298.

BROWN, M.B. (1976). Screening effects in multidimensional contingency tables. *Appl. Statist.*, 25, 37-46.

COX, D.R. (1970). *The Analysis of Binary Data.* Methuen, London.

CRAMER, H. (1946). *Mathematical Methods of Statistics*, Princeton Uni= versity Press.

DARROCH, J.N. (1962). Interaction in multi-factor contingency tables, *Biometrika*, 61, 207-214.

DARROCH, J.N. (1974). Multiplicative and additive interaction in con= tingency tables. *Biometrika*, 61, 207-214.

DARROCH, J.N. (1980) Book review of Gokhale and Kullback (1978). *Biometrics*, 36, 360.

DARROCH, J.N., LAURITZEN, S.L. and SPEED,T.P. (1980). Markov fields and log-linear interaction models for contingency tables. *Ann. Statist.*, 8, 522-539.

DARROCH, J.N. and RATCLIFF, D. (1972). Generalized iterative scaling for log-linear models. *Ann.Math.Statist.*, 34, 555-567.

DAVIS, J.A. (1971). *Elementary Survey Analysis.* Prentice-Hall, Engle= wood Cliffs, N.J.

FLEISS, J.L. (1973). *Statistical Methods for Rates and Proportions.* Wiley, New York.

FREEMAN, D.H. and JEKEL, J.F. (1980) Table selection and log-linear models. *J.Chron.Dis.*, Vol.33, pp 513-524.

GOKHALE, D.V. (1972). Analysis of log-linear models. *J.Roy.Statist. Soc.*, B, 34, 371-376.

GOKHALE, D.V. and KULLBACK, S. (1978). *The Information in Contingency Tables.* Marcel Dekker, Inc., New York and Basel.

GRIZZLE, J.E.,STARMER, C.F. and KOCH, G.G. (1969). Analysis of cate= gorical data by linear models. *Biometrics*, 25, 489-504.

GOODMAN, L.A. (1968). The analysis of cross-classified data : Inde= pendence, quasi-independence and interactions in contingency tables with or without missing entries. *J.Amer.Statist.Assoc.*,63,1091-1131.

GOODMAN, L.A. (1970). The multivariate analysis of qualitative data: Interaction among multiple classifications. *J.Amer.Statist.Assoc.*, 65, 226-256.

GOODMAN, L.A. (1978). *Analyzing qualitative/categorical data; Log-li= near model and latent structure analysis.* Abt Books, Cambridge, Massachusetts.

HABERMAN, S.J. (1972). Log-linear fit for contingency tables. *J.Roy. Statist. Soc.*, C, 21, 218-224.

HOLFORD, T.R. (1980). The analysis of rates and survivorship using log-linear models. *Biometrics*, 36, 299-305.

HUTCHINSON, T.P. (1979). The validity of the chi-square test when the expected frequencies are small : A list of recent research referen= ces. *Comm. Statist.* A: 8, 327-335.

IRELAND, C.T.,KU,H.H. and KULLBACK, S. (1969). Symmetry and marginal homogeneity of an rxr contingency table. *J.Amer.Statist.Assoc.*, 64, 1323-1341.

KATTI, S.K. (1973). Exact distribution for the chi-square in the one- way table. *Comm.Statist.*,2, 435-447.

KENDALL, M. (1977). *Multivariate Contingency Tables and some further problems in Multivariate Analysis. IV.* ed.P.R. Krisnaiah, North Holland Publishing Company, 483-494.

KILLION, R.A. and ZAHN, D.A. (1976). A Bibliography of Contingency Table Literature: 1900 to 1974. *Int.Stat.Rev.*,44, 71-112.

KOTZE, T.J.v.W. and GOKHALE, D.V. (1980). A comparison of the Pearson χ^2 and log-likelihood-ratio statistics for small samples by means of probability ordering. *J. Statist. Comput. Simul.*, 12, 1-13.

KULLBACK, S. (1959). *Information Theory and Statistics.* John Wiley and Sons, New York.

KULLBACK, S. (1973). *Program Manual. Dept. of Statistics.* The George Washington University, Washington, D.C.20052.

McCULLAGH, P. (1980). Regression models for ordinal data. *J.Roy.Statist. Soc.*,B, 42, 109-142.

NELDER, J.A. (1974). Log-linear models for contingency tables: a gene= ralization of classical least squares. *Appl.Statist.*, 23,323-329.

NELDER, J.A. and WEDDERBURN, R.W.M. (1972). Generalized linear models. *J.Roy.Statist.Soc.*, A, 135, 370-384.

PLACKETT, R.L. (1969). Multidimensional contingency tables: A survey of models and methods. *Bull.Inter.Statist.Inst.*,Vol.43, Book 1, 133-241.

SIMPSON, E.H.(1951). The interpretation of interaction in contingency tables. *J.Roy.Statist.Soc.*,B, 13, 238-241.

TATE,M.W. and HEYER,L.A. (1973). Inaccuracy of the χ^2 test of goodness of fit when expected frequencies are small. *J.Amer.Statist.Assoc.*, 63,836-841.

WHITTEMORE, A.S.(1978). Collapsibility of multidimensional contingency tables. *J.Roy.Statist.Soc.*,B,40,No.3,328-340.

SCALING A DATA MATRIX IN A LOW-DIMENSIONAL EUCLIDEAN SPACE

MICHAEL J. GREENACRE, *UNIVERSITY OF SOUTH AFRICA*
and
LESLIE G. UNDERHILL, *UNIVERSITY OF CAPE TOWN*

1. INTRODUCTION

The principles underlying data scaling originated in the field of psychology. Direct measurement of intelligence, aggressiveness, depression and other mental states are impossible, whereas it is fairly easy to observe various manifestations of these states. For example, a typical IQ test consists of a set of questions designed to test vari= ous aspects of the underlying quality which the researcher calls "in= telligence". Every person that undergoes such a test generates a vec= tor of responses and these are then converted into a single value which is called the IQ of that person. The process whereby this is achieved is called scaling, because each person becomes a point on a single di= mension, in this case the IQ scale. Geometrically, the original data vector may be considered a point in a high-dimensional space and the process of scaling maps this point to a point in a one-dimensional sub= space. These ideas can be generalized to mapping data vectors to points in a two-dimensional subspace so that the original vectors are "scaled" as points in a plane. Whether the dimensionality of the final subspace of representation be one, two or higher, the underlying principles are the same. Firstly, scaling transforms data points of very high dimen= sionality to points of much lower dimensionality. Secondly, the original data vectors are often points in a non-Euclidean space, that is dis= tances are not defined in the usual physical way. Whatever the origi= nal space of the data we would like the final low-dimensional displays to be Euclidean so that our view of the points is the same as our in= tuitive interpretation of physical space.

The problem of reducing the dimensionality of data and, in gene= ral, of displaying multidimensional data is one which permeates many applied disciplines today. Typically, a researcher who is studying the manifestation of a particular phenomenon, tries to record or measure all the aspects that he considers to be important in order to understand that phenomenon. Data can be collected with a view to quantifying a single underlying dimension, for example intelligence, or to describing several unknown dimensions which will be identified *a posteriori*, for example a survey of opinion on a controversial subject. At the end of this data collection process the researcher is faced with a vast amount of information from which he can extract certain simple summaries, but without any global view of what his data contain and how they inter= relate.

Scaling techniques provide a natural first stage in the explo= ration of a data matrix. Indeed, once the data are "seen", the resear= cher might not even find it necessary to perform further analyses. From a point of view of communication, a graphical description is very easily and quickly assimilated by the researcher. In particular, scaled data has the familiar appearance of a map of points and the proximi= ties, distances and groupings of the points are readily picked up by the eye.

We shall be dealing with two classes of scaling methods which, in our opinion, find the most widespread application in many varied contexts. These two types of method are depicted schematically in Fi= gure 1.1.

The first type of method deals with the scaling of profile or vector data, where the rows and/or the columns of the data matrix are vectors of description of the entities associated with the rows and/or columns respectively. For example, each can be vectors of measurements or observations on a person or an object, or the matrix can be a table of frequencies across income groups (columns) of the residents in dif= ferent suburbs (rows), where a row of frequencies characterizes the particular suburb. In general we call these techniques that lead to graphical displays of the rows and/or columns of such data matrices *vector* (or *profile*) *scaling methods*. We shall describe a specific, though very wide, class of vector scaling methods which we call *basic structure displays* (§2).

The second type of method deals with one set of points only, where the data form a square symmetric matrix of distances, dissimilari= ties or similarities between objects of a set, for example a matrix of genetic distances between human populations, a matrix of transport costs between towns or a matrix of similarity indices between regions based on their ecologies. The object here is to obtain a display of this set of points which concords as much as possible with the given data. In other words, if two regions, say, have relatively similar eco= logies then they should be represented by points which are quite close together in the Euclidean display. This type of scaling method is dis=

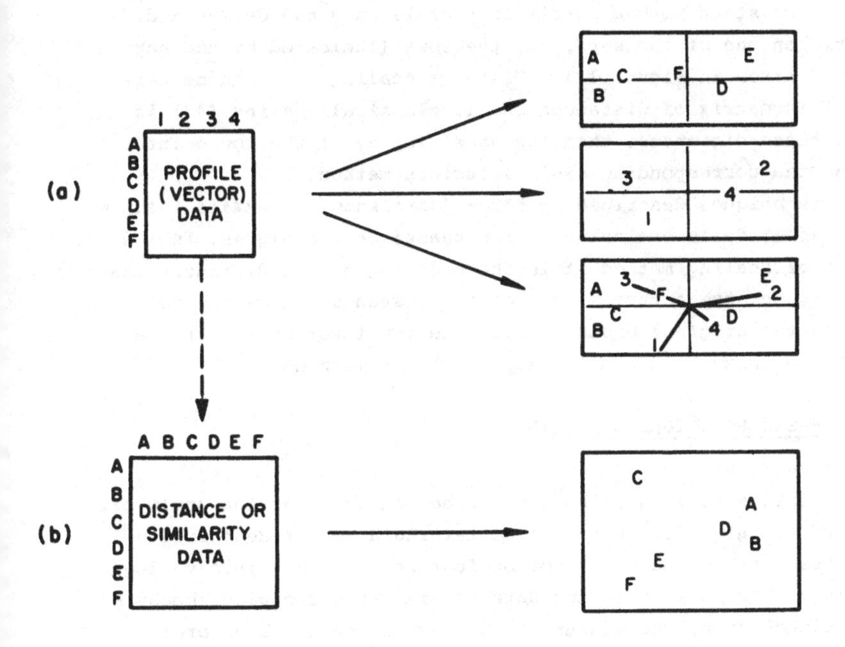
(a)
1 2 3 4
A B C D E F
PROFILE (VECTOR) DATA
A C F E B D
3 2 4 1
A 3 F E 2 B C 4 D 1
(b)
A B C D E F
A B C D E F
DISTANCE OR SIMILARITY DATA
C A D B E F

FIGURE 1.1

cussed in §§3, 4 and 5 and is often described in the literature simply as "multidimensional scaling". We use the more specific term *distance scaling*.

Often we start with a matrix of profile data and define a dis= tance metric on one of the sets, say the rows (indicated by the ver= tical dotted arrow in Figure 1.1). Distance scaling can then be per= formed on this matrix of distances and if classical scaling (§3) is applied to these distances, then the same display of the row points is obtained as the corresponding basic structure method.

The techniques described in §§2 - 4 are known as *metric* techni= ques, whereas §5 deals exclusively with *nonmetric* techniques. In the latter type of scaling method it is the ordering of the distances that is important, not the actual values of the distances. Here the relation= ship between the original distance data and the interpoint distances in the final display is a crucial aspect of the methodology.

2. BASIC STRUCTURE DISPLAYS

Introduction

The various scaling methods described in this section apply pri= marily to rectangular data matrices containing a set of description vectors. These methods have the common feature that they involve low rank matrix approximations to the data matrix. Traditionally, because of computational ease, the measure of approximation has been ordinary or generalized least squares. Because of the unifying role played by the *basic structure* (or *singular value decomposition*, as it is often known) in the least squares approximation of matrices, we term this class of graphical methods *basic structure displays*.

Eckart and Young (1936) and Householder and Young (1938) gave pioneering descriptions of matrix approximations in the context of ma= trix eigenstructure. Recent articles on matrix approximation particu= larly relevant to our approach are by Gabriel and Zamir (1976), Gabriel (1978) and Rao (1980).

2.1 Generalized basic structure

Basic structure

The concept of basic structure of a rectangular $n \times m$ matrix A of rank r is an extremely useful tool in matrix algebra, because it decomposes A into 3 elements of simple structure:

$$\underset{n\times m}{A} = \underset{n\times r}{U}\ \underset{r\times r}{D_\lambda}\ \underset{r\times m}{V^T} \tag{2.1.1}$$

where D_λ is a diagonal matrix of positive elements $\lambda_1, \ldots, \lambda_r$ and U and V have orthonormal columns: $U^TU = I = V^TV$.

If A is square symmetric, then U = V and the basic structure (2.1.1) is the usual eigenstructure of A.

Geometric interpretation

The column vectors in U and V form orthonormal bases for the columns and rows of A respectively (strictly speaking we should say: for the columns of A and A^T respectively). Thus the basic structure summarizes the structure of the matrix, with the *left* and *right basic vectors* (in U and V respectively) situating the subspaces in which the columns and rows are contained and the *basic values* (down the diagonal of D_λ) expressing the magnitude of the matrix in each of the r dimen= sions.

Low rank matrix approximations

The practical value of basic structure for our purposes is that it allows a data matrix A, usually of high rank r, to be approximated by a matrix $\hat{A}_{[p]}$, of low rank p. In fact if the basic values of A are ordered: $\lambda_1 \geq \lambda_2 \geq \ldots \geq \lambda_r > 0$, and then if $U_{(p)}$ and $V_{(p)}$ contain the first p vectors of U and V (correspondingly ordered) and $D_{\lambda(p)}$ the first p basic values down the diagonal, then we define:

$$\underset{n\times m}{\hat{A}_{[p]}} \equiv \underset{n\times p}{U_{(p)}}\ \underset{p\times p}{D_{\lambda(p)}}\ \underset{p\times m}{V^T_{(p)}} \tag{2.1.2}$$

$\hat{A}_{[p]}$ is an approximation of A in the sense that, of all matrices $A_{[p]}$ of rank p, it is the one which minimizes:

$$\|A - A_{[p]}\|^2 \equiv \mathrm{tr}[(A - A_{[p]})(A - A_{[p]})^T] \tag{2.1.3}$$

(equal to $\sum_i \sum_j (a_{ij} - \tilde{a}_{ij})^2$, where $A_{[p]}$ has elements $\tilde{a}_{ij}$).

The rank p of $\hat{A}_{[p]}$ is usually chosen to be much lower than that of A, that is the respective spaces of the rows and columns of $\hat{A}_{[p]}$ are of dimension p, and graphical exploration and interpretation is clear= ly easier in fewer dimensions.

Generalized basic structure and matrix appoximation

If the above concept is generalized we shall find that the (gene= ralized) basic structure underlies a large number of scaling techniques.

The matrix norm is generalized as follows:

$$\|A\|^2_{\Omega,\Phi} \equiv \text{tr}[\Omega A \Phi A^T] \text{ where } \Omega,\Phi \text{ are positive definite symmetric} \qquad (2.1.4)$$

(in (2.1.3) the norm was the special case $\Omega = \Phi = I$).

The positive definite matrices Ω and Φ can be considered as de= fining metrics in the column and row spaces of A respectively (cf. Gabriel, 1971, p.456), because of the following alternative expressions for (2.1.4):

$$\|A\|^2_{\Omega,\Phi} = \sum_{i=1}^{n} \sum_{i'=1}^{n} \omega_{ii'}\, a_i^T \Phi a_{i'} \qquad (2.1.5)$$

$$= \sum_{j=1}^{m} \sum_{j'=1}^{m} \phi_{jj'}\, a_j^T \Omega\, a_{j'} \qquad (2.1.6)$$

In (2.1.5) the squared norm is a weighted sum of all scalar products between the rows a_i^T of A in the metric Φ. Alternatively, in (2.1.6), it is a weighted sum of all scalar products between the columns a_j of A in the metric Ω.

Another notation for (2.1.4) which is particularly meaningful is to write $\text{tr}[\Omega A \Phi A^T]$ as $a^T(\Omega \otimes \Phi)a$, where $\Omega \otimes \Phi$ is the Kronecker product (or left direct product) of Ω and Φ, and a is the vector of elements of A, row by row. This notation shows clearly the analogy to vector quadratic forms, and the squared norm (2.1.4) could be called a "matrix quadratic form".

A special case which we shall often use is where either Ω or Φ is diagonal, for example if $\Omega = D_\omega$. Then (2.1.5) becomes:

$$\|A\|^2_{D_\omega,\Phi} = \sum_{i=1}^{n} \omega_i \|a_i\|^2_\Phi \qquad (2.1.7)$$

in other words, the weighted sum of squared norms of the rows of A (in the metric Φ).

The matrix approximation $\hat{A}_{[p]}$ of A in the sense of the norm $\|\cdots\|_{\Omega,\Phi}$ can be obtained from a correspondingly more general form of the basic structure of A:

$$A = N D_\alpha M^T \quad \text{where } N^T \Omega N = I,\ M^T \Phi M = I \qquad (2.1.8)$$

that is, where the left and right basic vectors are orthonormal in the metrics Ω and Φ respectively. $\hat{A}_{[p]}$ is then the appropriate $N_{(p)} D_{\alpha(p)} M^T_{(p)}$, as in (2.1.2), and it minimizes:

$$\|A - A_{[p]}\|^2_{\Omega,\Phi} = \text{tr}[\Omega(A - A_{[p]})\Phi(A - A_{[p]})^T] \qquad (2.1.9)$$

with respect to rank p matrices $A_{[p]}$.

Computation

In practice the basic structure is obtained by way of the eigen= structure. For the general problem (2.1.8) the square symmetric matrix $A\Phi A^T$ has the generalized eigenstructure:

$$A \Phi A^T = N D_\alpha^2 N^T \quad \text{where } N^T \Omega N = I \tag{2.1.10}$$

(since $M^T \Phi M = I$). The eigenvectors N are the left basic vectors of A and the eigenvalues are the squared basic values. Note that $A \Phi A^T$ is the matrix of scalar products between rows of A, in the metric Φ. The right eigenvectors M are obtained by using the *transition formula*:

$$M = A^T \Omega N D_\alpha^{-1} \tag{2.1.11}$$

which is a direct consequence of (2.1.8). The generalized eigenstruc= ture (2.1.10) is obtained by first computing the (ordinary) eigenstruc= ture of the matrix $\Omega^{\frac{1}{2}} A \Phi A^T (\Omega^{\frac{1}{2}})^T$:

$$\Omega^{\frac{1}{2}} A \Phi A^T (\Omega^{\frac{1}{2}})^T = U D_\lambda U^T \tag{2.1.12}$$

where the eigenvectors U have the usual normalisation $U^T U = I$; then $N = \Omega^{-\frac{1}{2}} U$ and $D_\alpha = +\sqrt{D_\lambda}$.

Quality and error of approximation

As mentioned before, the basic values express the magnitude of the matrix in each of its dimensions. In fact the squared norm of A in (2.1.8) is $\alpha_1^2 + \ldots + \alpha_r^2$. When approximating A by $\hat{A}_{[p]}$, the largest p basic values are retained and the squared norm of $\hat{A}_{[p]}$ is $\alpha_1^2 + \ldots + \alpha_p^2$. A measure of the quality of the approximation is thus:

$$\tau_{[p]} \equiv (\alpha_1^2 + \ldots + \alpha_p^2)/(\alpha_1^2 + \ldots + \alpha_r^2) \tag{2.1.13}$$

so that $0 \le \tau_{[p]} \le 1$.

The error of approximation can be measured in a complementary fashion by:

$$1 - \tau_{[p]} = (\alpha_{p+1}^2 + \ldots + \alpha_r^2)/(\alpha_1^2 + \ldots + \alpha_r^2) \tag{2.1.14}$$

2.2 Basic structure displays

The concept of generalized basic structure allows us to define a class of multidimensional techniques for displaying a rectangular data matrix in low-dimensional Euclidean space. Each technique typi= cally involves three phases (see Figure 2.1):

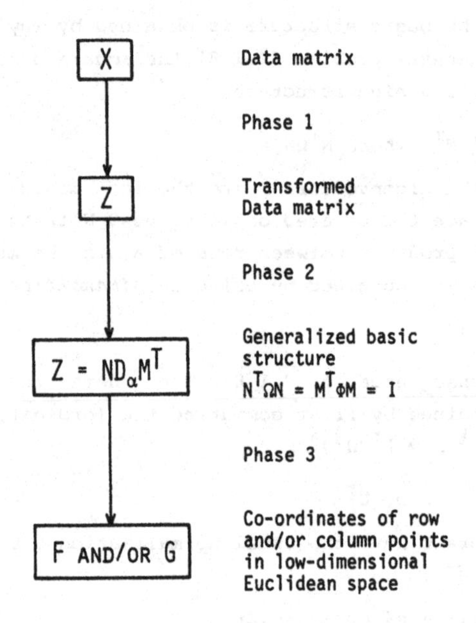

Figure 2.1

Three phases of a basic structure display

Phase 1...

The data matrix, X say, is pre-processed in some way, often by "centering" the data; for example, the observations on each variable are expressed as deviations from the variable's mean. This phase is peculiar to the type of data. Later we shall discuss the reasons for centering the data.

Certain types of data are often recoded to make them suitable for graphical analysis. For example, the data might be "heterogeneous", consisting of continuous and discrete variables, and the continuous data, say, is recoded is some way to be compatible with the discrete data.

Phase 2 ...

The transformed matrix, Z say, is decomposed into its generali= zed (Ω,Φ) basic structure, Ω and Φ again being peculiar to the type of data as well as the method used, so that a least squares approxima= tion of Z of any prescribed low rank p can be obtained, in the sense of the norm $\|\cdot\cdot\|_{\Omega,\Phi}$.

Phase 3 ...

The first p components of the basic structure provide a graphi= cal display in p-dimensional Euclidean space. There are a number of ways of obtaining a display from the basic structure, which we shall be discussing below.

An example: principal components

To illustrate these three phases let us take principal compo= nents analysis as an example. Suppose X(n x m) is a matrix of observa= tions of m continuous variables on n individuals or cases.

Phase 1 consists of subtracting the mean of each variable from its column of observations to obtain Z, the matrix of deviations. If $\bar{x}^T$ is the row vector of column averages, then:

$$Z = X - 1\bar{x}^T \qquad (2.2.1)$$

The (transposed) rows of Z are thus point vectors z_i in m-dimensional space and their centre of gravity, or mean vector, is the origin 0.

Let us suppose that p=2, for argument's sake. Then the first two principal axes form a plane which, often by definition (Pearson, 1901; Morrison, 1967), is the "closest" to all the points z_i (of all the planes through the centre of gravity of the z_i). The measure of close= ness of the plane to the set of points is the sum of squared distances from the points to the plane. If distance is the usual Euclidean dis=

tance and if the point closest to z_i in the (unknown) plane is deno= ted by $\tilde{z}_i$, then the problem is to minimize:

$$\sum_{i=1}^{n} (z_i - \tilde{z}_i)^T (z_i - \tilde{z}_i) = \| Z - Z_{[2]} \|^2 \qquad (2.2.2)$$

where $Z_{[2]}$ is a rank 2 matrix since all its rows $\tilde{z}_i^T$ lie in a 2-dimen= sional space. This is exactly the problem in (2.1.3), or (2.1.9) with $\Omega = \Phi = I$. Phase 2 thus consists of computing the ordinary basic structure of Z. If any other metric Φ was imposed on the rows of Z then phase 2 would consist of computing the generalized (I,Φ) basic structure of Z.

The actual matrix approximation $\hat{Z}_{[2]} = N_{(2)} D_{\alpha(2)} M_{(2)}^T$ is of little use to us, as its rows are still expressed relative to the original co-ordinate system. What we want are the co-ordinates of the points z_i relative to the principal plane, in other words the projections of the points onto this plane. Now the columns of $M_{(2)}$ are a basis for the rows of $\hat{Z}_{[2]}$ and therefore generate the principal plane. Be= cause they are an orthonormal basis, the projections of the points onto this plane are just the scalar products of the rows of Z with $M_{(2)}$ (denoting the matrix of co-ordinates by F(n x 2)):

$$F = Z\, M_{(2)} = N\, D_\alpha\, M^T M_{(2)} = N_{(2)}\, D_{\alpha(2)} \qquad (2.2.3)$$

since $M^T M = I$. (Again, if the norm had been generalized to Φ, then the co-ordinates would be in the rows of $F = Z\Phi M_{(2)}$.) Phase 3 thus consists of computing F and plotting its rows in two dimensions. Note that for the F of (2.2.3):

$$F\, F^T = \hat{Z}_{[2]} \hat{Z}_{[2]}^T \qquad (2.2.4)$$

This means that the set of scalar products of the rows of F, i.e. the points in the display, is exactly the set of scalar products of the rows of $\hat{Z}_{[2]}$. This is an obvious result since the rows of F are just the rows of $\hat{Z}_{[2]}$ expressed with respect to a different orthonormal basis, so they are different by a rotation which clearly preserves scalar products. The result corresponding to (2.2.4) when the metric in the m-dimensional space of the rows of Z is Φ, is:

$$FF^T = \hat{Z}_{[2]}\, \Phi\, \hat{Z}_{[2]}^T \qquad (2.2.5)$$

Here the set of (ordinary) Euclidean scalar products of the displayed points is the same as the set of generalized Euclidean scalar products of the rows of $\hat{Z}_{[2]}$ in the metric Φ.

Biplot

The object of principal components analysis is to replace the original set of m variables (dimensions) by a much smaller set of p variables which are uncorrelated (orthogonal) and which account for a large proportion of the total variance of the original variables. The rows of F (in (2.2.3)) may be plotted but attention is usually concentrated on the columns of $M_{(p)}$ which give the coefficients assigned to each original variable in the linear compounds which define the new variables.

It is also meaningful to plot the *rows* of $M_{(p)}$ as points representing the variables in the same display as the rows of F. If we suppose that p = 2 once again and write $G = M_{(2)}$ as the m x 2 matrix of the co-ordinates of these "variable points", then

$$\hat{Z}_{[2]} = N_{(2)}D_{\alpha(2)}M_{(2)}^T = F\ G^T \qquad (2.2.6)$$

and the interpretation of this joint display becomes clear. The (i,j)-th element of $\hat{Z}_{[2]}$ is equal to $f_i^T g_j$, the scalar product of the i-th row of F and the j-th row of G, that is the scalar product of the point vectors displaying the i-th case and the j-th variable. If $\hat{Z}_{[2]}$ is a good approximation of Z then this scalar product will be, on average, a good approximation of the element z_{ij} of the matrix Z. This joint display of two sets of points, where the between-set positions are interpretable in terms of scalar products, is called the *biplot* (Gabriel, 1971).

There is also a metric within-set interpretation of the above column points. Using the basic structure $Z = N\ D_\alpha M^T$ we can show that

$$Z^T(ZZ^T)^-Z = MM^T \qquad (2.2.7)$$

so that $GG^T = M_{(2)}M_{(2)}^T$ approximates $Z^T(ZZ^T)^-Z$, where $(ZZ^T)^-$ is a conditional inverse of ZZ^T. The scalar product of the column points (rows of G) thus approximate the scalar products of the columns of Z in the metric $(ZZ^T)^-$. This geometric interpretation is not particularly helpful in this present example but in other biplots, for example the covariance biplot (§2.3.4), this interpretation will be more meaningful.

Clearly there are an infinity of ways of performing phase 3 (Fig.2.1), that is of writing $\hat{Z}_{[2]}$ as $F\ G^T$ and thus obtaining a biplot display. For example F could be taken as $N_{(2)}$ and G as $M_{(2)}D_{\alpha(2)}$. The between-set interpretation remains the same, but the interpretation within the row points and within the column points changes. Various biplots will be described in §2.3.

2.3 Examples and applications of basic structure displays

Apart from principal components analysis and its associated bi=plot, described above, a number of other multidimensional techniques can be described in a similar fashion as basic structure displays: for example, various forms of biplot, canonical correlation analysis, canonical variate analysis and correspondence analysis. Furthermore, basic structure, in the special case of eigenstructure, will be shown to underlie classical scaling (Young-Householder scaling) and princi=pal co-ordinates analysis (§3). These methods differ only in the de=finition of the matrix "parameters" Φ and Ω (Phase 2) and in the way in which the co-ordinate matrices F and/or G are obtained (Phase 3) from the generalized basic structure of the (transformed) matrix of interest. We shall now describe some examples of these methods as well as actual applications .

2.3.1 *Principal components analysis and biplot*

First we shall discuss an application of principal components analysis, which is the easiest basic structure display to understand because it treats data in ordinary Euclidean space. Consider the data matrix (Table 2.3.1) of the marks of a class of 21 students in an exa=mination consisting of 5 questions, each question having a maximum mark of 20. The total mark (with a maximum of 100) is given and the names of the students have been replaced by their position in the class. The rows of the matrix are in descending order of class posi=tion and the codes 'a' and 'b' indicate a tied position.

Phase 1 of a principal components analysis consists of expressing all the marks as deviations from the class average for the particular question and the matrix of deviations is denoted by Z. For example, the first row of Z is [2.6 2.5 6.7 4.5 7.5].

Phase 2 is the computation of the basic structure of Z. Because the between-row metric is ordinary Euclidean (Φ=I) and because the num=ber of columns of Z is smaller than the number of rows, this amounts in practice to computing the eigenstructure $M\ D_\lambda M^T$ of the square, sym=metric 5 × 5 matrix Z^TZ. Then M contains the right basic vectors of Z, the basic values are the square roots of the eigenvalues in D_λ and $N = Z\ M\ D_\lambda^{-\frac{1}{2}}$ contains the left basic vectors (cf. the symmetric results (2.1.10) and (2.1.11)). Notice that Z^TZ is 20 times the (unbiased) covariance matrix S between questions so that M is the matrix of eigen=vectors of S and D_λ is 20 times the diagonal matrix of eigenvalues of S. This proportionality of the eigenvalues leads to the term "percentage

Table 2.3.1

STUDENTS (denoted by class position according to total mark)	QUESTIONS 1	2	3	4	5	TOTAL MARK	TOTAL "MARK" ACCORDING TO FIRST PRINCIPAL COMPONENT
1	20	20	20	14	12	86	82.4
2	20	20	20	14	11	85	81.0
3	20	19	15	14	12	80	76.5
4	20	20	20	12	7	79	73.4
5	19	18	14	13	12	76	72.9
6	20	20	15	12	7	74	68.1
7	20	19	15	12	7	73	67.5
8	20	20	15	11	5	71	64.3
9	18	19	13	11	5	66	59.9
10	18	18	13	11	5	65	59.3
11a	17	18	12	9	3	59	52.5
11b	17	18	13	9	2	59	52.1
13a	17	17	11	9	3	57	50.8
13b	18	18	13	8	0	57	49.1
15	16	16	11	9	3	55	49.3
16	16	17	11	8	0	52	44.8
17	16	16	10	7	0	49	41.9
18a	14	14	10	6	0	44	37.8
18b	15	14	10	5	0	44	37.5
20	14	13	10	5	0	42	36.1
21	10	14	8	0	0	32	25.7
Average	17.4	17.5	13.3	9.5	4.5	62.2	56.3

of variance" when referring to an eigenvalue of Z^TZ expressed as a percentage of the sum of the eigenvalues, because the sum of the eigen= values of S is the total variance of the 5 questions. The basic values of Z are calculated to be (in descending order): 31.23, 8.87, 6.75, 3.65 and 2.10. This gives the first two percentages of variance as 87.3% and 7.0% respectively, so that the quality of approximation of Z in two dimensions is $\tau_{[2]} = 0.943$ (or 94.3% as a percentage; see (2.1.13)).

Phase 3 defines the actual co-ordinates in a two-dimensional display as $F = N_{(2)}D_{\alpha(2)}$ and, if a biplot of the questions is required, $G = M_{(2)}$. The display is given in Figure 2.3.1.

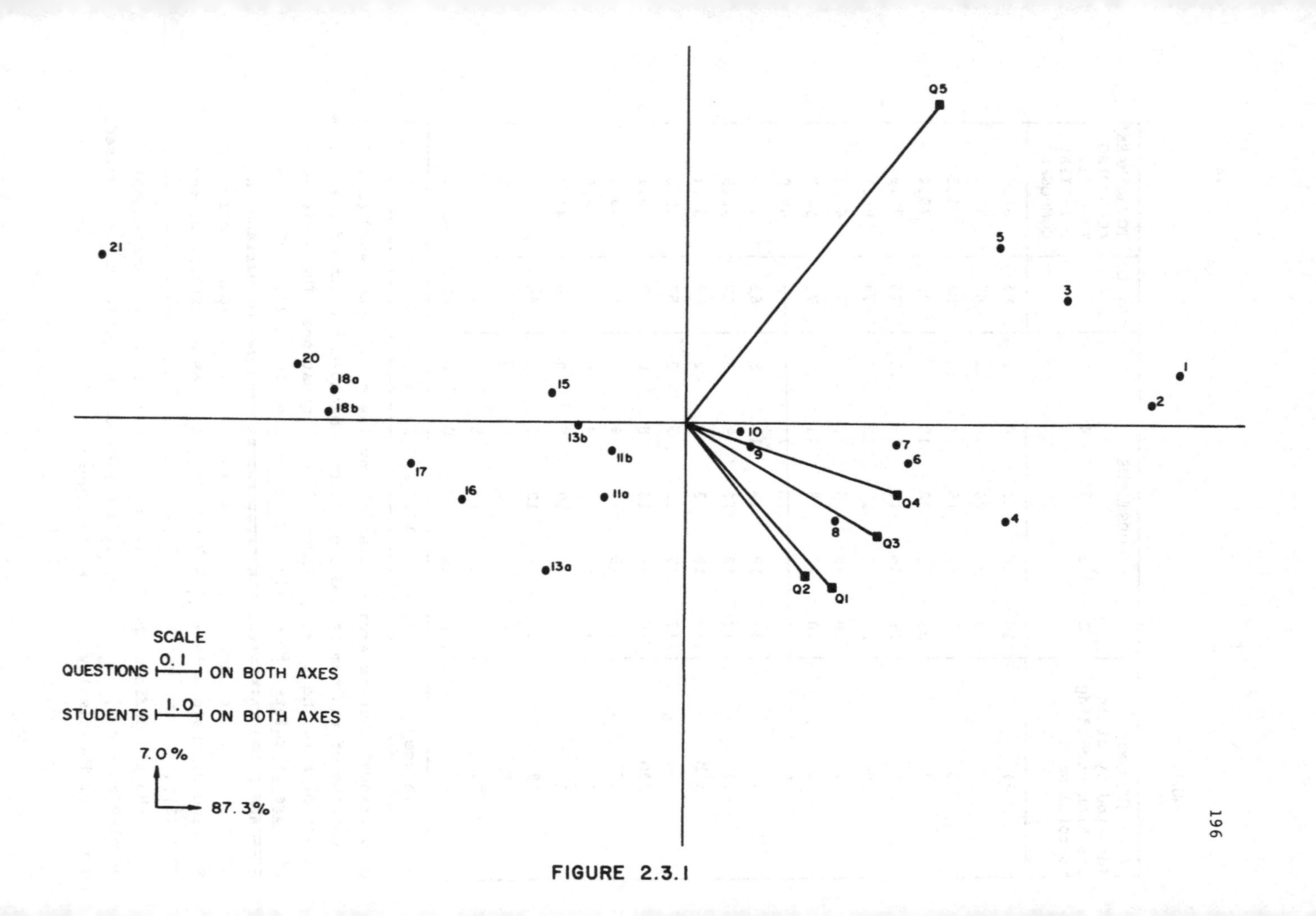
Q5
Q4
Q3
Q2
Q1
21
20
18a
18b
15
13b
11b
11a
13a
17
16
10
9
8
7
6
5
4
3
2
1
SCALE
QUESTIONS 0.1 ON BOTH AXES
STUDENTS 1.0 ON BOTH AXES
7.0%
87.3%

FIGURE 2.3.1

The matrix $G = M_{(2)}$ is actually:

$$G = \begin{bmatrix} 0.352 & -0.400 \\ 0.287 & -0.373 \\ 0.452 & -0.274 \\ 0.493 & -0.168 \\ 0.589 & 0.773 \end{bmatrix}$$

Notice that the scale of the display of the question points (rows of G) has been expanded so that these points are not bunched up near the origin relative to the display of the student points. The co-ordinate of a row (student) point on the first (horizontal) axis is just the scalar product of his vector of marks with the first column vector of G, up to a constant change in scale and origin. Because the elements of this latter vector are all positive the student's first axis co-ordinate (i.e. his first component *score*) can be considered a weighted sum of his five marks and it is a simple matter to scale the elements of the first column of G to add up to 5 so that the weighted sum, like the total mark, has a maximum of 100. These new weights are 0.81, 0.66, 1.04, 1.13 and 1.36 respectively, and these should be compared to the equal weights of 1 which are used to obtain the total mark. Questions 1 and 2 are down-weighted, while questions 4 and 5 are up-weighted. Question 5, having a large variance of marks across the class receives more than twice the weight of question 1, one of the easiest questions with relatively low variance. The total marks of each student accord= ing to this weighted sum are given in the last column of Table 2.3.1. The variance of these marks is calculated to be 258.1 and this is greater than the variance of the original total marks in the preceding column, namely 230.9. Working back we can show that the ordinary arith= metic total mark accounts for 78.2% of the total variance of the marks so that a further 87.3 - 78.2 = 9.1% of the total variance is included in the first principal component scores.

In practice this means that the first component scores are more discriminating between students than the total mark. This has interest= ing repercussions in discriminating between students, for example, the customary allocation of so-called "distinctions" to students who (in this study) receive a total mark greater than 80%. Student 4 misses this grade by only one mark in his total of 79, but has a difference from student 3 of over three "marks" in terms of their first component "marks", because of student 4's relatively poor mark for question 5. Similar reasoning explains why student 5, although 3 marks below student

4, only differs by 0.5 in his first component "mark". It seems that if student 4 is to be elevated to the grade "distinction" then so should student 5, although this is not clear from their original total marks. Notice that the first component "marks" are lower than the ordinary total marks and so new levels for awarding distinctions and passes should be decided upon.

The second axis in Fig. 2.3.1 provides a second set of scores on each student, a weighted average based on the second column of G. This axis opposes question 5 with the remaining questions and isolates the effect of this question, which on the face of it was the most difficult in the examination. In the two-dimensional display it appears as if student 5 is closer to the "distinction" group than student 4.

A variation of principal components analysis in common usage is to standardize the data initially, that is divide all data by the respective standard deviations of the variables, usually because of different units of measurement. This is equivalent to imposing the metric $\Phi = D_{s^2}^{-1}$ on the observation vectors (rows of data matrix X, say), where D_{s^2} is the diagonal matrix of variances of the variables (columns). This analysis is appropriate when the variables are observed on widely different scales, for example one variable in kilograms, another in millimeters etc.

2.3.2 *Symmetric biplot*

Sometimes attention is focussed on the structure of the matrix rather than the geometry of the row or column vectors, for example in identifying a model for the elements of the matrix (see Bradu and Gabriel, 1978). Here any biplot of the centered matrix is sufficient and to avoid different scales in the display of the row and column points the basic values are assigned "equally" to the left and right basic vectors:

$$F = N_{(2)} D_{\alpha(2)}^{\frac{1}{2}} \qquad\qquad G = M_{(2)} D_{\alpha(2)}^{\frac{1}{2}}$$

To illustrate this "symmetric" biplot we centred the data of Table 2.3.1 with respect to the overall mean of the data (which is 12.4; for discussion on centering, see Bradu and Gabriel, 1978), computed the basic structure of the centred matrix and then obtained the row and column co-ordinate matrices in two-dimensional space according to the above "symmetric" formulae. The display is given in Fig. 2.3.2.

Although interpretation of distances between points is not strictly permissible in this two-dimensional display one can interpret the

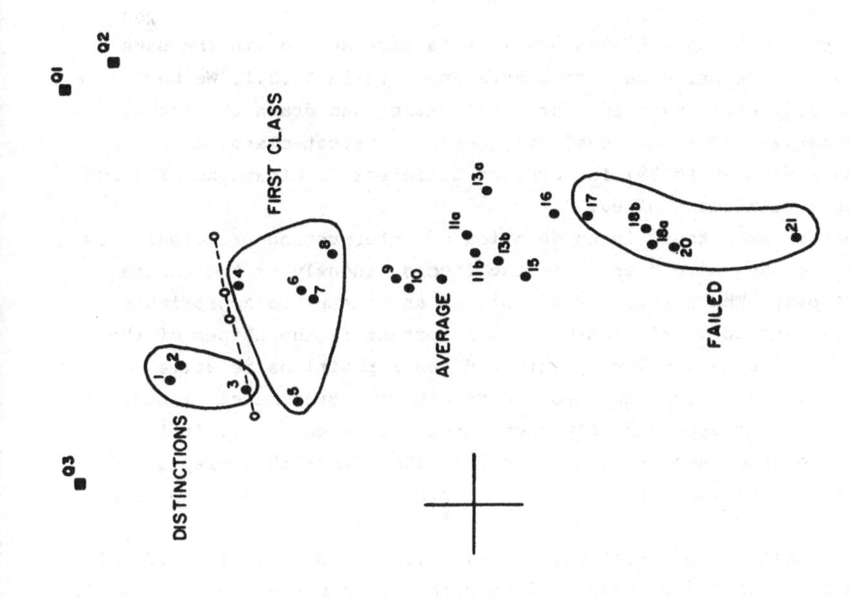

Q1
Q2
Q3
Q4
Q5
DISTINCTIONS
FIRST CLASS
AVERAGE
FAILED
1
2
3
4
5
6
7
8
9
10
11a
11b
13a
13b
15
16
17
18a
18b
20
21

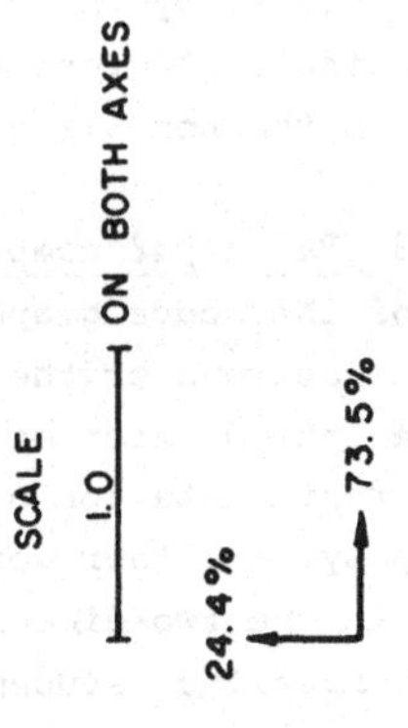

SCALE
1.0
ON BOTH AXES
24.4%
73.5%

FIGURE 2.3.2

relative positions of the students in this case and obtain the same conclusion as the principal components analysis in §2.3.1. We have also added the supplementary point for "full marks" and drawn the formal boundary between "distinctions" and so-called "first-class passes" (total mark from 60 to 79) in terms of different combinations of marks equalling a total mark of 80.

However here there is no question of interpreting principal axes and thus the axes have been indicated inconspicuously at the centre of the display. The display as a whole is an attempt to approximate the entire centred matrix and what is important is the shapes of the two clouds of row and column points and their positions relative to each other. In this case the two clouds are of approximately linear shapes and lie at approximately right-angles to each other. It is not difficult to show (see Bradu and Gabriel, 1978) that this display suggests that the original data matrix X approximately follows the simple additive model : $x_{ij} \approx \mu + \alpha_i + \beta_j$.

The additive "effects" α_i, $i = 1,...,21$ and β_j, $j = 1,...,5$ may be formally estimated or deduced less formally from the display itself. The row "effects" α_i will be interpreted as overall scores for the students and the column "effects" β_j as difficulty scores for the questions. Now a fixed straight line through the student points in the display will tend to lie to the right of the distinctions (students 1, 2 and 3) and student 5. Because this line represents the row effects and because an approximately perpendicular line represents the column effects, this means that students 1,2,3 and 5 have generally higher marks on the more difficult questions (especially question 5) than would be predicted by the additive model. (Student 5 again seems closer to the "distinction" group than student 4). A symmetric argument can be applied to the non-alignment of question 2 with the other questions.

2.3.3 *Principal components with weights*

Any of the above graphical analyses may be performed with weights assigned to the rows or the columns. In the context of the example in §2.3.1, the introduction of different row weights leads to improved quality of representation of the more highly weighted student points in the display. In other words, all the points are represented with some error in the two-dimensional subspace and if we choose to represent the "borderline" students, say, with greater accuracy (at the expense of the display of the other students), then we can assign them higher weight. The changes in the computation of the basic structure display are straightforward: in Phase 1, the matrix is centred with respect to the *weighted* row average (i.e. centre of gravity of weighted rows), and in Phase 2 the matrix Ω is defined as D_ω, the diagonal matrix of row weights. Usually a set of weights is scaled to have a sum of 1.

Weighting can be appropriate when sampling of subgroups of the population has not been proportional to their population frequencies and some kind of equalization of contribution is required for the different subsets of rows in the display. Another situation is where certain rows contain dubious data, in which case they can be down-weighted. If a weight of zero is assigned to a certain row, this means that the row becomes a "supplementary point" - it plays no role in the determination of the principal axes but its position relative to these axes is of interest.

2.3.4. *Covariance biplot*

In Phase 3 of the principal components analysis described in §2.3.1, the basic values are assigned to the left basic vectors to obtain the row co-ordinates: $F = N_{(2)} D_{\alpha(2)}$. If, instead, the basic values are assigned to the right basic vectors: $G = M_{(2)} D_{\alpha(2)}$, then the basic structure display defined by F ($=N_{(2)}$) and G is what we call a *covariance biplot*. In this display the relative squared lengths of the displayed column vectors approximate the column variances, while the scalar products between column vectors approximate the respective covariances between columns. The lengths of the displayed column point vectors are thus interpreted as approximate standard deviations, while the angle cosines between these vectors are approximate correlation coefficients between the columns of the data matrix. So whereas in principal components analysis the total variance (sum of the column variances) is decomposed along the principal axes, in the covariance biplot it is the individual variances and covariances that are displayed.

Furthermore, the distances between the displayed row points (the rows of $F = N_{(2)}$) may be interpreted as approximate Mahalanobis distances between the rows of the data matrix. This is described in detail by

Gabriel (1971, 1972).

An example of the covariance biplot applied to rainfall data is given by Gabriel (1972). Our following example also deals with rainfall data, namely measurements by five different types of rain-gauge of the rainfall at the same location on 42 "rain-days" (days on which at least one gauge recorded 1 mm. of rain). The covariances and correlations between the five gauges are given in the lower and upper triangles of the following matrix (the variances are in the diagonal):

77.7	*0.995*	*0.994*	*0.996*	*0.969*
76.1	75.1	*0.995*	*0.999*	*0.972*
74.3	73.1	71.9	*0.996*	*0..970*
77.0	75.9	74.0	76.9	*0.977*
76.1	75.1	73.2	76.3	79.3

Clearly there are very strong correlations between gauges 1 to 4, whereas these are all slightly less correlated with gauge 5, which is also the gauge with the highest variance. All this is summarized in the covariance biplot of Fig.2.3.3, along with points representing the 42 days. The vector G5 representing gauge 5 is longer and has a larger angle with the other vectors. The display of the days gives further information, for example the outlying day 21 has a large under-estimate of the rainfall by gauge 5. The spread of the days on the horizontal axis is just their ordering from low rainfall days (on the left) to high rainfall days (on the right). The vertical spread lines up the days in terms of under- or over-estimating the rain by gauge 5 compared to the other gauges. Gauge 4 is the most highly correlated with the mean rainfall vector. Notice too how the spread of the days on the second axis increases as rainfall increases, which is not unexpected. Figure 2.3.4 shows the five gauge vectors in the second and third dimensions of the covariance biplot (i.e. looking down the first axis). Although gauges 1 to 4 are very highly intercorrelated, gauge 1 is separated slightly from the other three.

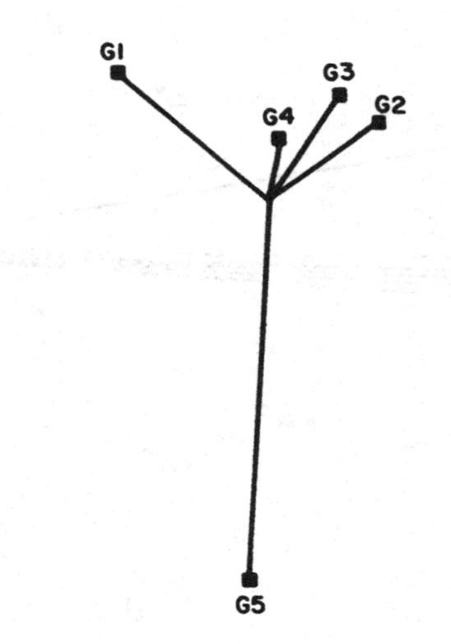

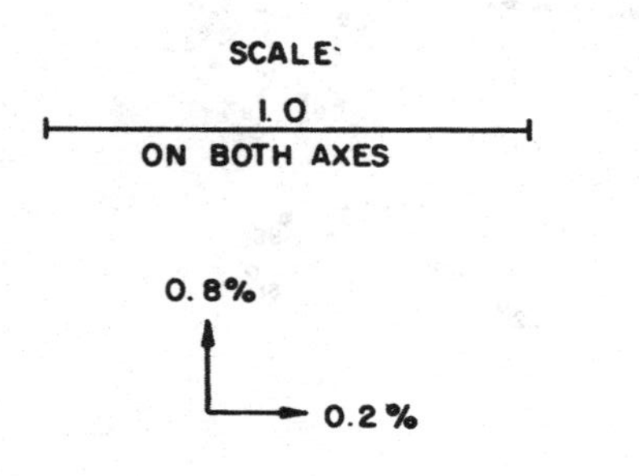

FIGURE 2.3.3

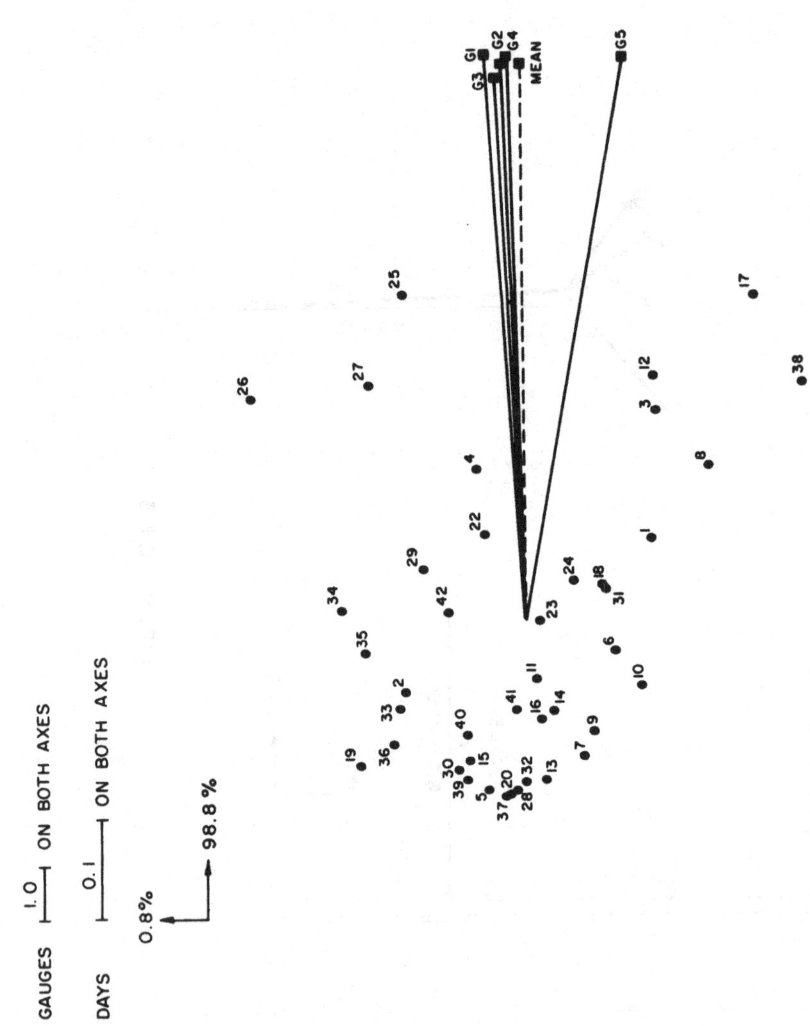
GAUGES 1.0 ON BOTH AXES
DAYS 0.1 ON BOTH AXES
0.8%
98.8 %
G1
G2
G3
G4
G5
MEAN

FIGURE 2.3.4

2.3.5 *Canonical variate analysis*

Canonical variate analysis is a dimension-reducing technique designed to show up differences between groups of points. It is a special case of canonical correlation analysis and the geometry of these two techniques is described fully by Falkenhagen and Nash (1978), who also give a detailed example in the context of forestry research.

If the n cases (rows) of X fall into g groups then we can compute $\bar{X}$, the $g \times m$ matrix of group means on the m variables, as well as W, the usual pooled within-groups sum-of-squares-and-cross-products (SSCP) matrix. The numbers of cases in each group $n_1, \ldots, n_g$ define the diagonal of the $g \times g$ diagonal matrix D_n. Canonical variate analysis is just a principal components analysis of the group mean vectors in the metric W^{-1} and with weights in D_n. In other words, it is a basic structure display where the matrix $\bar{X}$ is centred with respect to the overall mean row vector (Phase 1 : $Z = \bar{X} - 1\bar{x}^T$), where the generalized basic structure of the centred matrix is determined using $\Omega = D_n$ and $\Phi = W^{-1}$ (Phase 2 : $Z = N\, D_\alpha M^T$) and where the rows (groups) are displayed by assigning the basic values to the left basic vectors (Phase 3 , in two-dimensional space, say : $F = N_{(2)}\ D_{\alpha(2)}$).

Notice that if each case forms a "group" of its own (i.e. there are no formal groups), then $\bar{X} = X$, the original matrix, $\Omega = \Phi = I$, and the canonical variate analysis reduces to a principal components analysis. The idea of a canonical variate biplot is directly analogous to the principal components biplot and the rows of $G = M_{(2)}$ may be displayed in order to interpret the canonical axes.

2.3.6 *Correspondence analysis*

Correspondence analysis (Benzécri, 1973) can be described as a dual pair of basic structure displays.

If X is a matrix of nonnegative numbers, usually numerical frequencies, then P is defined as X "divided" by its total: $p_{ij} = x_{ij}/\Sigma\Sigma x_{ij}$. Let D_r and D_c be diagonal matrices of row and column sums of P respectively. $P^T D_r^{-1}$ is called the matrix of *row profiles*; c^T (the row vector of column sums of P) is the average row profile; D_r contains the weights assigned to the row profiles; and D_c^{-1} defines the metric (the "chi-squared metric") between the row profiles (for a detailed description in English of correspondence analysis, see Greenacre (1978,1981)).

A symmetric set of definitions applies to the column profiles PD_c^{-1}. The dual basic structure displays of the row and column profiles can be described as the single basic structure display of the matrix $Z = D_r^{-1} P D_c^{-1} - 11^T$ (Phase 1), where $\Omega = D_r$ and $\Phi = D_c$ (Phase 2) and $F = N_{(p)} D_{\alpha(p)}$ and $G = M_{(p)} D_{\alpha(p)}$ (Phase 3). Notice that the basic values of the generalized basic structure are assigned to both the left and right basic vectors, in contrast to previous displays. Thus the display of the rows of F and G is not a biplot of the matrix Z. Instead, the between-sets interpretation is in terms of the transition formulae:

$$F = D_r^{-1} P G D_{\alpha(p)}^{-1} \text{ and } G = D_c^{-1} P^T F D_{\alpha(p)}^{-1} \qquad (2.3.1)$$

(see Greenacre (1978)).

Applications of correspondence analysis are described by Benzécri (1973), Lebart *et al* (1977), and numerous articles in the French jour= nal *Cahiers de l'Analyse des Données* and (in English) by David *et al* (1974), Teil and Cheminee (1975), Hill (1974) and Greenacre (1978, 1981). The advantage of correspondence analysis over other basic struc= ture displays is its wide applicability to many different types of data. Although it is primarily defined on matrices of numerical frequencies most data matrices can be recoded in order to make them suitable for graphical display by correspondence analysis. The data in Table 2.3.1, for example, are unsuitable for correspondence analysis because a student's "profile" (his set of marks divided by their total) is not a sen= sible measure of that student's performance in the examination. The stan= dard approach to such data in correspondence analysis is to "double" the columns of the matrix, that is introduce a further 5 columns in this particular case where each new column is the maximum mark (20) minus the mark in the respective column. The total "weight" of 20 marks for each question is now distributed between the "positive" column (containing the actual mark) and the "negative" column (containing 20 minus the mark). The correspondence analysis of this doubled matrix can be shown to be closely related to a weighted principal components ana= lysis of the original data.

When the data matrix is a contingency table another basic struc= ture display of interest could be the symmetric biplot of the matrix of deviations $P - r c^T$ in the metrics $\Omega = D_c^{-1}$ and $\Phi = D_r^{-1}$. Since the squared norm:

$$\| P - r c^T \|_{D_r^{-1}, D_c^{-1}} = \text{tr} [D_r^{-1} (P - rc^T) D_c^{-1} (P - rc^T)^T] \qquad (2.3.2)$$

is equal to Pearson's mean-square contingency coefficient, this biplot displays approximately (in the form of scalar products) $f_i^T g_j$) the deviations $(p_{ij} - r_i c_j)/\sqrt{r_i c_j}$ whose squares are terms in the expansion of (2.3.2).

2.4 Stability of results and missing data

Stability of results

When faced with a basic structure display, a haunting question is whether a structure, which seems apparent in the display and which can be interpreted, is an indication of a true feature in the data or per= haps just a sampling fluctuation. Imposing statistical inference on the scaling problem is a new attempt to transform an almost purely geo= metric class of data-analytic techniques into statistical techniques where confidence statements can be attached to the observed dispersion of points.

A first approach in the context of basic structure displays is to ascertain whether a particular principal axis reflects nonrandom dis= persion within the cloud of points. Since the squared basic value (i.e. eigenvalue) is the "sum-of-squares explained" by the axis (cf. end of § 2.1), this question can be formalized as some significance test on the eigenvalues. For example, Anderson (1963) treats the hypothesis that an adjacent set of q eigenvalues of a covariance matrix Σ are equal. When this set consists of the last (smallest) q eigenvalues, the hypothesis is one of spherical dispersion of the principal compo= nents in the last q dimensions. There would thus be p "significant axes" in an m-variable problem if this hypothesis were accepted when $q=m-p$, but rejected when $q=m-p+1$.

Such inference is meaningful only when the original variables are on similar scales. When the data are standardized, it would be appro= priate to test whether the *percentages of variance* explained by the principal axes are "significant". For this case Krzanowski (1979) has investigated the exact percentage points for the first percentage of variance for multivariate normal observations on up to 4 variables. These give exact sampling intervals for the first percentage of variance due to chance alone.

Lebart *et al.* (1977) describe the history of the same problem in correspondence analysis. They give graphically the 5% significance points of the first five *percentages of inertia* $\tau_k \equiv \alpha_k^2/\sum_\ell \alpha_\ell^2$, for different size data matrices. For example, for a 40 × 20 contingency table, more than 15% inertia on the first axis is significant (at the 5% level), and more than 13% on the second axis is significant, etc ....

The distributional assumption made in establishing these significance curves is the normal approximation to the multinomial distribution.

Another possible field of inference in the context of scaling is to test whether a particular point deviates significantly from the centre of gravity or from certain other points in the display. In other words we want to place a confidence region around the point. Benzécri *et al.* (1973) have considered this problem in correspondence analysis, but the result is only valid under the strict assumption that the data form a true contingency table, i.e. the events (i,j) constitute a multinomial sampling.

Missing data

Missing data can be dealt with, but this complicates the computations involved. Gabriel and Zamir (1976) describe the situation in the context of the biplot and introduce weights w_{ij} on each datum, where $w_{ij} = 1$ for actual data and 0 for missing data. Bradu and Grine (1979) have applied this technique to skull data where many single measurements were missing.

Mutombo (1973) and Nora-Chouteau (1974) treat the situation in correspondence analysis in a similar fashion. Missing relative frequencies p_{ij} are estimated by the "zero-order" correspondence analysis, i.e. by $p_{ij} = r_i c_j$ (notation as in § 2.3.6), where r_i and c_j are the row and column relative frequencies using the available true data. Then we again get p_{ij} from the new $r_i c_j$ and iterate until p_{ij} stabilizes. Then the "first-order" correspondence analysis is used: $p_{ij} = r_i c_j (1+f_{i1}g_{j1}/\alpha_1)$, using the co-ordinates $f_{i1}(i=1,\ldots,n)$ and $g_{j1}(j=1,\ldots,m)$ on the first principal axis, again iterating until p_{ij} stabilizes. The process is repeated until a "p-th order" correspondence analysis is reached which reconstitutes the data and gives a suitable display. Greenacre (1974) describes an application of this technique, also in the craniometry of skulls.

3. CLASSICAL SCALING AND PRINCIPAL CO-ORDINATES ANALYSIS

3.1 Eigenstructure of a matrix of scalar products

When we dealt with the computation of the basic structure in §2.1, we showed that the left basic vectors, say, were eigenvectors of a square symmetric matrix (see (2.1.10)). With reference to the example of principal components analysis in §2.2, it is readily seen that the matrix F of co-ordinates in (2.2.3) can be obtained from the eigenstructure of the matrix ZZ^T:

$$ZZ^{T} = [(X - 1\,\bar{x}^{T})(X - 1\,\bar{x}^{T})^{T}] = N\,D_{\mu}\,N^{T} \quad (\text{where } N^{T}N=I) \tag{3.1.1}$$

The matrix F is then:

$$F = N_{(p)}\,D_{\mu(p)}^{\frac{1}{2}} \tag{3.1.2}$$

because the basic values α_k of Z are the square roots $\mu_k^{\frac{1}{2}}$ of the cor= responding eigenvalues of ZZ^{T}.

The matrix $C \equiv ZZ^{T}$ is the $n \times n$ matrix of scalar products between the rows of Z, i.e. scalar products between the rows of X with res= pect to their centre of gravity.

In practice this means that if we did not have the matrix $X - 1\,\bar{x}^{T}$ but instead the matrix of scalar products C (between rows, in this case), then the basic structure of C provides a graphical display of the row points, as shown above.

In general, if C is the matrix of scalar products between any set of point vectors with respect to their centre of gravity in any Euclidean or generalized Euclidean space, where each point is weighted by the respective element of the diagonal matrix D_{ω}, then the projec= tions of the points onto the "closest" low-dimensional subspace are obtained from the generalized eigenstructure of C:

$$C = N\,D_{\mu}\,N^{T} \quad (\text{where } N^{T}\,D_{\omega}\,N = I) \tag{3.1.3}$$

$$F = N_{(p)}\,D_{\mu(p)}^{\frac{1}{2}} \tag{3.1.4}$$

In practice the computation consists of constructing the symme= tric matrix $D_{\omega}^{\frac{1}{2}}\,C\,D_{\omega}^{\frac{1}{2}}$, finding the eigenstructure $U\,D_{\mu}\,U^{T}$, and then

$$F = D_{\omega}^{-\frac{1}{2}}\,U_{(p)}\,D_{\mu(p)}^{\frac{1}{2}} \tag{3.1.5}$$

In passing we should also remark that the matrix of scalar pro= ducts between columns of $X - 1\,\bar{x}^{T}$, i.e. $(X - 1\,\bar{x}^{T})^{T}(X - 1\,\bar{x}^{T})$, is just $(n - 1)$ times the covariance matrix S of the m variables. The co-ordinates G of the principal components biplot (§2.3.1) are thus obtainable from the eigenstructure of S, where $G = N_{(p)}$, the first p eigenvectors of S. (This analysis is often called the principal compo= nents analysis of the covariance matrix S.) The co-ordinates G of the covariance biplot (§2.3.4) are also obtainable from the eigenstructure of S, where G is now defined by $G = N_{(p)}D_{\mu(p)}^{\frac{1}{2}}$ (In this latter display the vector lengths approximate the actual sample standard deviations.)

From now on we shall be dealing with the scaling of one set of points and our attention will be focussed on the interpoint distances within this set. The indices i and j will now be used to indicate elements of the *same* set, e.g. δ_{ij} is the measured distance between two rows, say, i and j. Note too that the convention in the scaling literature is often that observed distances be denoted by the Greek letter δ, whereas the fitted or approximate distances in the display are denoted by d. (This convention is almost the opposite of that used elsewhere in statistics and in terms of which δ and d might be denoted d and $\hat{\delta}$ respectively.)

3.2 Classical (Young-Householder) scaling

If, instead of the set of scalar products between all the points, we had the set of all the between-point distances δ_{ij} $(i,j = 1,\dots,n)$, then we could still obtain the co-ordinates F. Using the result that the squared distances are related to the scalar products by:

$$\delta_{ij}^2 = c_{ii} + c_{jj} - 2c_{ij} \qquad (3.2.1)$$

it is not difficult to show (for example, see Greenacre (1978, pp. 65-66)) that the matrix C can be recovered from the matrix $\Delta^{(2)}$ of squared distances by the process of *double-centering*:

$$C = -\tfrac{1}{2}\,\Theta\,\Delta^{(2)}\Theta^{\mathsf{T}} \qquad (3.2.2)$$

where the idempotent matrix $\Theta(n \times n)$ is:

$$\Theta \equiv I - 1(1^{\mathsf{T}}\omega)^{-1}\,\omega^{\mathsf{T}} \qquad (3.2.3)$$

(ω is the diagonal of D_ω). C is thus the matrix of scalar products of the (weighted) points with respect to their centre of gravity, and then the display (3.1.3)-(3.1.4) applies. This display is what is generally known as *Young-Householder scaling*, after Young and House= holder (1938) or *classical scaling*. (see Torgerson, 1958).

The above technique can be applied to any symmetric matrix of squared interpoint distances $\Delta^{(2)}$, not necessarily Euclidean, to ob= tain a graphical display of the points in low-dimensional Euclidean space. If the double-centered matrix C is positive semi-definite then we have the assurance that the points do have a Euclidean representa= tion in some space of dimension at most $n - 1$.

3.3 Principal co-ordinates analysis

Finally, Gower (1966) showed that this same technique could be extended to display an $n \times n$ matrix S of interpoint similarities s_{ij}.

The crux of his argument is that if S is positive semi-definite then $s_{ii} + s_{jj} - 2s_{ij}$ is a true squared distance δ_{ij} in Euclidean space of dimension at most n-1. The matrix of these δ_{ij} can then be double-centered as described above in (3.2.2) and (3.2.3) and this leads to the scalar product matrix C with elements:

$$c_{ij} = s_{ij} - \bar{s}_{i.} - \bar{s}_{.j} + \bar{s}_{..} \qquad (3.3.1)$$

where $\bar{s}_{i.} \equiv \sum_j \omega_j s_{ij} / \sum_k \omega_k$

$\bar{s}_{.j} = \bar{s}_{j.}$ (since S is symmetric)

$s_{..} = \sum_i \sum_j \omega_i \omega_j s_{ij} / (\sum_k \omega_k)^2$

or, in matrix notation:

$$C = \Theta S \Theta^T \qquad (3.3.2)$$

where Θ is defined by (3.2.3). The graphical display (3.1.3)-(3.1.4) of C computed in this way from a similarity matrix S is called *principal co-ordinates analysis.*

Comparing (3.2.2) and (3.3.2) it is clear that if $S = -\frac{1}{2}\Delta^{(2)}$, then principal co-ordinates analysis reduces to classical scaling mentioned above. Classical scaling, in turn, can be used to find the single set displays of many of the basic structure displays described in §2, when the interpoint distance matrix is given as initial data. For example, if Euclidean distances are given, classical scaling yields a principal components analysis (representation of the rows only, described from (2.2.1) to (2.2.3)); if chi-squared distances between the row objects are given and the diagonal matrix D_ω of weights is equal to D_r, the diagonal matrix of row weights (see §2.3.6), then classical scaling yields the correspondence analysis of the row pro= files.

3.4 When is a "distance matrix" a distance matrix?

The technique of §3.2 can be applied to any symmetric matrix Δ of alleged interpoint distances. Such a distance matrix either may arise through direct measurement of the δ_{ij}, or may be computed by one of the many coefficients that measure dissimilarity or similarity. Non-Euclidean coefficients, and indeed coefficients that do not satis= fy the triangle inequality are frequently used and are justified by their protagonists on the grounds that, in practice, they give more intelligible results.

However, if our intuitive ideas of distance are not to be vio= lated, then the rows of F (3.1.5) which supply the co-ordinates in p dimensions must be real. This will happen if the first p eigen= values of the double-centered matrix C derived from Δ by (3.2.2) are positive. If we choose p = rank(C), then we have real co-ordinates if C is positive semi-definite. This provides the condition on Δ being a distance matrix rather than a "distance matrix". We use the term *dissimilarity matrix* to embrace both situations.

Direct measurement of the δ_{ij} may be done on an interval scale, rather than the ratio scale implicit in §3.2. A constant a then needs to be estimated so that δ_{ij} + a may be taken to be ratio data. This is known as the *additive constant problem*. Theoretical approaches have been considered by Messick and Abelson (1956) and by Torgerson (1958), but Cooper (1972) reports that these methods do not appear to work well in practice.

Two intuitive methods for estimating a are described by Kruskal (1977a). Other procedures that estimate the additive constant simul= taneously with the configuration fall into the category of least squares scaling (§4) (Cooper, 1972, Saito, 1978, and Bloxom, 1978).

3.5 Robustness of classical scaling

Introduction

If the n × n matrix Δ consists of Euclidean interpoint distances measured in p dimensional space without error, then C (3.2.2) has p positive eigenvalues and a zero eigenvalue with multiplicity n - p. In practice Δ is measured with error. This has the effect of "per= turbing" the eigenvalues so that some of those eigenvalues which ought to be zero are positive and some negative. (The double centering opera= tion (3.2.2) guarantees that one eigenvalue will be exactly zero).

In general p, the number of *genuinely positive* eigenvalues is unknown, and the smallest genuinely positive eigenvalue may not be obviously distinguishable from the largest *perturbed zero* eigenvalue.

Choice of dimensionality in classical scaling

Two guidelines for the choice of p are provided by Sibson (1979):

- 1. *Trace criterion*: choose p so that the sum of the genuinely positive eigenvalues is approximately equal to the sum of all the eigenvalues.

2. *Magnitude criterion*: accept as genuinely positive only those eigenvalues whose magnitude substantially exceeds that of the largest negative eigenvalue.

A robustness study

These results form part of a landmark paper (Sibson, 1979) in the theoretical error analysis of classical scaling as opposed to error analysis via simulation experiments. Sibson considered, amongst other things, perturbations of $D^{(2)}$, the matrix of true squared Euclidean distances, of the form

$$\Delta^{(2)}(\varepsilon) = D^{(2)} + \varepsilon F \tag{3.5.1}$$

with F symmetric, main diagonal zero. His main conclusion was that classical scaling is robust against errors that leave the observed dissimilarities approximately proportional to the true distances d_{ij}. Sibson found that even when $\|D^{(2)}\| \doteqdot \|\varepsilon F\|$ (i.e. when the magnitude of the errors in (3.5.1) approaches the magnitude of the d_{ij}^2) classical scaling performs tolerably well.

The problem that remains is that of recognizing when the error process generates δ_{ij} which are not proportional to the true distances, i.e. when the δ_{ij} are related to Euclidean distances by a monotonically increasing distortion other than a straight line through the origin. (It is also possible that the procedure for observing or computing the δ_{ij} is spurious, and that the δ_{ij} are little better than a collection of random numbers.)

If the distortion is such that it *bunches up* large distances (e.g. $\delta_{ij} = \log d_{ij}$) then classical scaling will need extra dimensions in order to accommodate the large number of nearly equal long "distances". This shows itself in extra positive eigenvalues that cannot be ignored. The apparent dimensionality of the configuration is increased.

On the other hand, if the distortion bunches up the small distances (e.g. $\delta_{ij} = d_{ij}^2$) then complex co-ordinates are required to accommodate the small distances. This shows itself in large negative roots that lead to the rejection as spurious of equally large positive roots. The apparant dimensionality is decreased.

In conclusion it is fair to state that if the suggested dimensionality according to the trace and magnitude criteria differ markedly then classical scaling is not appropriate.

The alternative actions then are either to use least squares scaling, estimating the distortion function, or to use nonmetric sca=

ling.

It would undoubtedly be better to estimate the distortion func= tion which would transform the observed dissimilarities to approxi= mate Euclidean distances. However the application of families of parametric monotonically increasing functions as distortion func= tions in scaling is poorly developed, and consequently nonmetric sca= ling, which fits a nonparametric function to overcome the distortion, is most widely used.

3.6 Examples of classical scaling

A study of dimensionality

A two dimensional confuguration of 50 points was generated (Fig. 3.6.1). Note the clustered pattern of the points. Almost all simula= tion studies in scaling can be faulted on the unrealistic assumption of generating a random configuration. The interpoint distances were computed and subjected to classical scaling. The largest and smallest eigenvalues of the double-centered matrix (see (3.2.2)) were 451.8, 365.0, 0.02, 0.02,...,-0.02, -0.02, -0.02. -0.02,..... leaving no doubt about the correct dimensionality by both the trace and magni= tude criteria. Since the dissimilarities are true distances, the exer= cise provides a test of the eigenstructure program which ought to return a zero eigenvalue of multiplicity 48.

The interpoint distances were distorted by adding a proportional error component so that

$$\delta_{ij} = d_{ij}(1+u) \qquad (3.6.1)$$

where u was uniformly distributed on the interval (-0.5, 0.5). The solution obtained by classical scaling of the distorted distances is essentially identical to the true solution (compare Figures 3.6.1 and 3.6.2), bearing out the comments of Sibson *et al* (to appear) on the robustness of classical scaling under this error model. The eigenvalues were 459.6, 391.8, 87.4, 80.3,...,-71.3, -72.3, -87.0, -101.6, so that by the magnitude criterion the dimensionality is clearly 2; this is confirmed by the trace criterion, the sum of the 48 smallest eigen= values being -23.7.

When, in addition, a logarithmic transformation of the distorted distances was made, the eigenvalues were 23.7, 19.5, 9.6, 4.9, 2.5, 2.2,...,-2.3, -2.4, -2.5, -2.9 and suggest dimensionalities of 5 and 3 by the magnitude and trace criteria respectively. When the distor= ted distances were exponentiated, the eigenvalues were 8615, 6365,

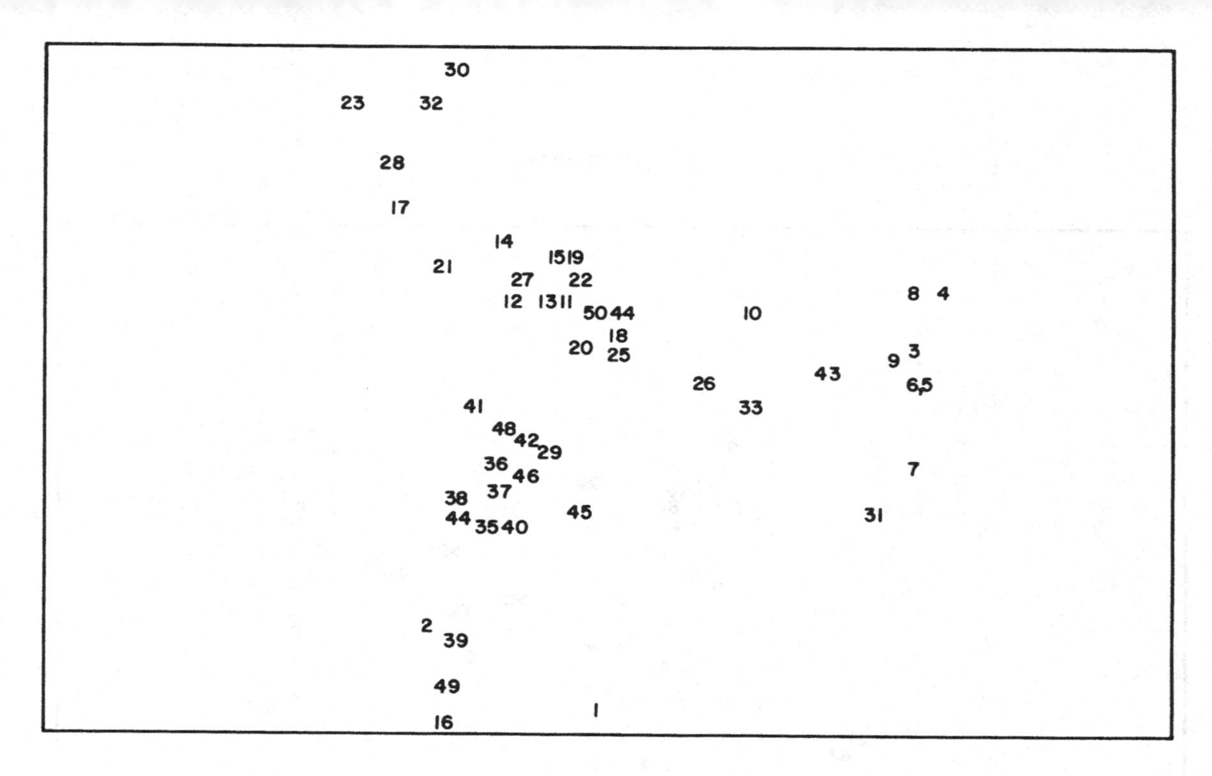

FIGURE 3.6.1

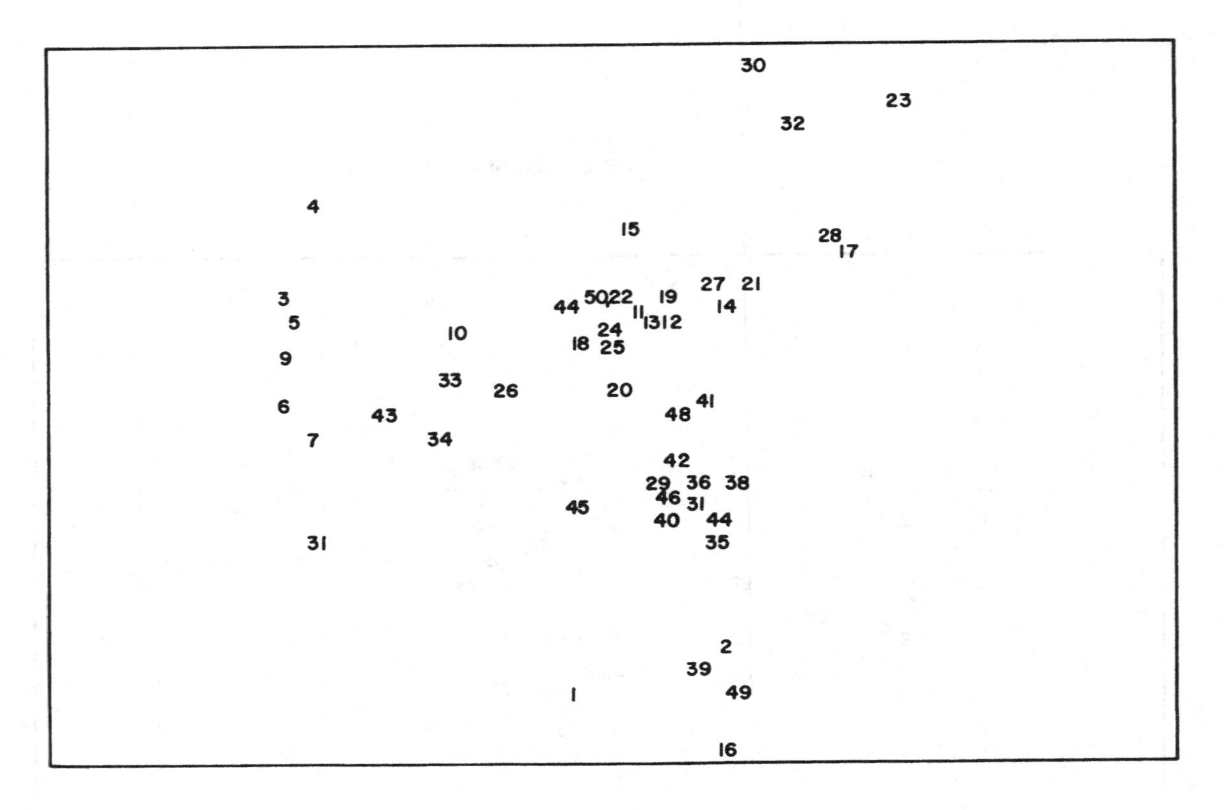
30
23
32
4
15
28
17
27
21
3
50
22
19
44
14
11
13
12
5
10
24
18
25
9
33
26
20
41
6
43
48
7
34
42
29
36
38
46
31
45
40
44
31
35
2
39
1
49
16

FIGURE 3.6.2

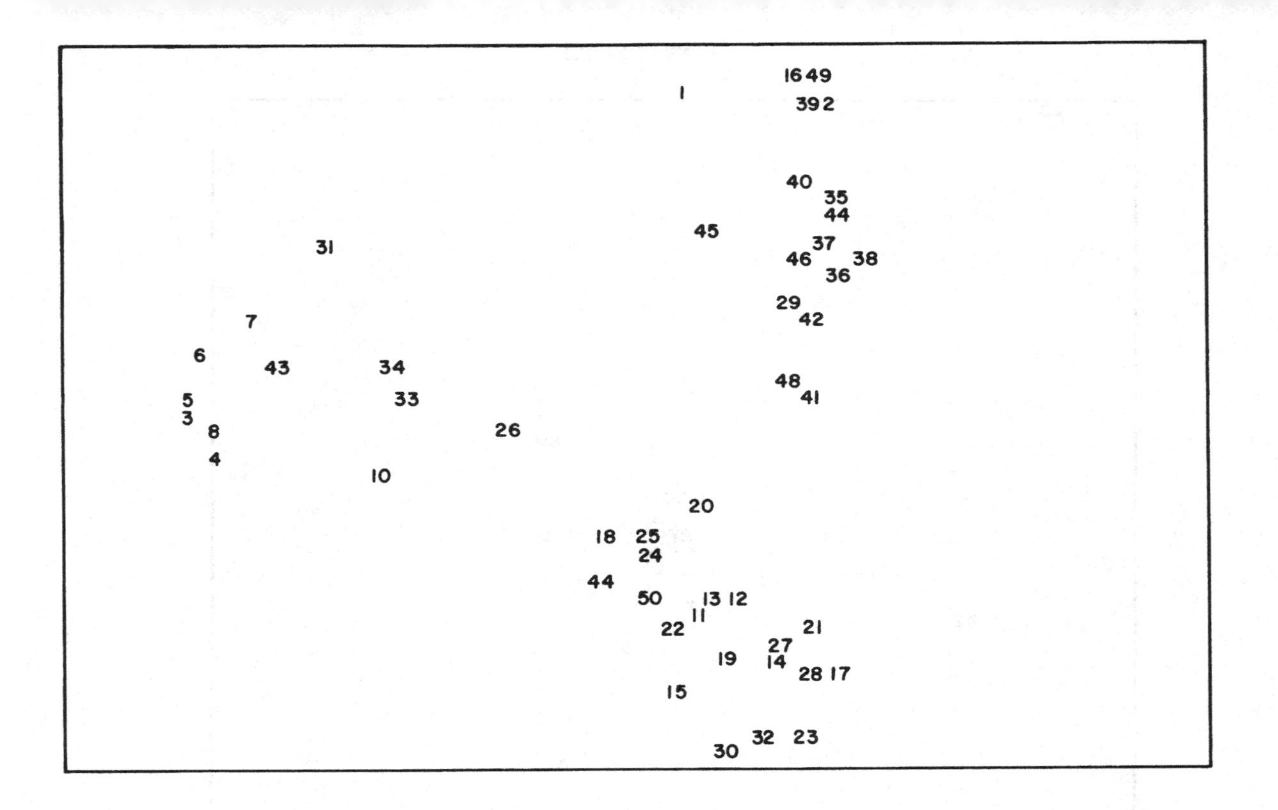

FIGURE 3.6.3

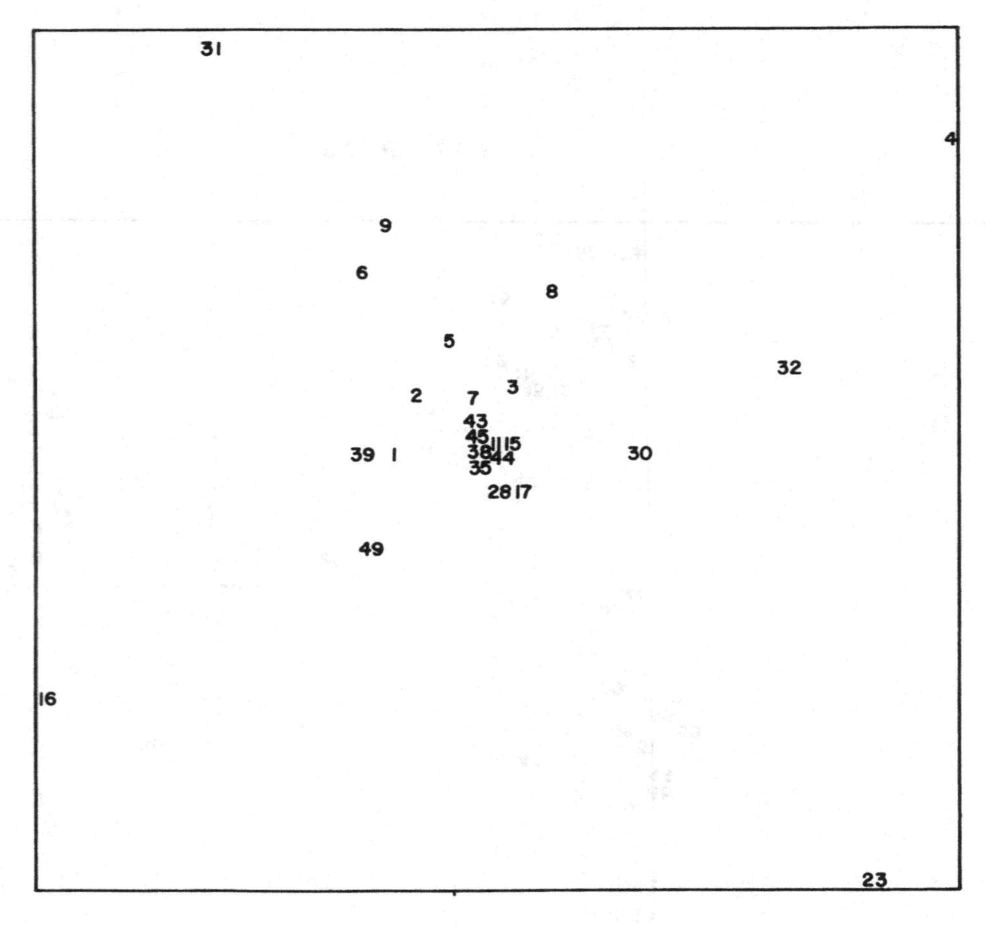

FIGURE 3.6.4

3977, 2626,...,-2480, -3705, -5889, -7231, suggesting dimensionali= ties of 1 and 0 by the magnitude and trace criteria. Note how, for these data, the logarithmic transformation, which makes the longer distances relatively more equal, produces a bunching up of the clusters (Figure 3.6.3), whilst the opposite trend occurs with the exponential transformation (Figure 3.6.4).

The minimum flying times between 13 southern African airports, taken from a South African Airways timetable, are shown in the upper triangular section of table 3.6.1, while the corresponding number of stopovers are shown in the lower triangular section. Application of classical scaling to the flying times yielded eigenvalues of 44910, 21114, 6312, 3600,...,-1137, -1343, -3589, -5682. A dimensionality of 2 is therefore appropriate - the solution (Figure 3.6.5) is fair re= presentation of the true locations of the airports (see Figure 3.6.6, on which the direct links between airports are plotted). That classi= cal scaling is successful is perhaps surprising, since the two and three stopover flying times frequently involve large detours, result= ing in a non-Euclidean structure. This example illustrates once again the robustness of classical scaling.

Two further examples, where the trace and magnitude criteria show classical scaling to be unacceptable, are given in §5.7.

4. LEAST SQUARES SCALING

4.1 Introduction

The scaling techniques considered here are linked in that their objective functions are variations on a theme of minimizing the sums of squares of differences between fitted and observed distances. They are similar to the classical scaling technique (§3) in that the fit= ted distances approximate the observed distances (or a parametric monotonic transformation of the observed distances). Classical sca= ling and least squares scaling are thus both metric scaling techni= ques.

Unlike classical scaling, there is here no algebraic solution to the minimization problem, and iterative methods are used to obtain solutions, with the attendant problems of local minima and sub-opti= mal solutions, choice of dimensionality and starting configuration. In this regard least squares scaling is akin to the non-metric tech= niques of §5. However the minimization problem is here better behaved. Sibson and Bowyer (to appear) and Sibson *et al* (to appear) have found

Table 3.6.1 *Flying times (in minutes) between southern African airports (upper triangular matrix) and number of stopovers (lower triangular matrix).*

	1	2	3	4	5	6	7	8	9	10	11	12	13
1 Johannesburg	–	60	50	50	85	95	125	110	125	120	85	125	115
2 Durban	0	–	60	90	50	65	95	110	115	135	105	145	175
3 Bloemfontein	0	0	–	30	90	125	145	130	120	80	45	85	125
4 Kimberley	0	1	0	–	60	95	125	135	125	85	40	80	120
5 East London	0	0	1	0	–	35	65	80	85	85	100	140	180
6 Port Elizabeth	0	0	1	1	0	–	30	45	50	65	135	145	195
7 Plettenberg Bay	2	2	1	2	2	0	–	15	20	65	135	145	195
8 George	0	1	1	1	1	0	0	–	15	50	120	130	180
9 Oudtshoorn	1	1	1	1	2	1	0	0	–	40	110	120	170
10 Cape Town	0	1	0	0	0	0	1	0	0	–	70	80	130
11 Upington	0	1	0	0	1	1	2	1	1	0	–	40	80
12 Keetmanshoop	1	1	1	1	2	1	2	1	1	0	0	–	50
13 Windhoek	0	1	1	1	2	2	2	2	2	1	0	0	–

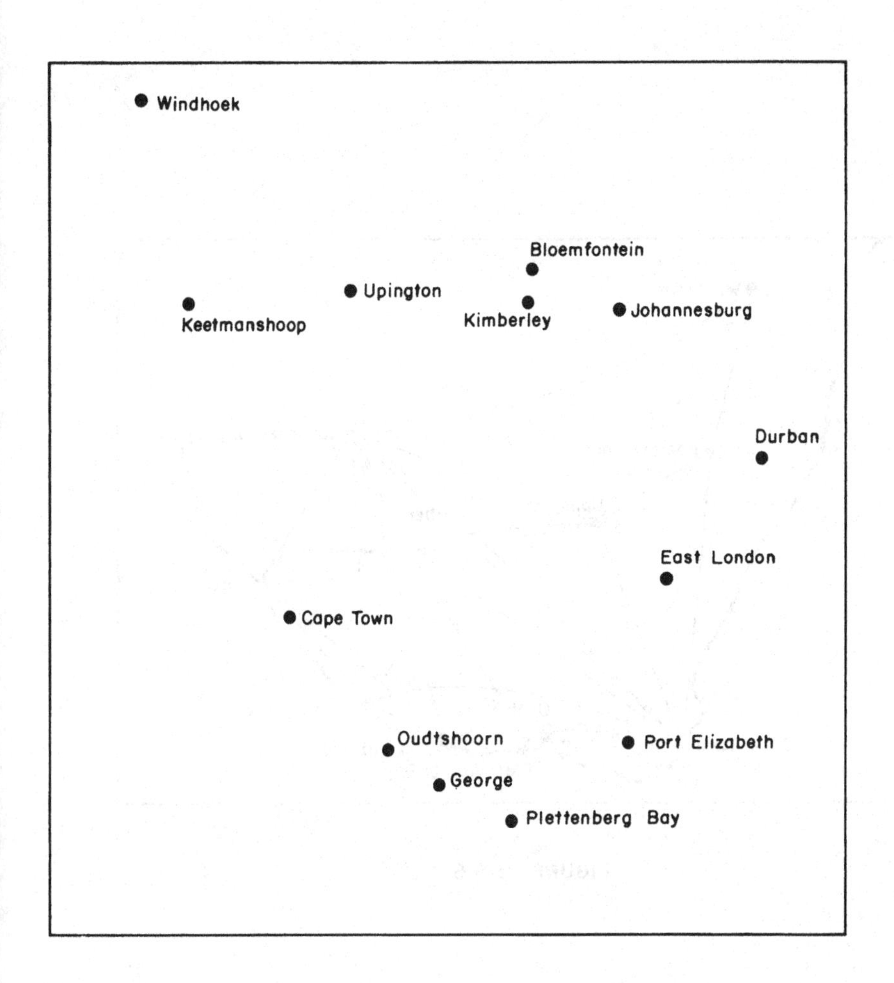
Windhoek
Bloemfontein
Upington
Keetmanshoop
Kimberley
Johannesburg
Durban
East London
Cape Town
Oudtshoorn
Port Elizabeth
George
Plettenberg Bay

FIGURE 3.6.5

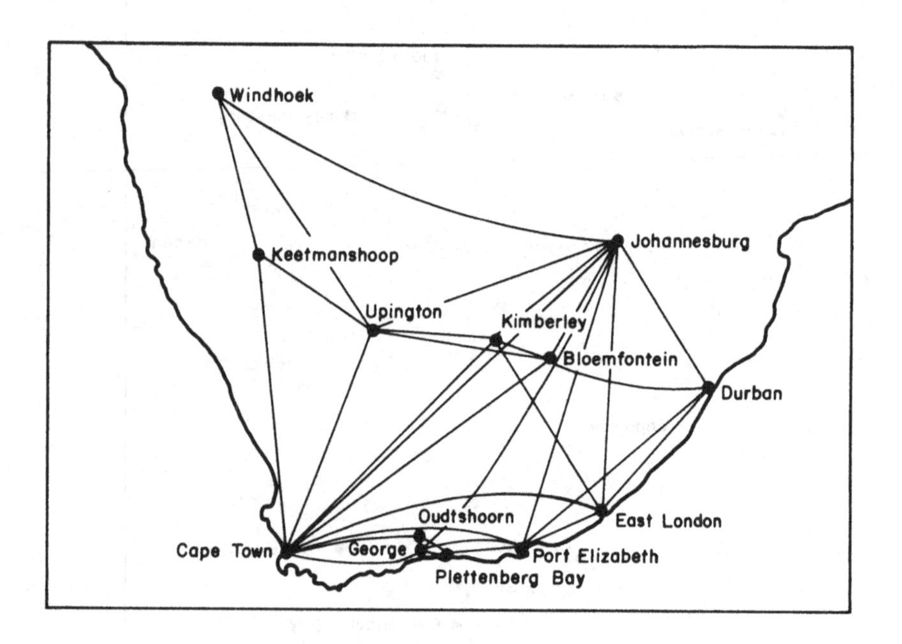
Windhoek
Keetmanshoop
Johannesburg
Upington
Kimberley
Bloemfontein
Durban
Oudtshoorn
East London
Cape Town
George
Port Elizabeth
Plettenberg Bay

FIGURE 3.6.6

that the conjugate gradient method (Fletcher and Reeves, 1964) works well. This method is available in several programme libraries (e.g. algorithm EO4DBF in the NAG libraries (NAG Library Manual (Mark 6) 1977) and algorithm VAO8A in the Harwell Program Library (Harwell Subroutine Library Manual 1978) and will not be considered further, except to note that it requires the availability of first order partial derivatives of the objective function.

It is of interest to note that least squares scaling was the only serious scaling method to be proposed by non-psychologists. Sammon's (1969) "nonlinear mapping" arose in an information retrieval context; Anderson (1971) suggested the "minimization of a quadratic loss function" as an alternative to the "crude method" of "polar ordination" pioneered in ecology by Bray and Curtis (1957); Rohlf and Archie (1978) developed "least squares mapping" to produce reasonably accurate maps of the location of trees in a forest in order to apply spatial analysis; Crippen (1977), in apparent ignorance of previous research, re-invented least squares scaling to find the conformation of a molecule by the "unusual means" of calculating atomic co-ordinates from a matrix of interatomic distances.

Objective functions for least squares scaling

Many of the proposed objective functions for least squares scaling are found by varying the parameters of

$$T_\alpha^2 = \sum_{i=1}^{n} \sum_{j=i+1}^{n} w_{ij} (d_{ij}^\alpha - \delta_{ij}^\alpha)^2 \qquad (4.1.1)$$

where the d_{ij} are fitted distances, the δ_{ij} are observed distances, the w_{ij} are positive weights and α, the weight in the exponent, is arbitrary, but non-zero. When a parametric monotonic function f is used as a distortion function to transform observed dissimilarities, the objective function

$$T_t^2 = \frac{\sum_{i=1}^{n} \sum_{j=i+1}^{n} w_{ij} (d_{ij} - f(\delta_{ij}))^2}{\sum_{i=1}^{n} \sum_{j=i+1}^{n} d_{ij}^2} \qquad (4.1.2)$$

may be used. The denominator is necessary to prevent degenerate solutions, i.e. to prevent all the points converging.

If we set $\alpha = 1$ in (4.1.1) we obtain the important family of objective functions

$$T_1^2 = \sum_{i=1}^{n} \sum_{j=i+1}^{n} w_{ij} (d_{ij} - \delta_{ij})^2 \qquad (4.1.3)$$

The simplest case, (4.1.3) with all $w_{ij} = 1$, seems to have been ne=glected, probably under the mistaken assumption that it is equiva=lent to classical scaling. It is proposed by Spaeth and Guthery (1969) and by Anderson (1971), but neither paper gives any indication that it was implemented. An extensive simulation study of its performance has recently been conducted (Sibson *et al*, to appear); it was found to be as robust as classical scaling, and in certain circumstances out-performed it.

Minimizing (4.1.3) makes sense from a theoretical point of view if it is assumed that the underlying error structure relating the observed distances (δ_{ij}) to the "true" distances (which, with non-confusing ambiguity, we will also denote d_{ij}) is $E(d_{ij} - \delta_{ij}) = 0$, $\mathrm{Var}(d_{ij} - \delta_{ij}) = c/w_{ij}$ and $\mathrm{Cov}(d_{ij} - \delta_{ij};\ d_{k\ell} - \delta_{k\ell}) = 0$. In fact, if $\delta_{ij} \sim N(d_{ij};\ c/w_{ij})$ then the minimum of (4.1.1) yields maximum likelihood estimates of the d_{ij}.

In particular choosing $w_{ij} = d_{ij}^{-1}$ is implied by assuming that the standard deviation of the errors in the δ_{ij} is proportional to δ_{ij}; choosing $w_{ij} = d_{ij}^{-\frac{1}{2}}$ assumes that the variance of the errors is proportional to δ_{ij}. (This is "nonlinear mapping" (Sammon, 1969), also investigated by Sibson *et al* (to appear) with favourable com=ments on its robustness, especially when this was the correct error model).

A useful family of weights to consider is

$$w_{ij} = 1/\delta_{ij}^{p} \tag{4.1.4}$$

where p should be chosen so that the variance of the errors is pro=portional to δ_{ij}^{p}. Putting p=0 implies homoscedasticity of errors. No serious investigation of the heteroscedasticity encountered in obser=ved dissimilarities has been attempted. In data given by Spuhler (1972), the variances of the observed dissimilarities are roughly proportional to δ_{ij}^{4}, which suggests a value of 4 for p.

When estimates of the variances of the observed dissimilarities are available, it would be appropriate to take

$$w_{ij} = 1/\mathrm{var}(\delta_{ij}) \tag{4.1.5}$$

Missing observations in the dissimilarity matrix can be dealt with by simply letting $w_{ij} = 0$ for unknown δ_{ij}.

Another way of exploiting the use of weights occurs if we wish to fix certain points with more accuracy than others (cf. §2.3.5). Let the vector ω reflect the relative weights we wish to give to the points. Then the appropriate weight for distance δ_{ij} is

$$w_{ij} = \omega_i \omega_j \tag{4.1.6}$$

Frequently, however, assumptions about the finer details of the error structure are of minor importance when considered alongside the far wider issue of whether it is appropriate at all to impose a Euclidean structure on the observed dissimilarities. In these situa= tions the justification for scaling is that "a picture is worth a thousand words" and scaling is used to produce a kind of visual sum= mary statistic of the data. The choice of weights is then governed by what features of the data we wish to emphasize. If it is important to us that small distances and the local structure are preserved (at the expense of poor representation of large distances) then positive values of p in (4.1.4) are called for: p=1 is Sammon's (1969) non-linear mapping, but values of p up to about 4 are useful. If preser= vation of the large distances is paramount (with loss of local struc= ture) negative values of p should be used. Anderson (1971) advocated $p = -1$.

The objective function (4.1.1) with $\alpha=2$, $w_{ij}=\delta_{ij}^2$ is shown by Gower (to appear) to be equivalent to the "parametric mapping" of Shepard and Carroll (1966). Assuming $d_{ij} \doteqdot \delta_{ij}$ it is easy to show that (4.1.1) is then roughly equivalent to (4.1.3) with $w_{ij} = 1/\delta_{ij}^4$. Thus parametric mapping gives little weight to the long distances and very large weight to the short distances.

Putting $\alpha=2$ and $w_{ij}=1$ in (4.1.1) was considered by Browne (1977). This approximates (4.1.3) with $p = -1$ and favours the representation of long distances. This particular choice of objective function has distinct computational advantages in that it drastically reduces the number of variables, and efficient numerical techniques which use second order derivatives become feasible (Browne, 1977).

In (4.1.2) any parametric monotonically increasing function may be used. The simplest choice is a straight line. Choosing

$$f(\delta_{ij}) = a + \delta_{ij} \tag{4.1.7}$$

is a useful strategy for the solution of the additive constant pro= blem (Cooper, 1972). Two popular computer packages, KYST and M-D-SCAL use polynomials up to degree four (Kruskal, 1977a), but there is no guarantee that they are monotonically increasing. This problem was circumvented by Browne (1979) who suggests the use of integrals of squared polynomials.

The most general family of always positive polynomials is given by

$$y = a \prod_{i=1}^{n} (x^2 + 2b_i x + b_i^2 + c_i^2) \qquad (4.1.8)$$

which has no real roots (Hawkins, pers. comm.). Integrating (4.1.8) yields the most general family of monotonically increasing polynomials of order 2n+1. The use of these polynomials as distortion functions is currently being studied.

Other families of monotonically increasing functions which may fruitfully be considered, are the four parameter transformations

$$y = \gamma + \delta \log \frac{x+\xi}{\xi+\lambda-x} \text{ and } y = \gamma + \delta \sin h^{-1}[\frac{x-\xi}{\lambda}] \qquad (4.1.9)$$

used by Johnson (1949) in his S_B and S_U distributions.

Shepard (1980) notes that the relationship between similarities and the distances in the configuration solution obtained by nonmetric scaling is frequently approximated by a one-parameter negative expo= nential curve. This suggests the use of

$$y = 1 - e^{-\lambda x} \qquad (4.1.10)$$

as another possible distortion function for least squares scaling.

A major problem is encountered with the fitting of distortion functions when they are near-vertical. Standard least squares methods, which minimize the sums of squares of vertical differences, are then unsatisfactory. It is recommended (M.W.Browne, pers.comm.) that, when distortion functions are fitted, perpendicular residuals are minimized.

4.2 Examples of least squares scaling

The inter airport flying time data (Table 3.6.1) were reanalysed using weighted least squares scaling. A weight of 1 was assigned to each direct flight, $\frac{1}{8}$ to one-stopover flights and $\frac{1}{20}$ to two-stopover flights. The solution is much the same as that obtained by classical scaling.

It seems reasonable that a fixed part of the flying time is al= lowed for take-off and landing; thus an additive constant model (4.1.7) is appropriate. This was further refined so that the additive constant was estimated from the direct flights only, and was applied twice to the one-stopover flying times and thrice to the two-stopover flying times. The additive constant was estimated to be 12,9 minutes. The

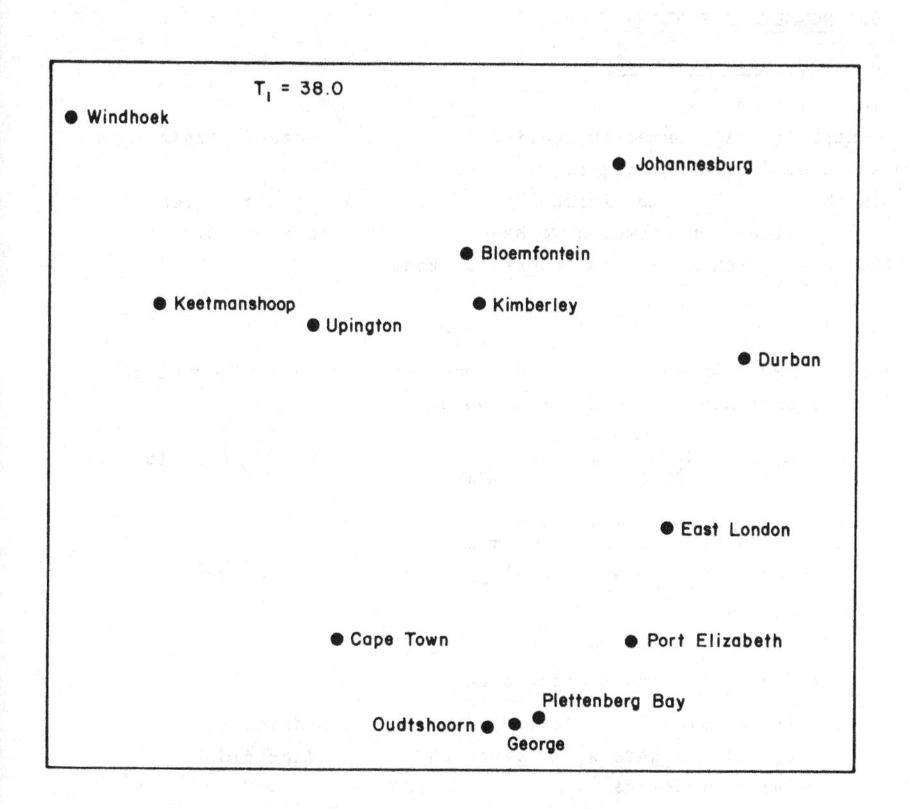
T_1 = 38.0
Windhoek
Johannesburg
Bloemfontein
Keetmanshoop
Upington
Kimberley
Durban
East London
Cape Town
Port Elizabeth
Plettenberg Bay
Oudtshoorn
George

FIGURE 4.2.1

solution (Figure 4.2.1) is markedly superior to the previous solu= tions. The remaining distortion is attributable to the fact that the nondirect flying times are frequently far from proportional to the true distances, due to the roundabout routes taken.

5. NONMETRIC SCALING

5.1 Monotonic regression

Introduction

Central to all nonmetric scaling methods is *monotonic regression*, also known as *isotonic regression*.

In the context of multidimensional scaling monotonic regression arises in a situation in which we have a set of n observed dissimi= larities δ_{ij}, which have been ordered so that

$$\delta_{i_1j_1} < \delta_{i_2j_2} < \ldots < \delta_{i_mj_m} \quad (5.1.1)$$

and a set of associated distances d_{ij} computed from a configuration in p dimensional space, and a set of weights w_{ij}:

$$d_{i_1j_1};\ d_{i_2j_2};\ \ldots;\ d_{i_mj_m} \quad (5.1.2)$$

$$w_{i_1j_1};\ w_{i_2j_2};\ \ldots;\ w_{i_mj_m}$$

It causes no confusion to adopt a simpler notation and to write

$$\delta_{i_kj_k} = \delta_k,\ d_{i_kj_k} = d_k \text{ and } w_{i_kj_k} = w_k \quad (5.1.3)$$

Definition of monotonic regression

The monotonic regression of d_i *on the ordered set* δ_i *with weights* w_i *is given by the function* $f(d_i)$ *which minimizes*

$$\sum_{i=1}^{m} (d_i - f(d_i))^2 w_i \quad (5.1.4)$$

subject to the monotonicity constraint

$$i < j \Rightarrow f(d_i) \leq f(d_j) \quad (5.1.5)$$

The function $f(d_i)$ that minimizes (5.1.4) has a simple graphical construction.

Plot points $p_1,\ldots,p_m$ with co-ordinates for p_i given by $(W_i;D_i)$ where

$$W_i = \sum_{j=1}^{i} w_j \text{ and } D_i = \sum_{j=1}^{i} d_j w_j \qquad (5.1.6)$$

Let $P_0 = (0,0)$. The line that joins these points is called the *cumulative sum diagram* (*CSD*). The slope of a line joining P_i to P_j is given by

$$\sum_{k=i}^{j} d_k w_k / \sum_{k=1}^{j} w_k \quad (D_j - D_i)/(W_j - W_i) \qquad (5.1.7)$$

In particular, the slope of the CSD immediately to the left of P_i is d_i. Now consider the graph formed by joining P_0 to P_m by means of a taut string constrained to lie below the CSD. This graph is called the *greatest convex minorant* (*GSM*).

Let $\hat{P}_0$ to $\hat{P}_m$ be the points on the GCM lying directly below $P_0,\ldots,P_m$. The points for which $\hat{P}_i = P_i$ are called "corner points"). Let the slope of the GCM immediately to the left of $\hat{P}_i$ be $\hat{d}_i$. Let $\hat{D}_i = \sum_{j=1}^{i} \hat{d}_i$. Clearly $\hat{P}_i = (W_i, \hat{D}_i)$.

Example of CSM and GCM

In Table 5.1.1, the first three rows contain the δ_i, d_i and w_i. In the fourth and fifth rows, W_i and D_i are computed by (5.1.6) and the CSD plotted (continuous line in Figure 5.1.1). The GCM is then plotted (dotted line) and the $\hat{D}_i$ and $\hat{d}_i$ computed (rows six and seven).

Table 5.1.1 *Example of CSM and GCM*

i	1	2	3	4	5	6	7	8	9	10
δ_i	0.2	0.7	0.8	1.3	1.7	1.9	2.0	2.3	2.6	3.0
d_i	0.6	1.8	1.4	0.1	0.9	3.6	1.0	4.0	0.6	3.0
w_i	1.0	0.5	0.5	2.0	1.0	0.5	0.8	0.2	1.5	1.0
W_i	1.0	1.5	2.0	4.0	5.0	5.5	6.3	6.5	8.0	9.0
D_i	0.6	1.5	2.2	2.4	3.3	5.1	5.9	6.7	7.6	10.6
$\hat{D}_i$	0.60	0.90	1.20	2.40	3.30	4.02	5.16	5.45	7.60	10.60
$\hat{d}_i$	0.60	0.60	0.60	0.60	0.90	1.43	1.43	1.43	1.43	3.00

Theorem of monotonic regression

The slopes $\hat{d}_i$ of the GCM yield the monotonic regression of d_i on the ordered set δ_i with weights w_i. The monotonic regression is unique.

For a proof of this theorem see Barlow *et al* (1972).

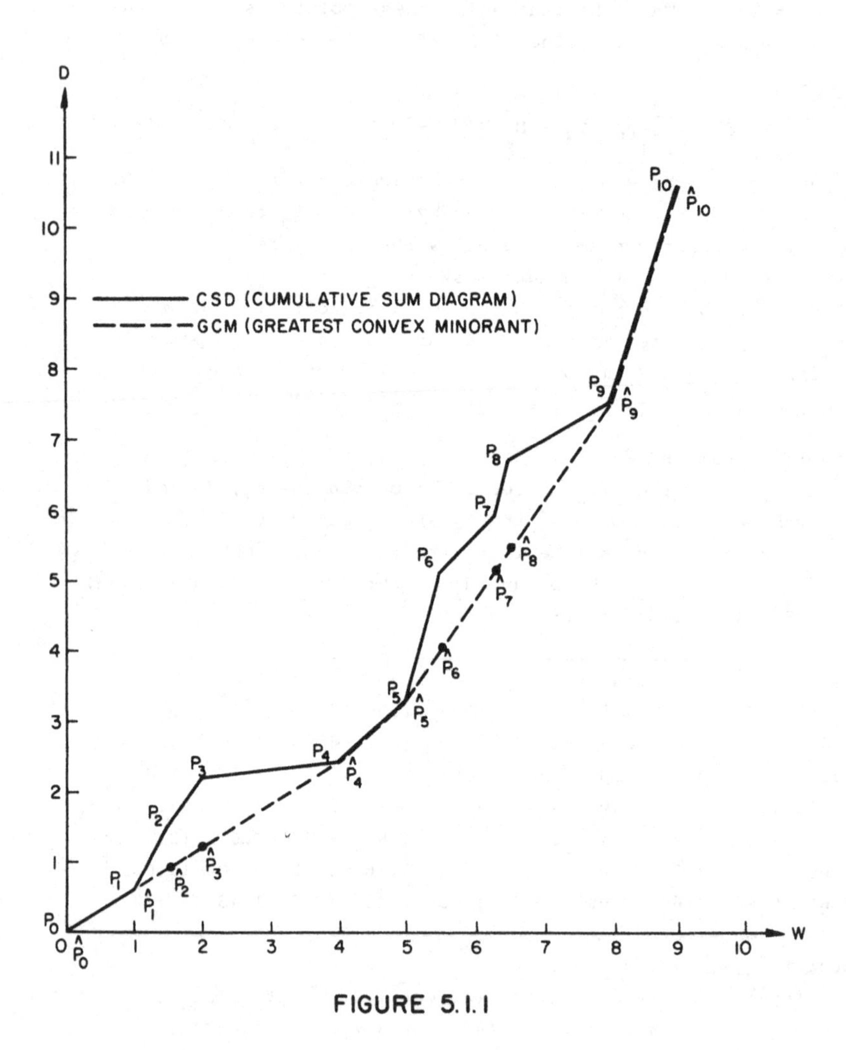
D
11
10
9
8
7
6
5
4
3
2
1
0
1
2
3
4
5
6
7
8
9
10
W
CSD (CUMULATIVE SUM DIAGRAM)
GCM (GREATEST CONVEX MINORANT)
P_0
$\hat{P}_0$
P_1
$\hat{P}_1$
P_2
$\hat{P}_2$
P_3
$\hat{P}_3$
P_4
$\hat{P}_4$
P_5
$\hat{P}_5$
P_6
$\hat{P}_6$
P_7
$\hat{P}_7$
P_8
$\hat{P}_8$
P_9
$\hat{P}_9$
P_{10}
$\hat{P}_{10}$

FIGURE 5.1.1

Pool-adjacent-violators algorithm

The validity of the algorithm hinges on the fact that it cal= culates the slopes $\hat{d}_i$ of the line segments that make up the GCM. The algorithm may be described in three steps:

Step 1:

Set $\hat{d}_i = d_i$, $\hat{w}_i = w_i$ and $n_i = 1 \quad i=1,\ldots,m$

(The n_i define a "block" structure. Block i is a block starting at $\hat{d}_i$ with n_i members and with $\hat{d}_i = \hat{d}_{i+1} = \ldots = \hat{d}_{i+n_i-1}$. We start with the finest block structure).

Step 2:

If $\hat{d}_1 \leq \hat{d}_2 \leq \ldots \leq \hat{d}_m$ we are finished. Otherwise continue to Step 3.

Step 3:

Choose any successive pair of blocks that "violates" the order= ing and "pool" them. Formally, if blocks i and j violate the ordering, we set

$$\hat{d}_i = \frac{\hat{w}_i\hat{d}_i + \hat{w}_j\hat{d}_j}{\hat{w}_i + \hat{w}_j} \tag{5.1.8}$$

and then set

$$\hat{d}_k = \hat{d}_i \qquad k = i + 1;\ldots;\ i + n_i + n_j - 1$$

$$\hat{w}_i = \hat{w}_i + \hat{w}_j, \qquad w_j = 0,$$

$$n_i = n_i + n_j, \qquad n_j = 0.$$

Return to Step 2.

Example of pool-adjacent-violators algorithm

In the example in Table 5.1.2 we use (5.1.8) first to pool blocks 2 and 3, and then to pool blocks 2 and 4. Figure 5.1.2 illu= strates graphically how the pool-adjacent-violators algorithm con= structs the GCM from the CSD.

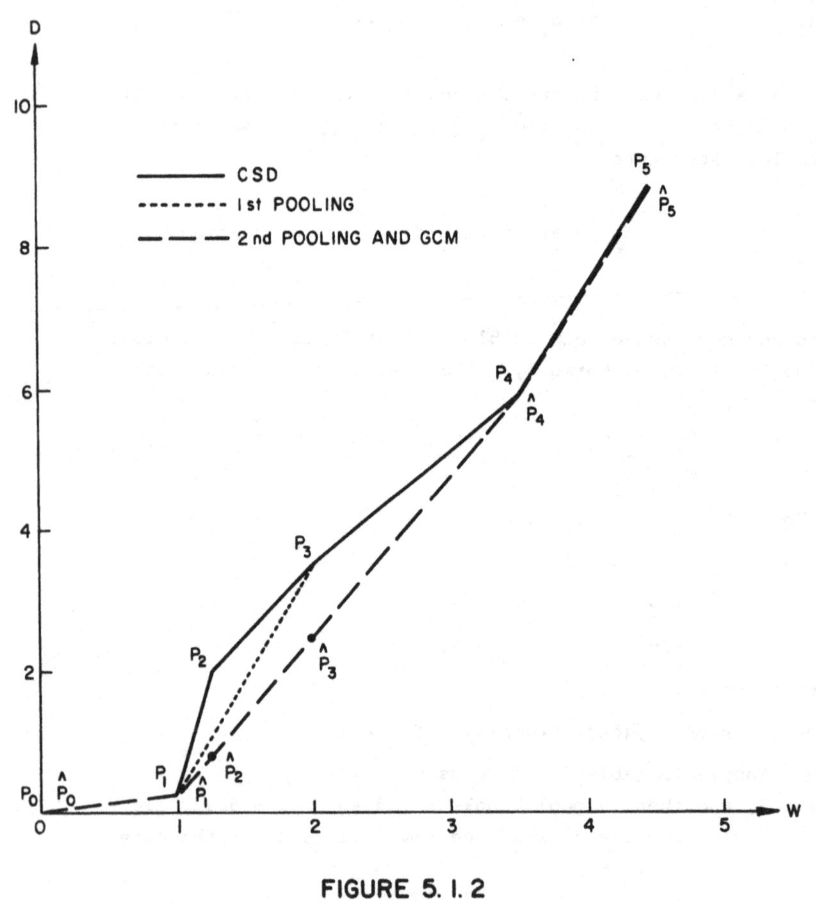
D
10
8
6
4
2
0
1
2
3
4
5
W
CSD
1st POOLING
2nd POOLING AND GCM
P_0
$\hat{P}_0$
P_1
$\hat{P}_1$
P_2
$\hat{P}_2$
P_3
$\hat{P}_3$
P_4
$\hat{P}_4$
P_5
$\hat{P}_5$

FIGURE 5. 1. 2

Table 5.1.2: *The pool-adjacent-violators algorithm*

	i	1	2	3	4	5
	δ_i	0.8	1.2	1.5	2.4	3.0
Step 1	$d_i = \hat{d}_i$	0.25	2.00	1.00	0.50	2.00
	$w_i = \hat{w}_i$	1.00	0.25	0.75	1.50	1.00
	n_i	1	1	1	1	1
Step 2			← pool →			
Step 3	$\hat{d}_i$	0.25	1.25	1.25	0.50	2.00
	$\hat{w}_i$	1.00	1.00	0	1.50	1.00
	n_i	1	2	0	1	1
Step 2			← pool →			
Step 3	$\hat{d}_i$	0.25	0.80	0.80	0.80	2.00
	$\hat{w}_i$	1.00	2.50	0	0	1.00
	n_i	1	3	0	0	1

Up-and-down-blocks algorithm

To program the pool-adjacent-violators algorithm efficiently, a rule is needed that decides which pair of adjacent violating blocks must be pooled at each stage. Kruskal's (1964b) algorithm achieves this.

It starts as in step 1 of the pool-adjacent-violators algorithm with each $\hat{d}_i$ forming a block. If we have three successive blocks i, j and k say (i < j < k) then block j is said to be "down-satisfied" if $\hat{d}_i \le \hat{d}_j$ and "up-satisfied" if $\hat{d}_j \le \hat{d}_k$. We define one block to be "active"; this block is "pooled" (as in step 3) with the block above if it is not up-satisfied and with the block below if it is not down-satisfied. When a block is both up-satisfied and down-satisfied, the block above becomes active. Initially block 1, the lowest block, is specified as the active block. It is necessary to make special provision to ensure initially that block 1 is down-satisfied and that block m is up-satisfied. This can be achieved by setting $\hat{d}_0 = 0$ and $\hat{d}_{m+1} = \hat{d}_m + 1$.

The flow chart (Figure 5.1.3) defines the Kruskal up-and-down-blocks algorithm.

Dealing with ties

In (5.1.1) the implicit assumption was that there were no ties amongst the δ_i. Ties can be dealt with in one of two ways; we can say either

$$\delta_i = d_j \Rightarrow \hat{d}_i \gtrless \hat{d}_j$$

or

$$\delta_i = \delta_1 \Rightarrow \hat{d}_i = \hat{d}_j$$

These have become known as the primary and secondary approaches to ties.

To implement these approaches into the algorithms it is neces= sary only to "preprocess" the d_i. For the pool-adjacent-violators algorithms we introduce either

Step 0 (primary approach)

Within each "tieblock", i.e. a block for which

$\delta_i = \delta_{i+1} = \ldots = \delta_j$, sort

$d_i, d_{i+1}, \ldots, d_j$ into ascending order.

Continue to step 1.

Step 0 (secondary approach)

Compute the mean $\bar{d}_{ij} = \frac{1}{j+i+1} \sum_{k=i}^{j} d_k$ of the d_k within each tie block from δ_i to δ_j and set

$$d_k = \bar{d}_{ij} \qquad k = i, \ldots, j$$

Step 1 (secondary approach)

Set $\hat{w}_i = \sum_{k=i}^{j} w_k \qquad w_k = 0 \qquad k = i+1, \ldots, j$

$$n_i = j - i + 1 \; n_k = 0 \qquad k = i+1, \ldots, j$$

We start not with the finest block structure, but with block structure determined by the tie blocks, noting that many tie blocks are of length 1.

5.2 Kruskal-Shepard scaling

The milestone papers in the development of nonmetric scaling are Shepard (1962a,b) and Kruskal (1964a,b). Shepard (1962a,b) demonstra= ted that low-dimensional configurations could be constructed from a rank-ordering of interpoint dissimilarities. The contributions of Kruskal (1964a,b) were to introduce the concept of "stress" of a con=

figuration to measure the departure from the required rank-ordering and to devise a successful algorithm to find a configuration with minimum stress. Shepard (1980) provides an interesting historic re= view of nonmetric scaling.

The *aim* of nonmetric scaling in p dimensions can be expressed as follows:

Given an $n \times n$ *matrix* $\Delta = (\delta_{ij})$ *of observed dissimilarities find an* $n \times p$ *configuration* X *with interpoint distances* $D = (d_{ij})$ *such that*

$$\delta_{ij} < \delta_{k\ell} \Rightarrow d_{ij} \leq d_{k\ell}. \tag{5.2.1}$$

We are thus aiming to find a nonparametric monotonic relation= ship between Δ and D.

In general (5.2.1) is not met for any given configuration X, and the need arises to assess the badness of fit of a configuration from the objective of monotonicity. Kruskal (1964a,b) used the theory of monotonic regression (§5.1) to develop such a badness of fit mea= sure, which he called stress.

Given a dissimilarity matrix $\Delta = (\delta_{ij})$ and a configuration X, the stress of X is found as follows: Compute $D = (d_{ij})$, the distance matrix for configuration X. (It is usual to use Euclidean distances, though this is not obligatory). Order the δ_{ij} as in (5.1.1) and relabel as in (5.1.3) so that $\delta_i < \delta_j$ for $i < j$. Let d_i be the dis= tance associated with δ_i. With weights $w_i = 1$ find $\hat{d}_i$, the monotonic regression of d_i on δ_i, using, say, Kruskal's up-and-down-blocks al= gorithm (§5.1). If there are ties among the δ_i, adopt either the pri= mary or secondary approach, and preprocess the d_i accordingly before computing the monotonic regression.

The "raw stress" of X is defined to be

$$S^{*2} = \sum_{i=1}^{m} (d_i - \hat{d}_i)^2 \tag{5.2.2}$$

By the theory of monotonic regression, S^{*2} is a minimum for the monotonic sequence $\hat{d}_i$. Raw stress is clearly invariant with respect to rotation, reflection and translation of X, i.e. transformation of X to XA + B, with A orthogonal and B arbitrary. To make the stress measure invariant with respect to dilation it is necessary to include a scale factor. It is easy to show that

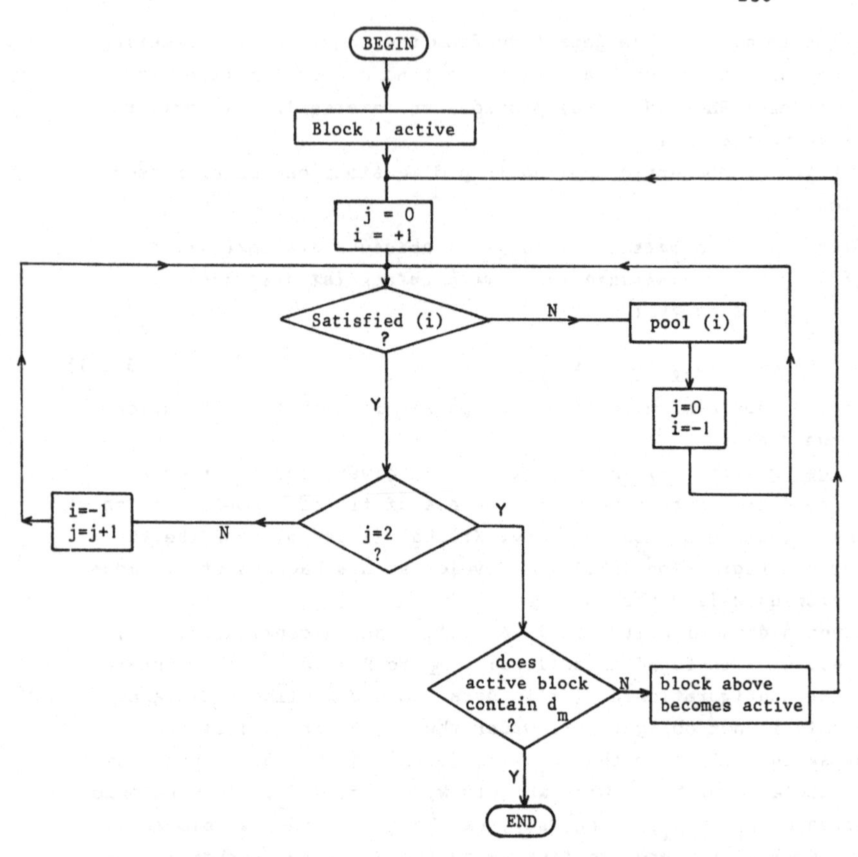

Figure 5.1.3 Flowchart for Kruskal's Up-and-down-blocks algorithm. In the flowchart it takes on the values +1 and -1, and we define

Satisfied (+1)	=	check if the active block is up-satisfied.
Satisfied (-1)	=	check if the active block is down-satisfied.
Pool (+1)	=	pool the active block with the block above: the pooled block becomes the active block.
Pool (-1)	=	pool the active block with the block below: the pooled block becomes the active block.

Note that when j=2 the active block is both up-and down-satisfied.

$$S_1^2 = \sum_{i=1}^{m} (d_i - \hat{d}_i)^2 / \sum_{i=1}^{m} d_i^2 \qquad (5.2.3)$$

and

$$S_2^2 = \sum_{i=1}^{m} (d_i - \hat{d}_i)^2 / \sum_{i=1}^{m} (d_i - \bar{d}_i)^2 \qquad (5.2.4)$$

$$(\text{where } \bar{d} = \frac{1}{m} \sum_{i=1}^{m} d_i)$$

are invariant with respect to dilation, i.e. transformation of X to kX, k a scalar. Without the denominators in (5.2.3) and (5.2.4) the raw stress can be made arbitrarily small by choosing k small, thus shrinking the configuration. In scaling applications, use of (5.2.3) or (5.2.4) makes little practical difference; we opt for (5.2.3).

A configuration that satisfies the objective (5.2.1) will have zero stress (the d_i then form a monotonically increasing sequence). Since the δ_i are observed dissimilarities measured with error, it is almost certain that the rank-ordering of the δ_i is incorrect, and it is highly improbable that a zero-stress configuration exists. We therefore search for a configuration X with minimum stress. So, thinking of stress as a function of the configuration, our solution X* in p dimensions is

$$S_1^2(X^*) = \min_X \{S_1^2(X)\} \qquad (5.2.5)$$

where X ranges over the set of all n x p matrices.

As with least squares scaling, there is no algebraic solution to the problem of choosing X to minimize the stress, and we resort to iterative methods. The most commonly used numerical procedure for finding the solution configuration is the method of steepest descent, which works surprisingly well, considering how ill-behaved the problem seems to be. Kruskal (1964b) produced a steepest descent algorithm to meet the specific requirements of the problem, and has subsequently "fine-tuned" the procedure still further (Kruskal 1977a). The overall operational procedure for Kruskal-Shepard scaling is shown in Figure 5.2.1.

Dealing with ties

As discussed in §5.1, there are two ways of dealing with ties among the δ_{ij}:

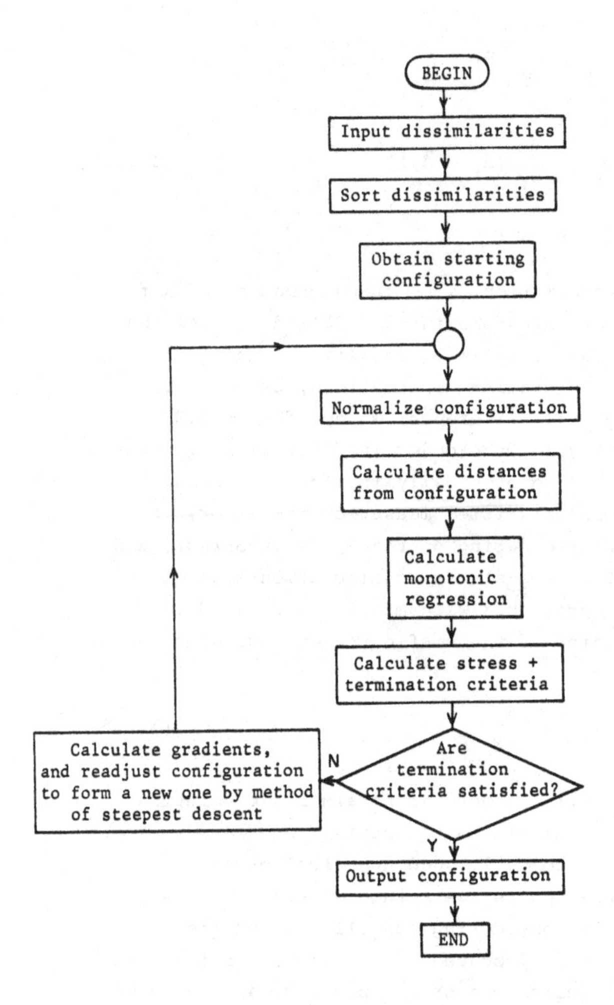

Figure 5.2.1

The operational procedure for Kruskal-Shepard scaling.

In the primary approach, we say

$$\delta_{ij} = \delta_{k\ell} \Rightarrow d_{ij} \gtrless d_{k\ell} \qquad (5.2.6)$$

thus not specifying the ordering of d_{ij} *and* $d_{k\ell}$. *The secondary approach to ties is restrictive and requires*

$$\delta_{ij} = \delta_{k\ell} \Rightarrow d_{ij} = d_{k\ell} \qquad (5.2.7)$$

The primary approach has been more widely used.

Kendall (1977) describes a compromise between the primary and secondary approaches, which he terms the *tertiary approach*. Certain tieblocks are handled by the secondary approach, and the remainder by the primary approach. This method proves valuable when the dissi= milarity coefficient takes on only a few values. For example, Kendall (quoted in Rivett, 1977) applies it to the reconstruction of maps from abuttal information in which the dissimilarity coefficient takes on only two values - $\delta_{ij} = 0$ implies localities i and j adjoin, $\delta_{ij} = \infty$ implies localities i and j do not adjoin. The tie block for $\delta_{ij} = 0$ receives the secondary approach, while the tie block for $\delta_{ij}=\infty$ receives the primary approach. Rivett (1977), who applies nonmetric scaling to multiple criterion decision making, has a dissimilarity coefficient taking on three values - two policies are alike ($\delta_{ij} = 0$), similar ($\delta_{ij} = 1$), or unlike ($\delta_{ij} = 2$). The tie blocks for $\delta_{ij} = 0$ and $\delta_{ij} = 1$ receive the secondary approach, the tie block for $\delta_{ij} = 2$ the primary approach.

A problem, still unresolved, suggested by Kendall (1971), is to extend the primary treatment of ties to dissimilarities which are nearly equal, i.e.

$$|\delta_{ij} - \delta_{k\ell}| < t \Rightarrow d_{ij} \gtrless d_{k\ell} \qquad (5.2.8)$$

Mimmack (1979) provides a sub-optimal solution. The difficulty lies with the modification of the monotonic regression algorithm to permit overlapping tie blocks.

5.3 Alternative stress functions for nonmetric scaling

A general comment on the method of least squares

Methods that minimize the sums of squares of residuals between data points and a function to be fitted do so by spreading the resi= duals as evenly as possible. The average size of the residuals tends to be more or less equal throughout the domain of the fitted function.

Implications for nonmetric scaling

In terms of Kruskal-Shepard scaling this implies that the *absolute* values of the residuals ($d_i - \hat{d}_i$) are of roughly the same order of magnitude for all $\hat{d}_i$, and that the *relative* absolute resi= duals, $|d_i - \hat{d}_i|/\hat{d}_i$ are larger for small values of $\hat{d}_i$ than for large values. This in turn implies that large d_i have more influence than small d_i in determining the final configuration, and that long dis= tances are relatively more accurately fitted than short distances. Ultimately it is fair to state that the long distances are fitted at the *expense* of the short distances.

The "folklore" of nonmetric scaling, and a countermeasure

The above paragraph explains what Kruskal (1977b) has called the "folklore" of Kruskal-Shepard scaling: "*scaling gives the infor= mation contained in the large dissimilarities... it is notorious that local features of the arrangement* (in the final configuration) *are not meaningful*". Graef and Spence (1979), in a simulation study, showed that omitting the smallest third of the dissimilarities made little difference to the recovered configuration. They concluded "*one is on much surer ground when considering the relative location of points that are far apart*", and cautioned users "*against inter= preting or attaching significance to the relative positions of points that are close together in the recovered space*".

This is not a fundamental problem of scaling *per se*, the problem lies in the choice of stress function. The essence of the countermea= sure to the problem has been in existence for as long as Kruskal-Shepard scaling. McGee (1966), in what he termed "*elastic distances multidimensional scaling*" introduced a stress function which is sen= sitive to *relative* residuals. McGee (1966) motivated his stress func= tion in physical terms as the "work" done to stretch or compress a spiral spring from length d_i to $\hat{d}_i$, and the significance of his method in terms of relative errors seems to have been lost.

Using a more direct motivation to generate a stress function similar to McGee's, we note that while ($d_i - \hat{d}_i$) measures the absolute difference between d_i and $\hat{d}_i$, $\frac{d_i}{\hat{d}_i} - 1$ measures their relative differ= ence. "*Using a time-honoured tradition of statistics*" the raw stress $\sum_{i=1}^{m} (d_i - \hat{d}_i)^2$ found by summing the squared differences measures the overall *absolute* deviation from monotonicity, and gives rise, with a scale factor included, to Kruskal-Shepard scaling. Using the same tra=

dition, but now summing over squared relative differences,

$$U_0^{*2} = \sum_{i=1}^{m} \left(\frac{d_i}{\hat{d}_i} - 1\right)^2 \tag{5.3.1}$$

measures the overall *relative* deviation from monotonicity, and is without modification scale invariant.

The folklore above does not apply to configurations obtained by minimizing U_0^{*2}. The relative errors in all distances are, on average, equal and as much significance can be placed on the interpretation of local arrangements within the configuration as can be placed on the overall arrangement.

Two families of stress functions

A family of stress functions that includes S_1^2 (Kruskal-Shepard stress) (5.2.5) and U_0^{*2} as special cases is

$$U_\alpha^2 = \sum_{i=1}^{m} \left(\frac{d_i - \hat{d}_i}{\hat{d}_1^{1-\alpha}}\right)^2 \Bigg/ \sum_{i=1}^{m} d_i^{2\alpha} \quad , \tag{5.3.2}$$

When $\alpha = 1$, $U_1^2 = S_1^2$; when $\alpha = 0$, $U_0^2 = U_0^{*2}/m$.

Compromise stress functions between U_0^2 and U_1^2 are obtained by choosing α from the interval [0,1]. This family of stress functions is a nonmetric scaling analogue of the objective function T_1^2 with $\sqrt{w_{ij}} = \delta_{ij}^{-(1-\alpha)}$. (See (4.1.3) and (4.1.4)).

A disadvantage of the stress family U_α^2 is that the $\hat{d}_i$, being the monotonic regression of d_i on δ_i, minimize $\sum_{i=1}^{m} (d_i - \hat{d}_i)^2$ and not $\sum_{i=1}^{m} ((d_i - \hat{d}_i)/\hat{d}_i^{1-\alpha})^2$

This can be overcome by iterating a weighted monotonic regression: on the first iteration use weights $w_i = d_i^{2(\alpha-1)}$, on the second and subsequent iterations use weights $w_i = \hat{d}_i^{2(\alpha-1)}$, where the $\hat{d}_i$ are ob= tained on the previous iteration. Iterate until the $\hat{d}_i$ on successive iterations are equal to within a specified tolerance.

A second family of stress functions, similar in spirit to the family (5.2.2), but without the difficulty of having to iterate the monotonic regression, may be motivated as follows. Since d_i and $\hat{d}_i$ are roughly equal,

$$\frac{d_i}{\hat{d}_i^{1-\alpha}} \doteq d_i^\alpha \quad , \tag{5.3.3}$$

we can consider

$$V_\beta^{*2} = \sum_{i=1}^{m} (d_i^\beta - \hat{d}_i)^2 \tag{5.3.4}$$

as a measure of raw stress, where the $\hat{d}_i$ are now obtained from the monotonic regression of d_i^β on δ_i with weights 1. Incorporating an appropriate scale factor to keep the stress scale invariant, we have the family of stress functions:

$$V_\beta^2 = \sum_{i=1}^{m} (d_i^\beta - \hat{d}_i)^2 / \sum_{i=1}^{m} d_i^{2\beta} \qquad (5.3.5)$$

If $\beta = 1$, $V_1^2 = S_1^2$, the Kruskal-Shepard stress. Since the role of a zero power is filled by the logarithm (Tukey, 1977) it is convenient to define

$$V_0^2 = \frac{1}{m} \sum_{i=1}^{m} (\log d_i - \hat{d}_i)^2, \qquad (5.3.6)$$

where the $\hat{d}_i$ are the monotonic regression of log d_i on δ_i. Note that V_0^2 is sensitive to relative errors in the same way as U_0^2.

The choice of α in (5.3.2) and β in (5.3.5) is ultimately governed (as it was in least squares scaling) by what the scaler desires to achieve. Choosing $\alpha = 0$ or $\beta = 0$ is neutral. Choosing $\alpha > 0$ or $\beta > 0$ emphasizes the accurate fitting of long distances at the expense of short distances. Choosing $\alpha < 0$ or $\beta < 0$ fits short distances with greater precision than long distances.

The popular ALSCAL package for multidimensional scaling (Takane *et al.*,1977) uses a computationally efficient method known as *alternating least squares* in place of the method of steepest descent. The price paid for increased efficiency is that alternating least squares demands that stress be measured by V_2^2. Users of ALSCAL should thus appreciate the fact that their solution configuration will suffer from the folklore of Kruskal-Shepard scaling, but to an even higher degree. (See de Leeuw and Heiser, 1980, p.505).

The changes that need to be made to computer programs that perform Kruskal-Shepard scaling are relatively minor. No doubt the stepsize procedure of Kruskal (1964b and 1977a) could be retuned, but the arrangements of Kruskal (1977a) seem to work efficiently enough. Mimmack (1979) attempted to adjust the stepsize procedure for a new stress function, but suggested only minor modifications.

Weighting points in nonmetric scaling

In both classical and least squares scaling we had the option of weighting certain points so that they are relatively more or less accurately fitted (see (3.1.4) and (4.7)). This can be achieved in nonmetric scaling by including weights in the monotonic regression.

If the weights are given by the n-vector ω then let weight

$$w^*_{ij} = \omega_i \, \omega_j \tag{5.3.7}$$

be attached to dissimilarity δ_{ij}. We reorder and relabel the w^*_{ij} as in (5.1.2-4) so that w^*_i is the weight attached to the ordered dis= similarity δ_i. Since the weights are relative it is convenient to standardize them so that they sum to m, the number of δ_i. The stress is now *weight invariant*. Thus we use

$$w_i = m \, w^*_i \Big/ \sum_{i=1}^{m} w^*_i \tag{5.3.8}$$

The Kruskal-Shepard stress with weights w (on the dissimilari= ties) is now found by computing the monotonic regression of d_i on δ_i with weights w_i:

$$S^2_{1,w} = \frac{\sum_{i=1}^{m} (d_i - \hat{d}_i)^2 w_i}{\sum_{i=1}^{m} d_i^2 \, w_i} \tag{5.3.9}$$

Similar changes are made to introduce weights into (5.3.5) to form $V^2_{\beta,w}$. For U^2_α (see (5.3.2)) we take the product of weights, using $d^{2\alpha-1}w_i$ as weight on the first iteration of the monotonic regression and $\hat{d}^{2\alpha-1}w_i$ on subsequent iterations.

An m-vector of weights w_i may also be used to reflect the ob= servers confidence in the accuracy of the observed dissimilarity δ_i.

5.4 Local order nonmetric scaling

Increasingly weaker assumptions

In classical scaling and least squares scaling it was assumed that the observed dissimilarities δ_{ij} were Euclidean distances with the addition of measurement error (or that the δ_{ij} were a parametric monotonic transformation of Euclidean distances plus error). Nonmetric scaling, so far, has used the ordering of the δ_{ij}; we have operated on the weaker assumption that the relationship between the recovered Euclidean distances and δ_{ij} is a nonparametric monotonic transforma= tion. We now weaken the assumptions still further, and assume that we can use only the "local ordering" (Sibson, 1972) of the δ_{ij}, that is, we can make order comparisons only between δ_{ij} with common first subscript. We now have as many nonparametric monotonic transformations as we have points in the configuration.

Formally, for local order nonmetric scaling, we replace the ob= jective (5.2.1) which states

$$\delta_{ij} < \delta_{k\ell} \Rightarrow d_{ij} \le d_{k\ell}$$

by

$$\delta_{ij} < \delta_{ik} \Rightarrow d_{ij} \le d_{ik} \qquad (5.4.1)$$

To assess how close a configuration X with interpoint distances d_{ij} comes to meeting the objective (5.4.1), we proceed as follows. Order each row of dissimilarities $\delta_{i1}, \delta_{i2}, \ldots, \delta_{in}$ so that

$$\delta_{(i1)} < \delta_{(i2)} < \ldots < \delta_{(in)} \qquad (5.4.2)$$

for i=1,...,n. Let $d_{(ij)}$ be the distance associated with $\delta_{(ij)}$. For each of the n row orderings (5.4.2) compute the monotonic regression $\hat{d}_{(ij)}$ of $d_{(ij)}$ on $\delta_{(ij)}$. We can now compute the stress for each row, and find the total stress by summing the row stresses. Any of the stress functions of §5.2 and §5.3 may be used. The analogue of Kruskal-Shephard stress is

$$S_L^2 = \sum_{i=1}^{n} \sum_{j=1}^{n} (d_{(ij)} - \hat{d}_{(ij)})^2 / \sum_{i=1}^{n} \sum_{j=1}^{n} d^2_{(ij)} \qquad (5.4.3)$$

Note that the matrix D = (d_{ij}) is symmetric, but that $\hat{D} = (\hat{d}_{ij})$ ($\hat{d}_{ij}$, the monotone regression value associated with d_{ij}) is asymmetric. The iterative procedure for local order scaling differs only in de= tails from "global order" scaling, and the Kruskal (1977a) stepsize procedure works efficiently.

Local order scaling is appropriate whenever it is not meaningful to compare the "unrelated" dissimilarities δ_{ij} and $\delta_{k\ell}$, and only dis= similarities δ_{ij} and δ_{ik} with common subscript (conventionally the first) are comparable. Or, put another way (Sibson 1973), "δ_{ij} *for* i *fixed and* j *varying represent the geography of the rest of the set of objects as seen from object* i".

A clear case for the use of local order scaling when the obser= vations are similarities σ_{ij} rather than dissimilarities and the σ_{ii} (the similarity of a point to itself) are not all equal (Sibson, 1973). This occurs frequently in practice, both when similarities are ob= served directly, and when they are computed from profile data by means of a "similarity coefficient" (Goodall 1973). In these cases the best procedure is to set $\delta_{ij} = -\sigma_{ij}$ and to use local order scaling.

Local order scaling may also be appropriate when the dissimilarity matrix is asymmetric.

Examples of the use of local order scaling are given by Sibson (1973), Rivett (1977) and Morgan and North (in press).

5.5 Local minima, starting configurations and choice of dimensionality

Local minima

In common with most iterative procedures, the method of steepest descent is prone to being trapped at local minima; a small perturba= tion of the current configuration produces a larger stress value, but the global minimum has not been found.

Starting configurations

The only reasonably certain way of finding the global minimum is to repeat the scaling analysis from a considerable number of dif= ferent starting configurations and to choose as final solution the configuration yielding minimum stress. Mimmack (1979) found, on the basis of simulation studies that if random starting configurations are used, roughly a third converged to the global minimum solution.

The method of steepest descent requires a starting configura= tion. A series of simulation studies (Spence 1972, Whitley 1971, Lingoes and Roskam 1973, amongst others) have shown that the problem of local minima is closely related to the starting configuration. Types of starting configurations that have been advocated are fixed L-shaped configurations (Kruskal 1964b), randomly generated confi= gurations, usually uniform in a sphere or a cube, or rational confi= gurations, (usually the solution obtained from classical scaling). The fixed L-shaped configurations are clearly inferior and the choice lies between random and rational starts. Arabie (1973) suggests that multiple random starts be used, unless there is sufficient evidence of linear relationship between dissimilarities and distances (in which case classical scaling or least squares scaling ought anyhow to be used). Arabie (1978a) concedes that "*ultimately a rational strategy... is preferable to taking the minimum over multiple random configurations*", but claims that the latter strategy may be more economical particularly if more than one rational start is required. The subject of starting configurations is by no means settled.

Nonmetric scaling ought to be used when classical scaling breaks down, in which case classical scaling cannot be expected necessarily to provide a good start. However, it ought always to provide a better-than-random start, in that the relative positions of the points are at least vaguely correct. A good compromise strategy (used by the authors' scaling program) is to use the classical start for the first iteration and then perturbing this configuration for subsequent starting configurations.

Another recommended technique (used, for example, by Gower(1971)) is to do the scaling in one dimension higher than what is finally required, to reduce the dimensionality by projecting onto the principal components, and using this configuration as the starting configuration.

Choice of dimensionality

Stress decreases as the dimensionality increases. The traditional approach to the choice of dimension for both nonmetric and least squares scaling is to plot the stress in p-dimensions against p and to say that the "right" dimensionality occurs at the "elbow" of the plot.

In practice, however, scaling is a summary and display tool, and the solutions that interest us most are usually those in two dimensions. A fairly poor solution in two dimensions, with an indication of how it is distorted, is of more practical use than the four dimensional configuration with low stress. Methods of assessing distortion will be given in §6.1.

5.6 Incomplete data and replicated data

Missing data, replicated data

One of the advantages that nonmetric scaling shares with least squares scaling over classical scaling is that it can take both missing data and replicated data in its stride. In both cases the stress is computed by summing over all available δ_i.

Designs for incomplete data

When n, the number of points, is large, then if all $m = \frac{1}{2}n(n-1)$ interpoint dissimilarities have to be observed, both the experimental task and the data analysis become a formidable problem. Spence and Domoney (1974) consider the problem of designing the experiment so that only a limited number of dissimilarities are observed. On the basis of simulation studies, they concluded that, with n = 48 points up to two-thirds of the dissimilarities can be omitted, provided the observation error is small. They deleted dissimilarities at random and in accordance with cyclical designs, but found little difference between the methods. When only a fixed proportion of the dissimilari= ties are to be observed, designs which avoid observing all three sides of triangles (i.e. all of δ_{ij}, δ_{jk} and δ_{ki}) are better (Rohlf and Archie, 1978). The proportion of dissimilarities that can be omitted increases with the number of points.

5.7 Examples of nonmetric scaling

Spuhler (1972) provides data on the genetic distances between 21 American Indian groups (Table 5.7.1). The trace and magnitude criteria suggest a dimensionality of about 16 for classical scaling (eigenvalues 155163, 74332, 41372,...,223, 88, 39, 0, -63, -158). It seems likely that the high dimensionality is caused by the high proportion of nearly equal large values amongst the genetic distan= ces. It is possible that a distortion function that "spreads out" the larger values would convert the genetic distances into Euclidean distances. To determine if a low dimensional solution could be ob= tained, nonmetric scaling was applied, and the two dimensional solu= tions obtained when the Kruskal-Shepard stress function S_1^2 is used (see (5.2.5)), is shown in Figure 5.7.1. The solution obtained when V_o^2, one of the alternative stress functions is minimized is shown in Figure 5.7.2. Comparison of the solutions is deferred to §6.1 when the "cluster loops" appearing in Figures 5.7.1 and 5.7.2 will be interpreted.

Data on the frequency distribution of 32 types of sentence end= ings in 45 works of Plato are given by Boneva (1971) who formed a similarity matrix using the coefficient

$$S_{ij} = \Sigma_{k=1}^{32} \min (P_{ik}, P_{jk})$$

where P_{ik} denotes the proportion of type k endings in the ith word.

The eigenvalues of the scalar product matrix, found by using the formula for principal coordinates analysis (3.3.1) were 3.19, 0.02, 0, -0.09,...,-0.84, -0.84, -0.85, -0.87, -0.88, -0.89. The magnitude cri= terion suggests a dimensionality of 1; the trace criterion is indetermi= nate, the sum of all 45 eigenvalues being -24.9. The nonmetric scaling solutions, using the stress measures S_1^2 and V_o^2, are shown in Figures 5.7.3 and 5.7.4. Comparison between the solutions will be made in §6.1.

Table 5.7.1 *Genetic distances between 21 American Indian Groups (Spuhler, 1972)*

	1	2	3	4	5	6	7	8	9	10	11	12	13	14	15	16	17	18	19	20	21
1 ESKIMO	-	-	-	-	-	-	-	-	-	-	-	-	-	-	-	-	-	-	-	-	-
2 KUTCHIN	.346	-	-	-	-	-	-	-	-	-	-	-	-	-	-	-	-	-	-	-	-
3 NAVAHO	.202	.292	-	-	-	-	-	-	-	-	-	-	-	-	-	-	-	-	-	-	-
4 NASKAPI	.199	.286	.132	-	-	-	-	-	-	-	-	-	-	-	-	-	-	-	-	-	-
5 CHIPPEWA	.331	.349	.275	.280	-	-	-	-	-	-	-	-	-	-	-	-	-	-	-	-	-
6 PIMA	.238	.333	.188	.208	.214	-	-	-	-	-	-	-	-	-	-	-	-	-	-	-	-
7 CORA	.306	.289	.238	.227	.198	.143	-	-	-	-	-	-	-	-	-	-	-	-	-	-	-
8 HUICHOL	.409	.415	.320	.322	.202	.223	.194	-	-	-	-	-	-	-	-	-	-	-	-	-	-
9 NAHUA	.259	.334	.148	.183	.289	.142	.201	.261	-	-	-	-	-	-	-	-	-	-	-	-	-
10 MARICOPA	.325	.321	.257	.278	.247	.122	.125	.201	.171	-	-	-	-	-	-	-	-	-	-	-	-
11 DIEGUENO	.273	.339	.221	.244	.314	.161	.199	.284	.110	.142	-	-	-	-	-	-	-	-	-	-	-
12 CHEROKEE	.300	.387	.269	.287	.237	.176	.254	.234	.218	.202	.218	-	-	-	-	-	-	-	-	-	-
13 TOTONAC	.303	.247	.288	.246	.217	.248	.183	.320	.298	.257	.288	.317	-	-	-	-	-	-	-	-	-
14 QUICHE	.253	.317	.162	.188	.266	.119	.164	.235	.054	.136	.095	.210	.271	-	-	-	-	-	-	-	-
15 MAM	.289	.304	.206	.222	.246	.126	.165	.231	.113	.113	.123	.197	.242	.093	-	-	-	-	-	-	-
16 TZELTAL	.278	.314	.166	.205	.298	.176	.222	.255	.102	.184	.144	.206	.314	.100	.140	-	-	-	-	-	-
17 TZOTZIL	.350	.334	.234	.280	.205	.166	.148	.179	.203	.146	.231	.247	.277	.178	.178	.210	-	-	-	-	-
18 CHOL	.225	.328	.211	.263	.262	.174	.188	.300	.212	.197	.188	.268	.266	.180	.220	.231	.221	-	-	-	-
19 HUASTEC	.328	.329	.220	.280	.248	.196	.197	.234	.199	.181	.214	.220	.313	.187	.216	.203	.145	.206	-	-	-
20 ZAPOTEC	.271	.365	.211	.278	.353	.212	.265	.338	.144	.217	.128	.276	.353	.144	.194	.183	.268	.176	.232	-	-
21 MAZATEC	.277	.401	.258	.323	.318	.203	.247	.320	.221	.213	.180	.262	.344	.201	.244	.244	.258	.107	.218	.149	-

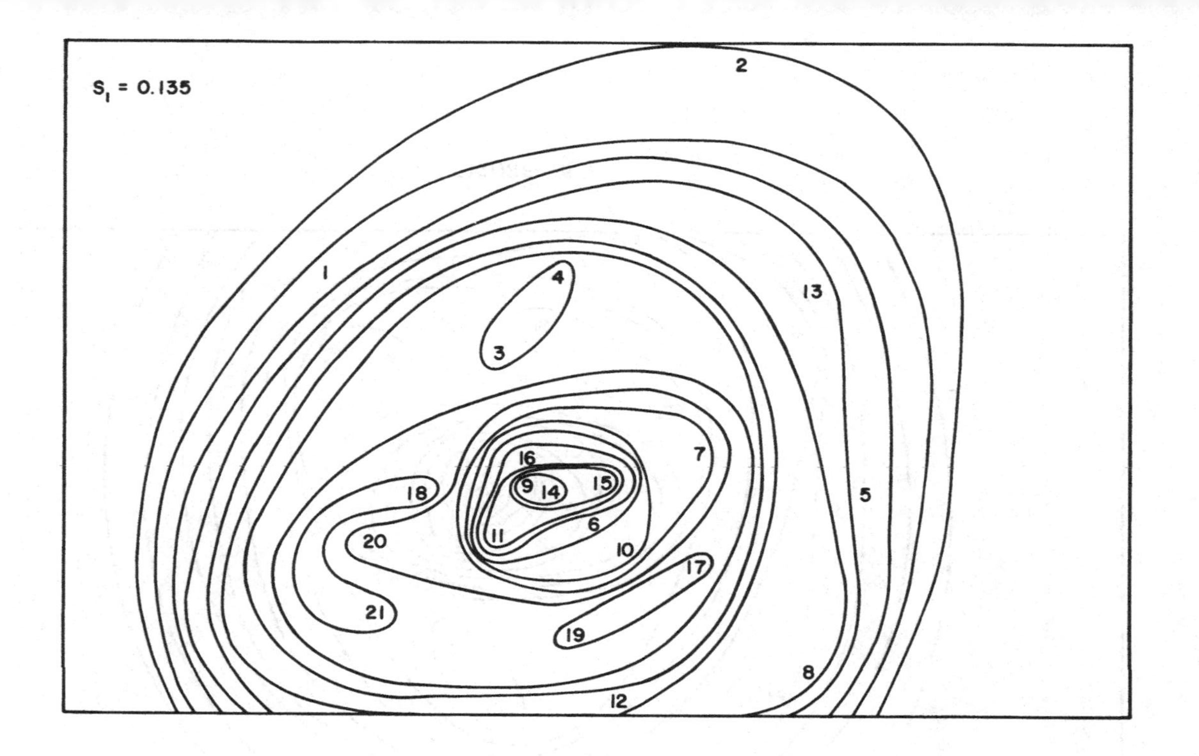

FIGURE 5.7.1

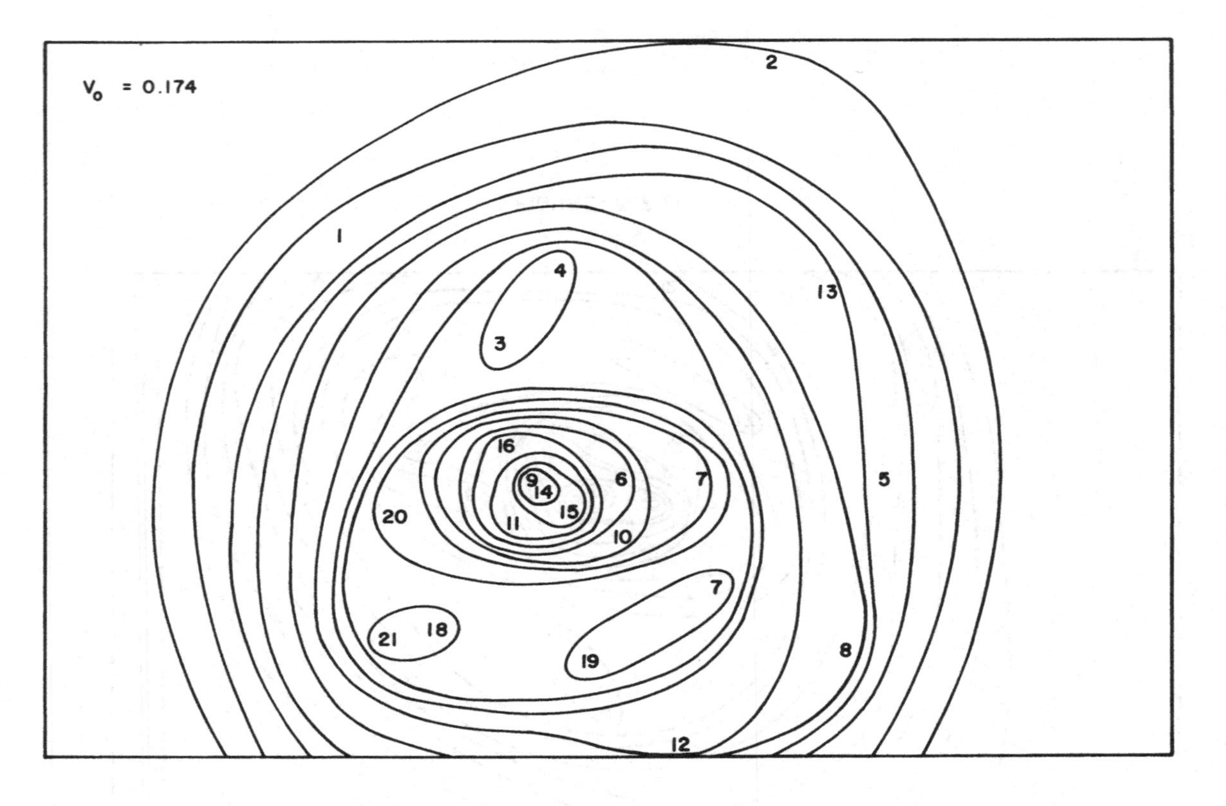

FIGURE 5.7.2

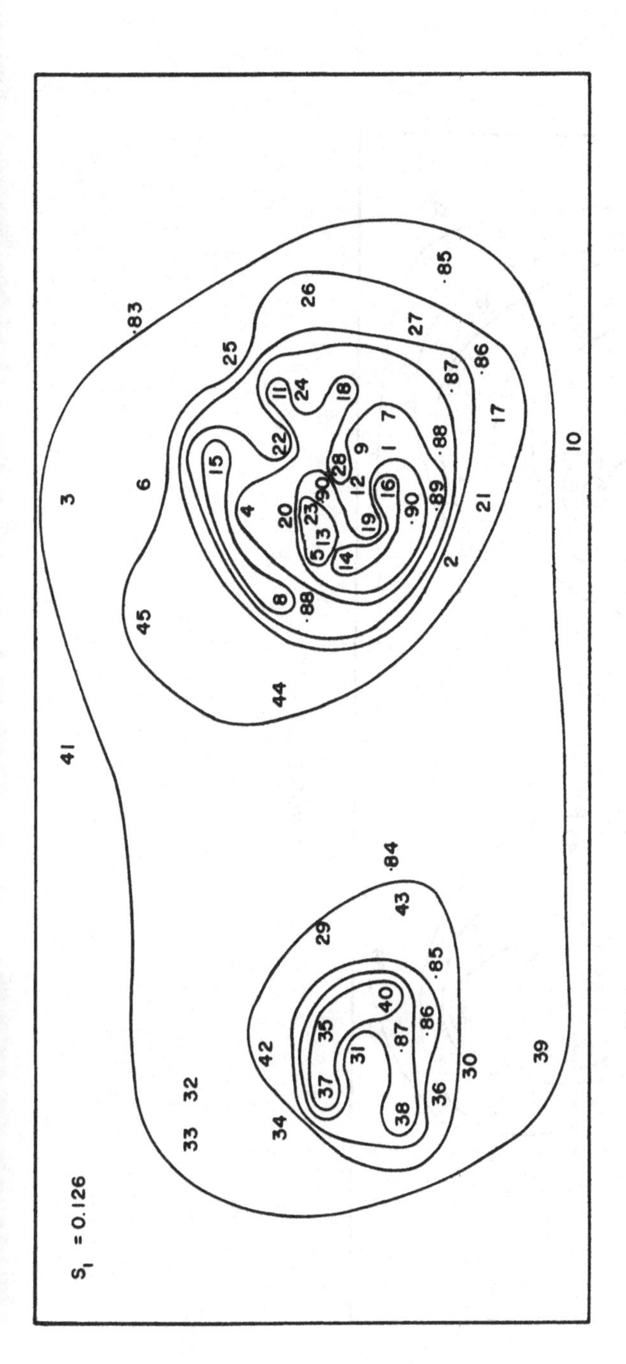
S1 = 0.126
3
6
41
.83
25
45
44
15
4
22
11
24
18
26
27
.87
.86
.85
7
9
1
.88
17
10
21
.89
.90
16
12
28
.90
23
20
13
5
14
19
2
8
.88
.84
43
29
.85
42
40
35
.86
.87
31
37
30
39
36
38
34
33
32

FIGURE 5.7.3

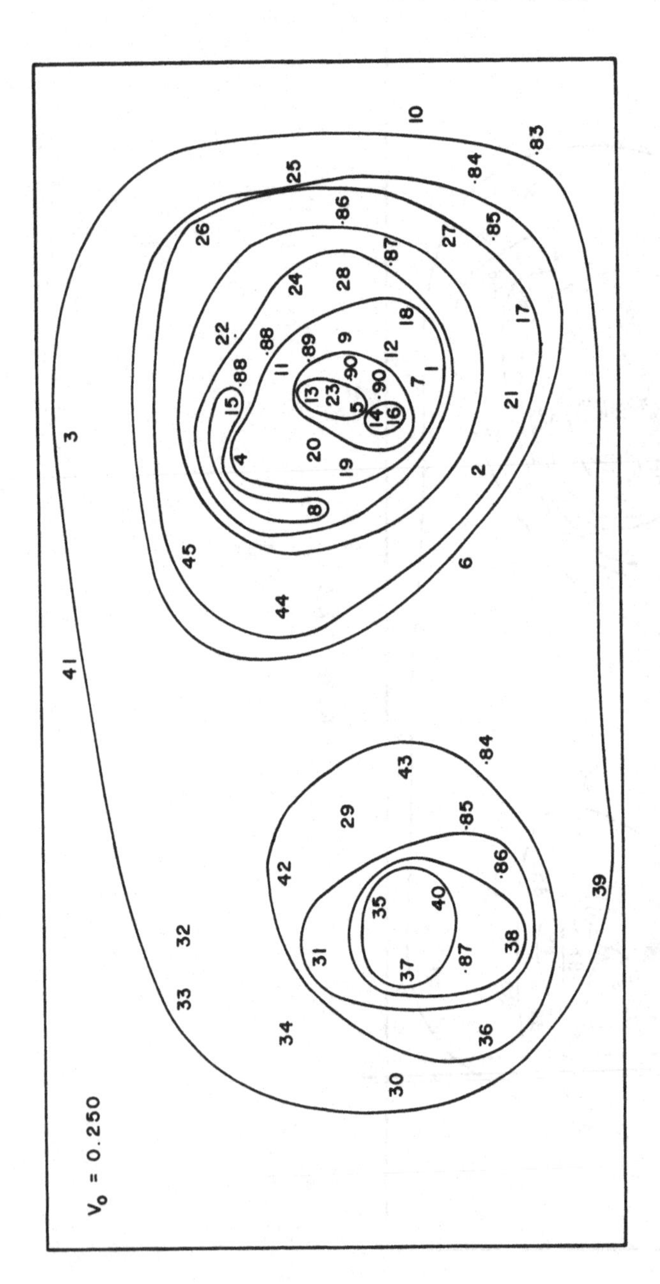

Vo = 0.250
3
41
10
.83
.84
.85
.86
.87
.88
.89
.90
25
26
27
28
24
22
18
17
9
12
11
1
7
13
23
5
14
16
15
4
20
19
8
21
2
6
45
44
32
33
34
30
42
29
43
.84
.85
.86
.87
35
40
31
37
38
36
39

FIGURE 5.7.4

6. MISCELLANEOUS SCALING TOPICS

6.1 Interpretation of scaling output

Interpoint distance interpretation

Intuitively we tend to interpret the solution display of any scaling method in the same way as a geographical map, that is in terms of interpoint distances in the display. However it should be remembered that for all the basic structure scaling methods as well as for classical scaling the within-set interpoint distances have not been directly fitted. In some basic structure displays, for example the symmetric biplot of §2.3.2, the within-set distance interpretation is invalid. In all basic structure displays where row and column points are displayed simultaneously, the between-set interpoint distance interpretation is invalid. This latter interpretation is only valid in multidimensional unfolding (§6.4).

Representation of distortion

In all basic structure displays the axes of the display are independent in the sense that, for a particular problem, the first two axes, say, of a three-dimensional display are exactly the same as the two axes of the two-dimensional display. Poorly represented points in two dimensions can thus be identified by examining the co-ordinates of all the points in third and higher dimensions; in other words we examine the vector residuals.

For classical scaling we would look at the distance residuals $\delta_{ij} - d_{ij}$, where the d_{ij} are the displayed interpoint distances. Because these residuals are always nonnegative ($d_{ij}^2 \leq \delta_{ij}^2$) and because interpretation is usually in terms of relative distances, it might be more meaningful to look at the residuals $\delta_{ij} - \hat{\alpha} d_{ij}$, for example, where $\hat{\alpha}$ is the regression coefficient when δ_{ij} is regressed on d_{ij} through the origin, i.e. the display is scaled up to fit the δ_{ij} (cf. Greenacre, 1978, §4.2.2).

For least squares and nonmetric scaling this implies a consideration of the individual residuals which form the stress function in use.

Distortion in the solution is indicated by large residuals. For Kruskal-Shepard scaling a positive value of $d_{ij} - \hat{d}_{ij}$ indicates that points i and j ought to be closer together, while a negative value implies they should be further apart. It is convenient to consider those residuals which exceed some threshold value t. If the configuration has been normalized so that $\|X\| = 1$, a value of $t = 0.7$ has proved useful in highlighting the gross distortions. The interpoint distances whose residuals exceed the threshold can be indicated on the configuration plot as follows:

i ——→——←—— j $d_{ij} - \hat{d}_{ij} > t$ (the points i and j ought to be closer together)

i ——←——→—— j $d_{ij} - \hat{d}_{ij} < -t$ (the points i and j ought to be further apart)

It is a common experience when solutions from multiple random starts are compared, that the majority of the points are stable from solution to solution whilst a few points differ radically. These are the points which the scaling algorithm has had the most difficulty in placing, and for which the fit is poor. Generally these points make the largest relative contribution to the overall stress, and may be regarded as "outliers" in the sense that they do not fit into the required dimensionality. Alternatively, one or more of the dis= similarities associated with an ill-fitting point may be in error, and this needs to be checked.

For Kruskal-Shepard scaling, the percentage contribution of point i to the overall stress is given by

$$S^2(i) = \frac{\sum_{i=1}^{n} (d_{ij} - \hat{d}_{ij})^2}{2\, S^{*2}} \times 100\% \qquad (6.1.1)$$

where S^* is the raw stress (5.2.4). The factor two is in the denomi= nator since each residual is added to the stress contributions of two points.

Scaling and cluster analysis

The dissimilarity matrix Δ is frequently subjected also to clus= ter analysis (Hartigan, 1975, Everitt, 1974 and Chapter 6 of this volume). A useful technique (Shepard, 1974) is to place contour loops around points that have been grouped together by the cluster analysis. A discussion of the relationship between cluster analysis and scaling is given by Kruskal (1977b).

It is sometimes recommended (e.g. Everitt, 1978) that the mini= mum spanning tree (Gower and Ross, 1969) be plotted onto a configu= ration. Finding the minimum spanning tree is equivalent to doing a single linkage cluster analysis. Plotting these cluster loops onto the configuration conveys a lot more information about the distortion at little extra expense.

It is generally easier to plot cluster loops onto a configura= tion that has been found using U_0^2 and V_o^2 stress ((5.3.2),(5.3.5)). Clusters obtained by single linkage cluster analysis of the American Indian groups data (Table 5.7.1) were plotted as cluster loops on the solutions obtained by minimizing Kruskal-Shepard stress (Figure 5.7.1) and by minimizing V_o^2 (Figure 5.7.2.) Similarly, the single linkage clusters obtained from the simi= larities between the works of Plato are depicted in Figures 5.7.3 and 5.7.4.

The complex shapes of the cluster loops for the small clusters bear out the warnings of Kruskal (1977b) and Graef and Spence (1979), quoted in §5.3, that significance should not be attached to points that are close together in Kruskal-Shepard scaling. On the other hand, the relatively circular or elliptic cluster loops of Figures 5.7.2 and 5.7.4 bear out the value of the alternative stress measures in overcoming the limitations of Kruskal-Shepard scaling. The visual clusters, even the small ones, correspond closely with those found by cluster analysis.

6.2 Simulation studies

In the vast majority of simulation studies in scaling the pro= cedure has been to generate a "true" configuration (usually a fixed number of points within the unit cube or sphere) and then to intro= duce errors into the "observed" dissimilarities by perturbing either the co-ordinates or the distances. Typical examples of these methods are those used by Graef and Spence (1979):

$$\delta_{ij} = \sum_{k=1}^{p} (x_{ik} + \varepsilon_i - x_{jk} - \varepsilon_j)^2 \qquad \varepsilon_i \sim N(0;\sigma^2) \qquad (6.2.1)$$

(a different "error" is added each time each co-ordinate is used) and

$$\delta_{ij} = d_{ij} \times Z \qquad\qquad Z \sim N(1;\sigma^2). \qquad (6.2.2)$$

Ramsay (1969) advocated (6.2.1), Wagenaar and Padmos (1971) origina= ted (6.2.2). Using (6.2.1) the errors introduced into the δ_{ij} are roughly proportional to the square root of the d_{ij}, with (6.2.2) the errors are proportional to the magnitude of d_{ij}.

Both error models are unrealistic in that they assume that the errors are independent. Sibson *et al.* (to appear) have devised seve= ral methods which introduce dependent errors. Since this is the way errors arise in practice it is strongly recommended that future simu= lation studies should use methods based on those of Sibson *et al.* (to appear).

A development of these methods, which has been used in ecolo= gical studies, generates an $n \times m$ profile matrix A with underlying two dimensional structure X, in the following way:

Step 1: Generate m points $y_k = (y_{1k};y_{2k})$ at random in the unit square to act as "variable centres".

Step 2: For each variable centre, generate ℓ_k points $y_{ki} = (y_{1k} + d_i \cos \theta_i, y_{2k} + d_i \sin \theta_i)$ $i = 1 \ldots \ell_k$, where the distance d_k has (say) exponential distribution with parameter λ_k, and θ_k is uniformly distributed on $(-\pi;\pi)$. These points form overlapping clusters around their variable centres.

Step 3: Generate a configuration of n points $x_j = (x_{1j};x_{2j})$ to form the "true" configuration X. $(k = 1,\ldots,m)$.

Step 4: Let a_{jk} be the number of points y_{ki} that lie within a distance r_j of point x_j; r_j having (say) an exponential distribution with parameter $\rho_j (j = 1, \ldots, m)$.

The matrix A generated in this way may be interpreted as an ecological abundance matrix. Each species k prefers a certain habi= tat type centered at y_k, is common close to y_k and rare far from it. An observer at x_j counts all individuals of species k within a radius r_j. Varying the parameters ℓ_k, λ_k and ρ_j, and the underlying distri= butions used, enables realistic looking abundance data to be simula= ted.

Any dissimilarity coefficient may be used to convert A into a dissimilarity matrix $\Delta = (\delta_{ij})$, to which scaling methods (§§3-5) are applied to recover X. A simulation study conducted along these lines would be able to test the appropriateness of both the dissimi= larities coefficient and the scaling method. Application of corres= pondence analysis (§2.3.6) directly to the profile matrix A should recover both the X and Y.

6.3 Comparison of solutions

Cophenetic correlation

If we apply two scaling techniques to a dissimilarity matrix, and obtain two solutions X and Y it is natural to ask how similar is X to Y. Or, alternatively, if we have done a simulation study, we have a "true" configuration X and a recovered configuration Y (or several recovered configurations, obtained by different methods) and we want to ask how close are the true and recovered configuration (or which recovered configuration is closest to X). The problem arises because scaling methods recover the interpoint relationships only, and produce solution configurations which are arbitrary with respect to rotation, reflection, translation and, in the case of nonmetric scaling, a scale factor.

A longstanding approach has simply been to compute the so-called *cophenetic correlation coefficient* between the sets of distances $d_{ij}(X)$ and $d_{ij}(Y)$ (Sokal and Rohlf, 1962, Kruskal and Carroll, 1969). The principal objection to this procedure is that for n points, the $\frac{1}{2}n(n-1)$ interpoint distances are not independent, and a specious cor= relation may occur (Gower, 1971). Sibson *et al.* (to appear) bluntly state that the cophenetic correlation coefficient "fails to relate properly to either the geometrical or the probabilistic aspects of the problem".

Procrustes statistics

The alternative approach, developed chiefly by Gower (1971) and reviewed by Sibson (1978) is the direct one - to translate, rotate, reflect and dilate the configuration Y so that it matches most closely to X.

The measure used to evaluate the distance between X and Y is

$$G(X,Y) = \mathrm{tr}((X-Y)(X-Y)^T) = \sum_{i=1}^{n} \sum_{j=1}^{p} (x_{ij}-y_{ij})^2 \qquad (6.3.1)$$

the sums of squared Euclidean distances between corresponding points (given by the row of X and Y). The most general form of the Procrus= tes statistic is $G_S(X,Y)$, defined to be

$$G_S(X,Y) = \inf(F(X,Y\Sigma),\Sigma \in S) \qquad (6.3.2)$$

where S is the set of $p \times p$ matrices that generate rotations, re= flections and dilations.

It is easy to see that the best translational fit occurs when X and Y have the same centroid:

$$G(X,Y) = G(X_0,Y_0) + G(\bar{X},\bar{Y}) \qquad (6.3.3)$$

where $\bar{X} = \frac{1}{n}\, 1\, 1^T X \qquad \bar{Y} = \frac{1}{n}\, 1\, 1^T\, Y$

and $X_0 = X - \bar{X} \qquad Y_0 = Y - \bar{Y}$.

It is convenient, as in (6.3.3), to have the centroid at the origin, for this is then unaltered by subsequent rotation, reflection and dilation.

After translating X and Y to centroid-at-origin, the best ro= tation/reflection fit is found by

$$G_H(X,Y) = \min\{G(X,YH),\ HH^T = I\}. \tag{6.3.4}$$

To show that the minimum is reached and to find the associated orthogonal matrix H, consider

$$G(X,YH) = tr(XX^T) + tr(YY^T) - 2\ tr(YHX^T) \tag{6.3.5}$$

Let the basic structure X^TY be $U\ D_\lambda V^T$. Then

$$\begin{aligned} tr(YHX^T) &= tr\ (HX^TY) \\ &= tr\ (V^THUD_\lambda) \le tr\ (D_\lambda). \end{aligned} \tag{6.3.6}$$

because V^THU is orthogonal and thus none of its entries exceed 1. Equality in (6.3.6) and the minimum in (6.3.4) is achieved only when

$$H = VU^T. \tag{6.3.7}$$

Finally to match under dilation, we note that G(X,kYH) is a quad= ratic in k with minimum

$$k = tr(YHX^T)/tr(YY^T). \tag{6.3.8}$$

so that substituting (6.3.7) and (6.3.4) into G(X,kYH) we have

$$G_S(X,Y) = tr(XX^T) - \frac{\{tr(YVU^TX^T)\}^2}{tr(YY^T)}. \tag{6.3.9}$$

Sibson (1978) recommends division by $tr(XX^T)$, a standardization which removes both the effect of the number of points and of the scale of X. Thus the solution for the problem, how close are X and Y, is given by

$$\gamma_S = 1 - \frac{\{tr(Y\ V\ U^TX^T)\}^2}{tr(XX^T)\ tr(YY^T)} \tag{6.3.10}$$

It is recommended that this statistic be used, and that the cophenetic correlation coefficient be abandoned.

6.4 Multidimensional unfolding

The term "unfolding" was originally introduced by Coombs (1950) in the context of graphical representation of preference data in the form of ranks. Here the objects are considered a set of m *stimuli* and the subjects a set of n *individuals* who have ranked the stimuli in different orders of preference. Coombs' idea was that the stimuli and the individuals might be represented on the same *scale* (a straight line) where each individual's preference order could be obtained by *folding* the scale at that individual's point. For example, consider the hypothetical case of four presidential candidates A, B, C and D (m=4) which each of n voters (i=1,...,n) has ranked in order of pre= ference. *A posteriori* it might be found that the following scaling of the candidates is appropriate:

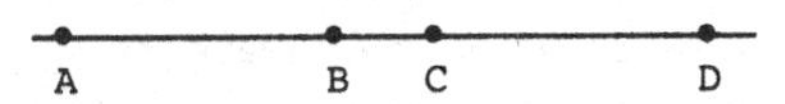

By representing a voter i=1 in the following position:

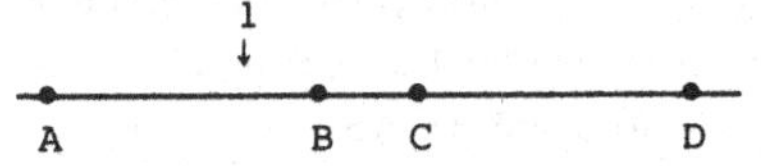

we would indicate graphically that voter no.1 has a preference order= ing of the candidates B C A D (B first, C second, etc.).
This order can be obtained by folding the scale at point 1:

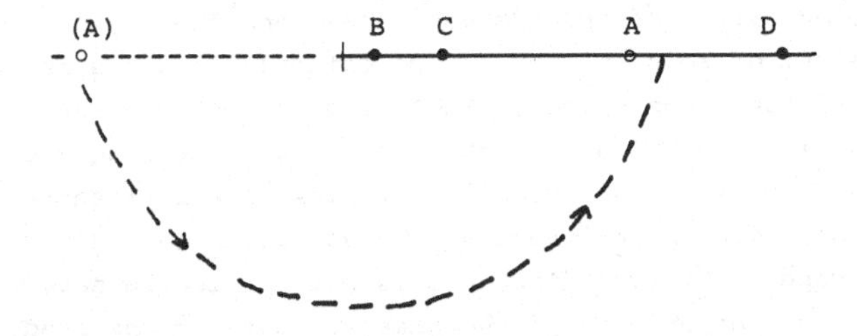

Similarly voters 2 and 3 with preference orders C B D A and C B A D respectively could be represented by points in the following positions on the scale:

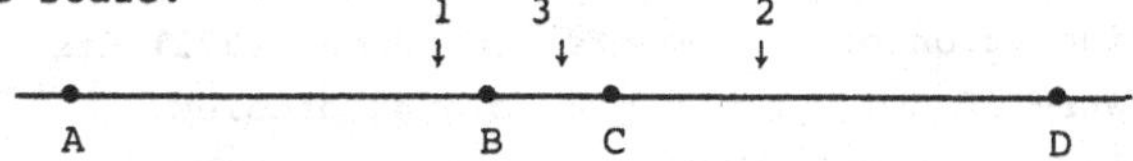

Two remarks about this type of graphical display are immediately apparent. Firstly it is clear that the actual representation provides more information than the original data. The data here are simple orderings (cf. nonmetric scaling, §5) while the representation neces= sarily displays specific *metric* information, although it *should* be interpreted only as far as its properties of *ordonnance* (or *ordered metric*) are concerned. Secondly it can be seen in the particular configuration of the candidates above that no point exists for a vo= ter who ranks C first and A second. In fact any configuration of the 4 candidates will exclude a certain number of possible orderings which can then only be displayed with a certain "error", just as in nonmetric scaling. Clearly there are regions on the scale within which identical orderings are generated, and these regions are bounded by the midpoints of all pairs of candidates on the scale.

Coombs' proposed model can be thought of as effectively *unfold= ing* the original preferences of each individual in a way as to obtain a scaling of the stimuli on which the individuals are displayed at the points of unfolding. In the light of what we have said above it is conceivable that there is a unique configuration which minimizes some form of "error", where the error may be defined by one of the stress functions, for example, described in §5.

The unfolding analysis applies not only to preference data but also to any data which reflects dissimilarities (or, inversely, simi= larities) between the elements of two different sets. The data thus consist of a rectangular matrix of "pseudo-distances" and the analysis attempts to scale the set of rows and the set of columns as points in a joint low-dimensional Euclidean space so as to fit the pseudo-dis= tances. The question arises, as it did in the case of scaling a square symmetric matrix (§3.4), whether the "pseudo-distances" are represent= able exactly in some high-dimensional Euclidean space. Clearly the triangle inequality property of true distances is not applicable here because the distances between objects of the same set are not observed. Schönemann and Wang (1972) discuss this question in the context of metric unfolding. Bennett and Hays (1960) and Hays and Bennett (1961) have generalised the nonmetric problem, as described by Coombs in the one-dimensional case, to a multidimensional space of representation. These papers respectively describe the determination of the dimension of the space and a configuration of the points. Davidson (1972) dis= cusses further the geometry of the nonmetric unfolding problem.

Other articles relevant to unfolding are those by Coombs and Kao (1960), Ross and Cliff (1964), Schönemann (1970), Wang, Schönemann & Rusk (1975) and Ramsay (1980). Ross and Cliff (1964) showed the pro=

blem to be more tractable mathematically when squared distances are treated. This is a direct consequence of the simplification of the Euclidean distance when it is squared. Greenacre (1978) follows Browne's (1977) metric scaling approach and gives the algorithm which finds the least-squares fit to squared "distances", and demonstrates the danger of multiple solutions. The problem of local minima is more acute in unfolding, both metric and nonmetric, than in scaling a single set of objects based on their inter-object distances. This is due to our observing only the between-set distances and none of the within-set distances, as if only a rectangular $n \times m$ submatrix of a larger $(n + m) \times (n + m)$ square symmetric matrix is being taken into account when scaling the $n + m$ points.

An unfolding analysis produces a joint display of the rows and columns of the data matrix, as do many of the basic structure methods described in §2.3. However it should be remembered that in unfolding we should only interpret distances between points of different sets, unless the representation of the between-set distances is of high quality, in which case the within-set displayed distances are reliable measures of inter-object dissimilarities in the context of the data.

REFERENCES

ANDERSON, A.J.B. (1971) Ordination methods in ecology. *Journal of Ecology* 59,713-726.

ANDERSON, T.W. (1963) Asymptotic theory for principal component ana= lysis. *Ann. Math. Statist.* 34, 122-148.

ARABIE, P. (1973) Concerning Monte Carlo evaluations of nonmetric multidimensional scaling algorithms. *Psychometrika*, 10, 607-608.

ARABIE, P. (1978) Random versus rational strategies for initial con= figurations in nonmetric multidimensional scaling. *Psychometrika*, 43, 111-113.

BARLOW, R.E., BARTHOLOMEW, D.J., BREMER, J.M. and BRUNK, H.D. (1972) *Statistical Inference under Order restrictions*. New York, Wiley.

BENNETT, J.F. and HAYS, W.L. (1960) Multidimensional unfolding determining the dimensionality of ranked preference data. *Psychometrika*, 25, 27-43.

BENZÉCRI, J.P. and collaborateurs (1973) L'Analyse des Donneés (Tome 2) *L'Analyse des correspondances*. Dunod, Paris.

BLOXOM, B. (1978) Constrained multidimensional scaling in N spaces. *Psychometrika*, 43, 397-408.

BONEVA, L.I. (1971) A new approach to a problem of chronological seriation associated with the works of Plato. *Mathematics in the archeological and historical sciences*, (Hodson, C.R., Kendall, D.G. and Tâuti, L, eds.), 173-185. Edinburgh Univer= sity Press.

BRADU,D.and GABRIEL, K.R. (1978): The biplot as a diagnostic tool for models of two-way tables. *Technometrics*, 20, 47-68.

BRADU,D.and GRINE, F.E. (1979): Multivariate analysis of Diademodon= tine Crania from South Africa and Zambia. *South African Journal of Science*, 75, 441-448.

BRAY,J.R.and CURTIS, J.J. (1957): Ordinations of the upland forest communities of southern Wisconsin. *Ecological Monographs*. 27, 325-349.

BROWNE, M.W. (1977): Graphical display of similarity data. Paper at South African Statistical Association Conference, University of Port Elizabeth.

BROWNE, M.W. (1979): Graphical display of similarity matrices. Talk at Medical Research Council, Tygerberg.

COOMBS, C.H. (1950): Psychological scaling without a unit of measure= ment. *Psychological Review*. 57, 148-158.

COOMBS,C.H.and KAO, R.C. (1960): On a connection between factor ana= lysis and multidimensional unfolding, *Psychometrika*, 25, 219-231.

COOPER, L.G. (1972): A new solution to the additive constant problem in metric multidimensional scaling, *Psychometrika*, 37, 311-322.

CRIPPEN, G.M. (1977): A novel approach to calculation of conforma= tion: distance geometry. *Journal of Computational Physics*, 24, 96-107.

DAVID. M., CAMPIGLIO, C. and DARLING, R. (1974) Progress in R- and Q-mode analysis: correspondence analysis and its application to the study of geological processes. *Can. J. Earth Sci.*, 11, 131-146.

DAVIDSON, J.A. (1972) A geometrical analysis of the unfolding model: nondegenerate solutions. *Psychometrika*, 37, 193-211.

DE LEEUW, J. and HEISER, W. (1980) Multidimensional scaling with restrictions on the configuration. In: *Multivariate Analysis V* (P.R.Krishnaiah, ed.) Amsterdam, North Holland Publishing Company.

ECKART, C. and YOUNG, G. (1936) The Approximation of one matrix by another of lower rank. *Psychometrika*, 1, 211-218.

EVERITT, B.S. (1974): *Cluster Analysis*. London : Heinemann.

EVERITT, B.S. (1978) *Graphical Techniques for Multivariate Data*, London, Heinemann.

FALKENHAGEN, E.R. and NASH, St.W. (1978). Multivariate classification in provenance research. *Silvae Genetica*, 27, 14-23.

FLETCHER, R. and REEVES, C.M. (1964). Function minimization by conju= gate gradients. *Computer Journal*, 7, 149-153.

GABRIEL, K.R. (1971). The biplot graphical display of matrices with application to principal component analysis. *Biometrika*,58,458-467.

GABRIEL, K.R. (1972). Analysis of meteorological data by means of cano= nical decomposition and biplots. *J.Appl.Meteor.*,11, 1071-1077.

GABRIEL, K.R. (1978). Least squares approximation of matrices by addi= tive and multiplicative models. *J. Roy. Statist. Soc. B*, 40, 186-196.

GABRIEL, K.R. and ZAMIR, S. (1976). Lower rank approximation of ma= trices by least squares with any choice of weights. *Mimeograph, University of Rochester, Rochester, N.Y.*

GOODALL, D.W. (1973). Sample similarity and species correlation. In: *Handbook of Vegation Sciences, Part V, Ordination and Classifi= cation of Communities* (R.H. Whittaker, ed.) The Hague, Dr. W. Junk.

GOWER, J.C. (1966). Some distance properties of latent root and vec= tor methods used in multivariate analysis. *Biometrika*, 53, 325-338.

GOWER, J.C. (to appear). Comparing multidimensional scaling configu= rations.

GOWER, J.C. (1971) Statistical methods of comparing different multi= variate analyses of the same data. In: *Mathematics in the archeological and historical sciences* (Hodson, C.R., Kendall, D.G. and Tăutu, P. eds.) 138-149 (Edinburgh, University Press.

GOWER, J.C. and ROSS, G.J.S. (1969) Minimum spanning tree and single linkage cluster analysis. *Applied Statistics*, 18, 54-64.

GRAEF, J. and SPENCE, I. (1979) Using distance information in the design of large multidimensional scaling experiments. *Psycho= logical Bulletin*, 86, 60-66.

GREENACRE, M.J. (1974) L'analyse des correspondances des donneés sur des crânes d'ours de cavernes. Rapport de stage (D.E.A.), Université Pierre et Marie Curie, Paris.

GREENACRE, M.J. (1978) *Some Objective Methods of Graphical Display of a Data Matrix* Doctoral thesis, Université Pierre et Marie Curie, Paris, published as a special report by the University of South Africa, Pretoria.

GREENACRE, M.J. (1981) Practical correspondence analysis. In: *Interpreting Multivariate Data* (Ed. Vic Barnett), Wiley, Chi= chester, to appear.

HARTIGAN, J.A. (1975) *Clustering Algorithms* New York, Wiley.

HARWELL Subroutine Library Manual (1978) Computer Science and Systems Division, A.E.R. Establishment, Harwell, Oxfordshire, England.

HAYS, W.L. and BENNETT, J.F. (1961) Multidimensional unfolding : determining configuration from complete rank order preference data. *Psychometrika*, 26, 221-238.

HILL, M.O. (1974) Correspondence analysis : a neglected multivariate technique. *Appl. Statist.* 23, 340-354.

HILLS, M. (1969) On looking at large correlation matrices. *Biometrika*, 56, 149-153.

HOUSEHOLDER, A.S. and YOUNG, G. (1938) Matrix approximations and la= tent roots. *Am. Math. Monthly*, 45, 165-171.

JOHNSON, N.L. (1949) Systems of frequency curves generated by methods of translation. *Biometrika*, 36, 149-176.

KENDALL, D.G. (1971) Seriation from abundance matrices. In: *Mathematics in the archeological and historical sciences*, (Hodson, C.R., Kendall, D.G. and Tăutu, P. eds.), 215-252. Edinburgh, Univer= sity Press.

KENDALL, D.G. (1977) On the tertiary treatment of ties. Appendix to Rivett, B.H.P. (1977). Policy election by structural mapping. *Proc. R. Soc. Lond. A.*, 354, 407-423.

KRUSKAL, J.B. (1964a) Multidimensional scaling by optimizing goodness-of-fit to a non-metric hypothesis. *Psychometrika*, 29, 1-27.

KRUSKAL, J.B. (1964b) Nonmetric multidimensional scaling : a nume= rical method. *Psychometrika*, 29, 115-129.

KRUSKAL, J.B. (1977a). Multidimensional scaling and other methods for discovering structure. In: *Statistical methods for digital computers*. (A.J. Ralston, H.S. Wilf and K. Enslein, eds.). 296-339. New York, Wiley.

KRUSKAL, J.B. (1977b). The relationship between multidimensional scaling and clustering. In: *Classification and Clustering*. (J. van Ryzin, ed.) 17-44, New York, Academic Press.

KRUSKAL, J.B. and CARROLL, J.D. (1969). Geometric models and badness-of-fit functions. In: *Multivariate Analysis II* (P.R. Krishnaiah, ed.) 637-671. New York, Academic Press.

KRZANOWSKI, W.J. (1979). Some exact percentage points of a statistic useful in analysis of variance and principal component analysis. *Technometrics*, 21, 261-263.

LEBART, L., MORINEAU, A. and TABARD, N. (1977) *Méthodes de Description Statistique*, Dunod, Paris.

LINGOES, J.C. and ROSKAM, E.E. (1973) A mathematical and empirical study of two multidimensional scaling algorithms. *Psychometrika Monograph Supplement*, 38:4, part 2.

McGEE,V.E. (1966). The multidimensional scaling of "elastic" dis= tances. *The British Journal of Mathematical and Statistical Psychology*, 19, 181-196.

MESSICK,S.J.and ABELSON, R.P. (1956): The additive constant problem in multidimensional scaling, *Psychometrika*, 21, 1-15.

MIMMACK, G.M. (1979). An investigation into aspects of nonmetric multidimensional scaling. Unpublished masters thesis, Univer= sity of Cape Town.

MORGAN, B.J.T. and NORTH, P.M. (to appear). On using cluster-analysis and multidimensional scaling for describing animal movements and bird territories.

MORRISON, D. (1967). *Multivariate Statistical Methods*, McGraw-Hill, New York.

MUTOMBO, F.K. (1973). *Traitement des Données Manquantes et Rationali= sation d'un Réseau de Stations de Mesures*, Doctoral Thesis, Université Pierre et Marie Curie, Paris.

NAG Library Manual (Mark 6) 1977. Oxford.

NORA-CHOUTEAU, (1974). Une Méthode de Reconstitution et d'Analyse de Données Incomplètes. Doctoral thesis, Université Pierre et Marie Curie, Paris.

PEARSON, K. (1901). On lines and planes of closest fit to a system of points. *Philosophical Magazine* Ser. 6 2, 559-572.

RAMSAY, J.O. (1969). Some statistical considerations in multidimen= sional scaling. *Psychometrika*, 34, 167-182.

RAMSAY, J.O. (1980). The joint analysis of direct ratings, pairwise preferences, and dissimilarities. *Psychometrika*, 45, 149-165.

RAO, C.R. (1980) Matrix approximation and reduction of dimensionality in multivariate statistical analysis. In: *Multivariate Analysis V* (P.R. Krishnaiah, ed.) Amsterdam, North-Holland.

RIVETT, B.H.P. (1977) Policy selection by structural mapping. *Proc. R. Soc. London. A.*, 354, 407-423.

ROHLF, F.J. and ARCHIE, J.W. (1978) Least-squares mapping using interpoint distances. *Ecology*, 59, 126-132.

ROSS, J. and CLIFF, N. (1964) A generalization of the interpoint distance model. *Psychometrika*, 29, 167-176.

SAITO, T. (1978) The problem of additive constant and eigenvalues in metric multidimensional scaling. *Psychometrika*, 43, 193-201.

SAMMON, J.W. (1969) A nonlinear mapping for data structure analysis. *IEEE Transactions on Computers*, 18, 401-409.

SCHÖNEMANN, P.H. (1970) On metric multidimensional unfolding, *Psychometrika*, 35, 349-366.

SCHÖNEMANN, P.H. and WANG, M.M. (1972). An individual difference model for the multidimensional analysis of preference data. *Psychometrika*, 37, 275-309.

SHEPARD, R.N. (1962a) The analysis of proximities: multidimensional scaling with an unknown distance function. I. *Psychometrika*, 27, 125-140.

SHEPARD, R.N. (1962b) The analysis of proximities: multidimensional scaling with an unknown distance function II. *Psychometrika*, 27, 219-246.

SHEPARD, R.N. (1974) Representation of structure in similarity data: problems and prospects, *Psychometrika*, 39, 373-421.

SHEPARD, R.N. (1980) Multidimensional scaling, tree-fitting and clustering. *Science*, 210, 390-398.

SHEPARD, R.N. and CARROLL, J.D. (1966) Parametric representation of nonlinear data structures. In: *Multivariate Analysis Proceed= ings of an International Symposium* (P.R. Krishnaiah, ed.) 561-592. New York, Academic Press.

SIBSON, R. (1972) Order invariant methods for data analysis. *Journal of the Royal Statistical Society, Series B (Methodological)* 34, 311-349.

SIBSON, R. (1973): Local order multidimensional scaling. *Bulletin of the International Statistical Institute*, 45, 207-218.

SIBSON, R. (1978) Studies in the robustness of multidimensional scaling : procrustes statistics. *Journal of the Royal Statistical Society, Series B (Methodological)* 40, 234-238.

SIBSON, R. (1979): Studies in the robustness of multidimensional scaling : Perturbational analysis of classical scaling. *Journal of the Royal Statistical Society, Series B (Methodological)*, 41, 217-229.

SIBSON, R. and BOWYER, A. (to appear) Trilateration and scaling methods in surveying and photogrammetry.

SIBSON, R.,BOWYER, R. and OSMOND, C. (to appear) Studies in the robustness of multidimensional scaling : Euclidean models and simulation studies.

SOKAL, R.R. and ROHLF, F.J. (1962) The comparison of dendrograms by objective methods. *Taxon*, 11, 33-40.

SPAETH, H.J. and GUTHERY, S.B. (1969) The use and utility of the monotone criterion in multidimensional scaling. *Multivariate Behavioral Research*, 4, 501-515.

SPENCE, I. (1972) A Monte Carlo evaluation of three nonmetric multidimensional scaling algorithms. *Psychometrika*, 37, 461-486.

SPENCE, I. and DOMONEY, D.W. (1974) Single subject incomplete designs for nonmetric multidimensional scaling. *Psychometrika*, 39, 469-490.

SPUHLER, J.N. (1972) Genetic, linguistic and geographical distances in native North America. In: *The assessment of population affinities in man.* Oxford, Clarendon Press.

TAKANE, Y., YOUNG, F.W. and DE LEEUW, J. (1977) Nonmetric individual differences multidimensional scaling : an alternating least squares method with optimal scaling features. *Psychometrika*, 42, 7-67.

TEIL, H. and CHEMINÉE, J.L. (1975) Application of correspondence factor analysis to the study of major and trace elements in the Erta Ale Chain (Afar, Ethiopia). *Math. Geol.*, 7, 13-30.

TORGERSON, W.S. (1958) *Theory and methods of scaling*. New York, Wiley.

TUKEY, J.W. (1977) *Exploratory Data Analysis* Reading, Massachusetts, Addison-Wesley.

WAGENAAR, W.A. and PADMOS, P. (1971) Quantitative interpretation of stress in Kruskal's multidimensional scaling technique. *British Journal of Mathematical and Statistical Psychology*, 24, 101-110.

WANG, M.M., SCHÖNEMANN, P.H. and RUSK, J.G. (1975) A conjugate gra= dient algorithm for the multidimensional analysis of preference data. *J.Multivar. Behav. Res.*, (1975) 45-79.

WHITLEY, V.M. (1971) Problems in multidimensional scaling. Unpublished doctoral dissertation. University of California, Irvine.

YOUNG, G. and HOUSEHOLDER, A.S. (1938) Discussion of a set of points in terms of their mutual distances. *Psychometrika*, 3, 19-22.

AUTOMATIC INTERACTION DETECTION

DOUGLAS M. HAWKINS, *COUNCIL FOR SCIENTIFIC AND INDUSTRIAL RESEARCH*
AND
GORDON V. KASS, *UNIVERSITY OF THE WITWATERSRAND*

1. INTRODUCTION

Automatic Interaction Detection (AID) is a family of methods for handling regression-type data in a way that is almost free of the usual assumptions necessary to process the data using linear hypothesis me= thods.

In AID, one has a dependent variable Y which one wishes to pre= dict, and a vector of predictors X from which to predict Y. The predic= tors are all categorical (i.e. either nominal or ordinal), and generally take on only a few possible values. Interval predictors may be reduced to this form by grouping their possible values into classes, and then using the (ordinal) class variable as the predictor.

Various different methods within the AID family have been devised for situations in which the dependent variable Y is: (a) a scalar inter= val variable, (b) a scalar nominal variable, (c) a vector of interval variables. Other possibilities such as an ordinal Y or a vector of no= minal Y are easy to fit into the general conceptual framework of AID.

The name AID suggests that the function of the technique is to discover whether the linear hypothesis model predicting Y from X con= tains only main effects, or whether interactions also occur. This is indeed one of the things that AID can do, but it has a number of other uses as well, which we consider overshadow this use in importance.

Before going into a detailed study of the aims and methods of AID, it may help to consider a simple example. Suppose one had a set of data with a factor A at two levels and factors B and C each at three levels. Suppose that a very large sample gave the following cell means (assumed to be effectively without error)

		C	
	1	2	3
A=1 B=1	10	11	10
2	10	12	10
3	10	12	10
A=2 B=1	9	11	9
2	9	10	9
3	9	10	9

The conventional method of analysis of these data would be by a three-way analysis of variance. As the three-way interaction exists, the predictor of all cell means would require the values of 48 parame= ters (of which 18 are essential, and the further 30 linearly dependent on them but advisable for simplicity). The linear model would also be very difficult to interpret, as is commonly the case when interactions are present.

Analysis using AID might produce the following *dendrogram*

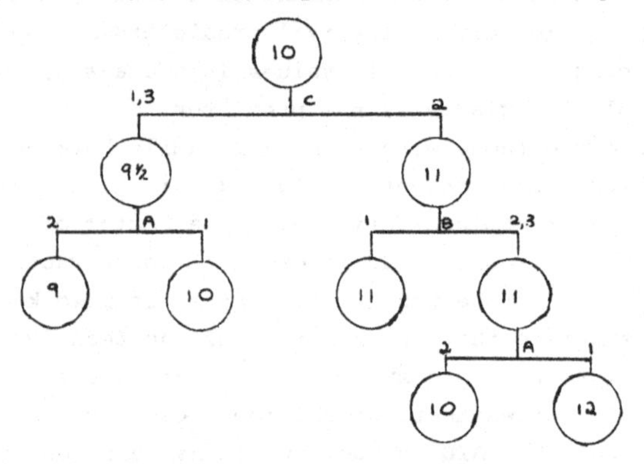

This dendrogram serves a number of purposes. First, its unbalan= ced structure shows that, in linear hypothesis terms, there is an inter= action between A, B and C, so that if one wishes to set up a regression-type model, interactions must be included.

Second, it gives a parsimonious summary of the data so that one could well scrap the original 18 element table and replace it with the 5 terminal nodes of the dendrogram.

Third, the dendrogram may be used to predict the Y corresponding to a future unknown with given X. Furthermore, this prediction involves no arithmetic whatever - merely a tracing of the unknown down the bran= ches of the dendrogram until one reaches a terminal node.

The first and second of these uses correspond to AID as a preli=

minary to a regression or analysis of variance; the third uses the AID as the complete analysis. We have found all three of these uses to be very valuable, though the third (surprisingly) often meets with resistance from statisticians, who prefer an explicit mathematical formula, even one that is not easy to calculate, to the more graphic and easily in= terpretable dendrogram. This latter use however is familiar in compu= ter science as constituting a decision tree, from which it differs in having a stochastic element.

2. MODE OF OPERATION

Applied to a set of data, AID works through a number of stages, each having the following form.

The categories of one of the predictors are investigated, and grouped into classes by merging categories which do not differ signi= ficantly in their Y values. (This merging is governed by some restric= tions on which categories are allowed to be lumped together:- these restrictions will be set out below). At the end of this reduction, the original c categories on that predictor will have been grouped toget= her into $k \le c$ classes.

For example, we may start with a predictor with 7 categories labelled a,e,i,o,u,h,y. After the reduction, these categories may be reduced to 3: {y}, {a,u,o} , {e,i,h} , with significant differences in Y between the 3 classes, but not between the categories within a class.

Each of the p predictors is investigated in this way, and that predictor which best (in a suitably defined sense) partitions the data is actually used.

The data are then physically split into k sets, one set for each of the classes in the reduction.

Finally, each of these k subsets in turn is analyzed in the same way. The process continues until it is no longer possible to split any of the available subsets.

As a further illustration of the dendrogram produced by AID, but with a nominal dependent variable, we show in Figure 1 the results of an analysis of the performance of 798 first year Commerce students in a mathematical course at the University of the Witwatersrand. Their term-end performance is to be predicted from information available at the time of admission.

The nodes of the dendrogram represent groups of students isolated during the analysis. Each node contains the percentage breakdown of its members into the three classes of this dependent variable. The node is

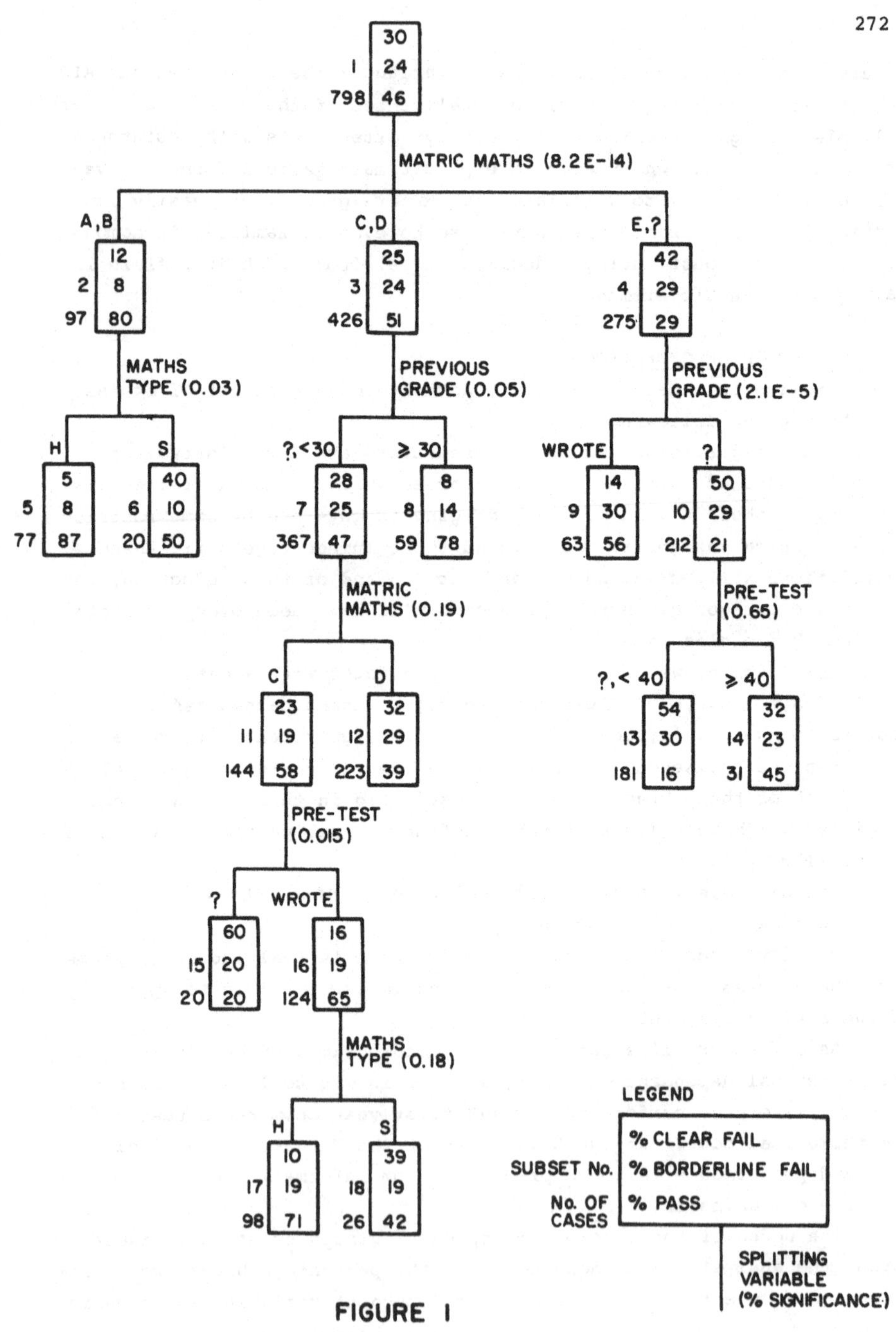

FIGURE 1

DENDROGRAM FROM CHAID

also subscripted with the number of cases it contains, and superscrip=
ted with the class of the split defining it. The arcs define splits,
and are labelled with the predictor defining them.

Later we shall say more about the interpretation of the dendro=
gram; for the present however it will suffice to note that the origi=
nal 798 students have been partitioned into 10 terminal groups with
mean pass rates going from 16% to 87%, and that the likely future per=
formance of a student is easily predicted by simply finding to which
terminal node he corresponds. This use of the dendrogram is remini=
scent of the use of identification keys (Payne and Preece, 1980),
which are often represented in tree form.

2.1 Types of predictor

As noted in the introduction, the predictors in an AID must be
categorical. Often they are dichotomous (eg. Sex:Male/Female) and
usually have relatively few categories. More than ten categories, for
example, are seldom found in practice, and, as we shall see below, are
also undesirable in theory.

For the puposes of AID a continuous predictor must be divided
into groups, either following conventional groupings (eg. Age into
10-year groups or the cruder Young/Middle Aged/Old) or arbitrarily as
one would form a histogram (perhaps a Type II histogram assigning 20%
say to each group).

AID recognises a number of types of predictor. We have had ex=
perience with the first three; the last two have appeared in the li=
terature but are not widely used.

(i) Monotonic predictors. For these predictors, the categories
are considered *ordered*. AID will then only form groups of contiguous
categories. A typical example is that of a high school Mathematics
symbol. Since we would expect a positive relation between this predic=
tor and the performance on a Mathematically oriented university sub=
ject, we would like to disallow a subset formed by Mathematics symbols
"A" and "C" without including "B" in the same subset. (Our logic would
suggest that such an unfortunate division that excluded "B" could only
arise from sampling fluctuations). Declaring a predictor to be mono=
tonic precludes such an unfelicitous division. While a monotonic pre=
dictor is practically by definition,ordinal rather than nominal, the
converse is not necessarily true. An ordinal predictor should only be
regarded as monotonic if one has good reason to expect that its rela=
tionship with the dependent variable is monotonic. Thus while age is

ordinal, if one were predicting, say, monthly income from age, one would not specify age as a monotonic predictor.

A dichotomous predictor, which could logically be regarded as having any of the types we list, is normally and most conveniently regarded as monotonic.

(ii) Free predictors. Here *no* restrictions are placed on the possible groupings of the predictor's categories. An example of this would be (in the previous context) the degree for which the student was registered eg. B.Com (Full Time General), B.Com (Full Time Legal), B.Com (Part Time General), B.Acc, etc. Here it is impossible to order the types of degree, so we treat the various categories as purely nominal.

If we are unsure about the monotonicity of a given predictor it clearly would be safer to treat it as free to allow possible non-monotonicities to be exploited. We have seen many cases of osten= sibly monotonic predictors whose regression is not monotonic and ad= vise that if one is in any doubt about the regression, one should re= gard the predictor as free.

Note, however, that since there are many more alternative ways to group the categories of a Free predictor than a Monotonic predictor, the computational effort is correspondingly more tedious.

Sometimes one has a predictor which, on prior grounds, may be expected to be monotonic, but one would like a "safety net" in case it is not. For example, in predicting mortality from age, one would generally expect the effect of age to be monotonic, but this would not necessarily be the case. The "safety net" can be provided by creating two predictors from this suspect predictor and specifying that one of the derived predictors is monotonic (or floating) and the other free. Correct inferential procedures (see Section 3) will ensure that the monotonic version of the predictor will be used automatically unless there is strong evidence of non-monotonic response.

(iii) Floating predictors. These predictors contain categories that are ordered except for one single category, which we allow to float up or down the ordered scale. The best example (and typical use) of a floating predictor arises with a predictor which one believes to be monotonic except for one particular category corresponding to cases for which the predictor is not measured - either fortuitously or because it is logically not measurable (for example the grade in a course not taken by a student). Usually there are no good logical grounds for putting this missing information category at any prefixed position in the monotonic scale, and one would prefer the data them= selves to locate this class in the otherwise monotonic scale. This may be done by specifying the predictor as floating, with the "missing information" category being the one that is permitted to float across the other categories and position itself wherever its Y values cor= respond best with those of the monotonic categories.

Many sets of data contain missing information, either because it is unavailable (as in this case) or because the respondent refuses to supply the requested information (if it involves opinions he may be undecided). If there is a single homogeneous reason for missing in= formation, this group can be classified into a category of its own. If the remaining categories are monotonic, then the missing category becomes a floating category.

Taking this idea a bit further, the absence of an item may re= sult from more than one discernible cause - for example undecided, re= fuse to answer, illegible, inapplicable. The reason for non-appearance of the predictor may itself have some predictive power. This possibility may be tested in AID by adding a free predictor, one of whose catego= ries indicates that the item is present while the other categories indicate the reason for its absence.

As a matter of convenience a 3 category free predictor is often entered as a floating predictor, as floating predictors are faster to process than free ones.

(iv) Partially constrained predictors. A generalization of the floating predictor is a partially constrained predictor in which there are some constraints on the possible grouping of predictors, but these constraints are not as severe as those assumed in monotonicity. A good example of this class is an interleaved predictor, which consists of K different monotonic scales, the i-th having C_i categories. The mono= tonic, free and floating types are all special cases of an interleaved predictor.

As a concrete example, consider the outcome of an examination in a subject written at one of two levels - say Higher and Lower. In ana= lyzing such a predictor, one would impose monotonicity of the Higher level symbols, and of the Lower level, but would not be able to specify the relative positions of, say, a Higher level D and a Lower level B. The partial constraints would be:

HA < HB < HC < HD < HE

LA < LB < LC < LD < LE

and in this example, it would also be sensible to specify

Hx > Lx $\qquad$ x = A,B,C,D,E.

though this additional constraint would not always be applicable.

We are not aware of any computer implementation of such partially constrained predictors, nor of any formal inferential results for them corresponding to the results of Section 3 (though some of the latter could be extended quite easily).

In the absence of such an implementation, one is compelled to use other and less satisfactory solutions. For example, in the two-level examination, one might use two predictors - a monotonic one for the grade and another for the level; or impose some assumed monotoni= city, or ignore the partial constraints and treat the predictor as free. The first of these possibilities is probably the least objec= tionable.

(v) Cyclic predictors. We consider such a predictor as having categories that lie on a circle, and permit grouping of adjacent cate= gories. Equivalently, a Cyclic predictor is a Monotonic predictor with the first and last categories being considered adjacent.

The implicit assumption that the regression is circular would make this model suitable for orientation data, but other situations where it would be appropriate are very uncommon, and generally little is lost by declaring such predictors to be free.

As a matter of historical interest, AID as set out so far differs considerably from that originally defined by Morgan and Sonquist (1963). There, only monotonic, free and cyclic predictors were permitted, and only binary (rather than k-way) splits on a predictor were permitted. There were also differences in the order in which subgroups were selec= ted for analysis, and the criteria for making a split. We shall say more about these points below.

It is often stated that AID is related to cluster analysis. While true, this point may well mystify the uninitiated. A situation in which one has n data points and a single predictor with n classes, so that there is one observation per class, provides the analogy. If the pre= dictor is free, then the AID is equivalent to a regular cluster analysis. If the predictor is monotonic, then the AID is equivalent to a segmen= tation. Within the AID context, a categorical predictor with one obser=

vation per category would be absurd; fortunately the analogy between the distribution theory of the two models extends to more realistic situations as well, and allows fruitful interchange of theoretical results between the two application areas.

2.2 Analysis of a set of cases

As the preceding remark implies, the reduction of the categories of a predictor to k groups which differ from one another but are in= ternally homogeneous, is equivalent to a cluster analysis of catego= ries, except that in an AID one normally has many cases within each category. These cases provide information about the 'within cluster' distribution of Y, and so are able to simplify the analysis into clus= ters.

There are two sources of difficulty in carrying out an AID - com= putational and inferential. We shall consider the computational as= pects first.

In Chapter 6 , we discuss a number of methods of clustering data, many of which could easily be adapted to the reduction of categories. Most of them however suffer from the defect of requiring heavy compu= tations. Since AID involves very many such clusterings (one on each of the variables on every subset of cases produced in the entire ana= lysis), we prefer to sacrifice the guarantee of optimality in the final clustering that is attainable (e.g. Fisher 1958, Hawkins 1976) using dynamic programming in favour of a not necessarily optimal but very much faster method related to the semi-hierarchic stepwise procedure of Chapter 6 . This method operates in two stages: Initially, all cate= gories are kept distinct. The *merge* phase consists of carrying out suitable two-sample tests to compare all categories or previously mer= ged groupings of categories that could legitimately be merged. (The rules of legitimacy restrict, for example, the merging of monotonic pre= dictors to adjacent categories). When all such tests have been carried out, the least significantly different categories (or groupings of categories) are merged into one provided they are not significantly different. This merging continues so long as there are categories or groupings of categories that may legally be merged, and are not signi= ficantly different.

The other phase, the *splitting* phase, comes into operation every time a composite of 3 or more categories is created. This grouping is investigated by creating all legal binary divisions of the composite into two subsets of categories and a two-sample statistic for identity of the two-group split so created is computed. If the most signifi= cant such binary split is significant at a specified significance level, then the composite of categories is redivided accordingly.

This repeated splitting and merging is considerably faster than a full optimization by dynamic programming, but generally produces an optimal or near-optimal grouping of categories.

The significance levels for splitting and merging of categories are specified by the user. As with other such stepwise procedures, the user must specify a more stringent splitting than merging criterion to prevent an infinite loop occurring.

As the above description implies, it is possible that in the end, all categories of a predictor will be merged into one, implying that that predictor is not suitable for splitting that set of data.

Morgan and Sonquist's original AID carried out only a single splitting of the fully merged set of categories, thus inexorably lead= ing to only binary splits. This simplification persists in a number of current implementations of AID. It reduces the computational load and simplifies the statistical inference associated with the procedure, but has the defect of requiring multiple stages to produce a k-way split, and this can lead to clumsy or misleading dendrograms.

2.3 AID test statistics

As is clear from the preceding discussion, the AID procedure re= quires a suitable test statistic for the comparison of two groups to establish whether splitting or merging is needed. A statistic is also needed, once the grouping of categories has been completed to establish the overall significance of the resultant k-way classification of the data. It is generally sensible and convenient (but not essential) to select for the two-group statistic the specialization to k=2 of the overall statistic for the k-way classification.

Within this very broad specification, certain particular statis= tics are indicated for particular types of dependent variable.

First (and first historically), if Y is an interval variable which one hopes can be modelled reasonably by the normal distribution with constant variance, then the test statistic of choice for the k-way split is analysis of variance. Its specialization to k=2 is the Student's t statistic.

An implementation of the resulting AID procedure is described in Heymann (1981). We shall refer to this procedure as XAID.

The model supposition of constant variance is much less restric=

tive than it is in the corresponding analysis of variance situation. When analyzing a particular subset, one uses only information about σ^2 within that subset, and it is of no import if some subset in the XAID has a different σ^2.

Next, if one has an interval dependent variable, but is unhappy about making an assumption of normality, then one could base the pro= cedure on a non-parametric equivalent - for example using the Kruskal-Wallis Statistic for the k-way split, and its two-group specializa= tion - the Wilcoxon-Mann-Whitney test.

This approach is described in Worsley (1977).

At the next level of complexity, if Y is a p component vector of interval data for which a normal model is appropriate, then one could use any of the criteria suitable for the corresponding model in clus= ter analysis. Taking the one-way Manova layout resulting from a k-way grouping of the categorical variable, the test statistics include:

1) $Tr(W)/tr(W + B)$
2) $|W|/|W + B|$
3) Trace (BW^{-1})
4) Largest latent root of (BW^{-1})
5) $\prod_{i=1}^{k} \{|W_i/n_i|/|W + B|\}^{n_i}$

where W_i refers to the sum of squares and cross products matrix with= in the i-th grouping of categories; W the sum of the W_i, and B the between groupings sum of squares and cross products matrix.

The conditions for applicability of the criteria are sketched more fully in Chapter 6 . Criterion 1 assumes an error covariance matrix of the form $\sigma^2 I$. Criteria 2 to 4 are applicable to homoscedastic data with a general error covariance matrix (but, as in the univariate case, the assumption of homoscedasticity is less restrictive than it appears to be). Criterion 5 is suitable for heteroscedastic data.

MAID, a multivariate AID package uses the first or third of these criteria (Gillo and Shelly 1974), but we are not aware of any package implementing the other criteria.

At present, we are investigating and implementing a further ex= tension in which the data will also include q covariates. The case q=0 includes the multivariate setup discussed above, while the further restriction to p=1 reduces to conventional Anova-based XAID. The ex= tension allows one to set up hybrid models in which some of the pre= dictors (i.e. the covariates), are treated using conventional general linear model techniques, while others (which may also include some of the covariates) are treated in the more assumption-free AID way. The covariates will be fitted using the general linear model, so that this too will fall within the ambit of our generalized AID.

Turning now to a scalar nominal dependent variable, two criteria are in common use: "the proportion of the sample classed correctly when using the optimal-prediction-to-the-mode strategy" (Morgan and Messenger 1973) forms the basis of the THAID program. The conventional χ^2 test for a contingency table forms the basis of CHAID (Kass 1980). While the χ^2 test is theoretically inferior to the MDI statistic dis= cussed in Chapter 3, its use is advisable in AID because of the very large number of such tests that must be performed in an AID.

The extension to a vector of nominal dependent variables follows naturally from the latter criterion. For this extension, one would use as test statistic the MDI test for association between predictor and the set of dependent variables. This likelihood ratio follows from, and is discussed more fully in Chapter 3 of this volume.

When sample size permits, it is very much in accordance with the minimal-assumption philosophy of AID that an interval dependent variable be reduced to an ordinal grouping and analyzed by CHAID. While the usual XAID criterion detects only a change in location, and sup= poses that the data are normal and homoscedastic, the more assumption-free CHAID analysis is able to detect any shift in distribution and makes practically no model assumptions. This analysis will detect, for example, significant changes in spread, which XAID cannot do, and is not affected by skew or heavy tailed distributions as XAID is.

A further, more subtle advantage of so categorizing an interval dependent variable is that, for some purposes, its scale may not really be interval *for the user's purposes*. Thus a 5 percentage point inter= val spanning the pass mark in an examination may have much greater im= portance than an interval of the same width elsewhere in the scale, thus implying that, for the real-world problem, the ostensibly inter= val scale is not in fact interval.

Another advantage of reducing the dependent variable to a group= ing is that it gives one the opportunity to include a category for cases for which the dependent variable is missing. For example, in an educational context one might want to predict a student's year-end mark. The very fact of the mark's being missing might be an important measure of failure. In a parametric analysis, one might set all such missing values to some predetermined constant. The much more satisfac= tory categorized analysis, however, allows the data themselves to deter=

mine the importance of a missing value from its predictability.

The categorization of the dependent variable does require a much larger sample than would be adequate for XAID, however, and so is not always possible. We shall have more to say later about such sample size requirements.

3. FORMAL INFERENCE IN AID

The early implementations of AID included no formal inferential mechanisms. Splits were made on the basis of the proportion of vari= ance explained by a split, without regard to the null distribution of this proportion. Since this null distribution depends on the sample size (which changes from one stage to another in AID); on the type of predictor (free variables can "explain" more variation when the null hypothesis is true); and on the number of categories in the predictor (with more categories, more variation is explained by chance), it is not surprising that no control could be exercised over the probabili= ties of Types I and II errors. In the event, Morgan and Sonquist's rules of thumb on the explained variation needed to justify a split led to excessively frequent Type I errors, and this gave AID a poor reputation as a method "guaranteed to find structure in almost any set of data".

This defect has since however, been alleviated considerably, and today there can be no justification for continuing to use Morgan and Sonquist's original recommendation.

There are two inferential problems in the AID analysis - the inference associated with reducing the original c categories to the final k categories; and the assessment of the significance of the fi= nal k-way split. Although it is logically, the later problem, we shall consider the second problem first.

3.1 Exact and asymptotic theory

Exact distribution theory for the AID criteria is very difficult to find. We will sketch out some of the known exact results first, and then show a perfectly general, but conservative procedure applicable for any criterion.

The exact results are for an AID using the explained variation criterion and performing a single binary split using a single predictor. Kass (1975) obtains an asymptotic permutation distribution, for the explained variation when a binary split is made on a monotonic predic= tor, and when it is made on a free predictor. This distribution is exactly the same as that arising if the data are assumed $N(\xi,\sigma^2)$.

Thus if σ is known, or if the total sample size is so large as to estimate σ with negligible error (as is commonly the case), then from this exact distribution of the explained variation, one can de= duce that of the ratio of explained to residual variation to an ade= quate degree of approximation.

Related results which are asymptotically exact are due to Scott and Knott (1976) and Engelman and Hartigan (1969). Both pairs of au= thors derive the asymptotic test theory for a free predictor, assum= ing, respectively, known and unknown variance.

Some asymptotic theory arising from the segmentation problem for a special case of the nonparametric AID, again restricted to bi= nary splits is given in Pettitt (1979). His results are for the change-point problem - an AID with a monotonic predictor and one observation per category - but could be adapted to cover several cases per cate= gory.

Note that all of these exact theoretical results are restricted to: a binary split; a single monotonic or free predictor; and a single stage of the analysis.

3.2 Conservative theory

Let us now turn to the more general k-way split with a predictor of arbitrary type. We shall continue to restrict attention to a single stage of the analysis and a single predictor.

We have available a suitable test statistic (see the earlier general definitions of the AID procedure) for testing the association between the predictor and the dependent variable. It would be very wrong, however, to simply apply the usual theory of that predictor for a k-way categorization of the data. The k groups have been formed by "data-snooping" on the original c categories, and so some allowance must be made for this fact when obtaining the overall significance of the association. Exact theory appears unattainable, and we are driven to use some other method. We know of two approaches, both of which give a conservative estimate of the significance of the association.

(i) Bonferroni inequality. Let B be the number of ways in which k groups can be formed from the initial c categories. B depends on the type of predictor, since the type has a strong effect on which group= ings are permissible. From Kass (1980) we obtain the following:

For a monotonic predictor,

$$B = \binom{c-1}{k-1} .$$

For a floating predictor,

$$B = \frac{k - 1 + k(c - k)}{c - 1} \binom{c - 1}{k - 1}$$

For a free predictor,

$$B = \sum_{i=0}^{k-1} (-1)^i \frac{(k - i)^c}{k!(k - i)!}$$

The test statistic T is the most significant of B possible statis= tics $T_1, T_2, \ldots, T_B$ that could be obtained by making all B groupings on the predictor. Thus by Bonferroni's inequality

$$\text{Significance of } T \le \sum_{i=1}^{B} \text{Significance of } T_i$$

$$= B. \text{ Significance of } T_i.$$

Thus by testing T at the α/B level of significance of a random T_i, we would obtain an overall significance $\le \alpha$. In practice, we use this procedure slightly differently by computing the tail area of T under the theory for a random T_i, and multiplying this by B to obtain the "Bonferroni significance" of the predictor. (This may exceed 1 very considerably - a fact that confuses the inexperienced user, but really involves no paradox.)

The Bonferroni approach cannot but be conservative. Its conser= vatism is slight when one has a monotonic predictor and k is near 2 or c. If k=2 or k=c, the degree of conservatism is negligible (Worsley 1979). It is most conservative when one has a free predictor, c is large and $k \doteq \frac{1}{2}c$. Generally, with free predictors and any c larger than about 5, the Bonferroni approach is extremely conservative.

(ii) Multiple comparisons. This approach may not be applicable to all test statistics. It is perhaps best illustrated by specific examples. Consider first an XAID, in which the overall statistic is an analysis of variance using the quantities S_b - the sum of squares between the k groups - and S_e - the sum of squares within groups. The Bonferroni approach would give the overall significance by computing

$$F = (n - k)S_b/\{(k - 1)S_e\}$$

and reporting B times the tail area of F under the F distribution with k - 1 and n - k degrees of freedom, B depending on the type of the predictor.

The multiple comparison approach observes that S_b cannot exceed, and S_e must exceed the corresponding statistics for the original c category classification. A conservative significance for the associa= tion may thus be obtained by referring

$$(n - c)S_b/\{(c - 1)S_e\}$$

to the F distribution with c - 1 and n - c degrees of freedom.

An exactly analogous approach applies to the multivariate ex= tensions of this homoscedastic anova.

In the case of the χ^2 test which provides the basis for CHAID, asymptotic additivity results show that the χ^2 obtained on the k-group classification cannot, in sufficiently large samples, exceed that of the c category classification from which it was collapsed. If the dependent variable has r categories, then the Bonferroni approach would use B times the significance under a χ^2 distribution with $(r-1)(k-1)$ degrees of freedom. The multiple comparison approach would refer the χ^2 value to a χ^2 distribution with $(r-1)(c-1)$ degrees of freedom.

Neither of these conservative bounds is uniformly tighter than the other. For two examples of this suppose first that in an XAID $c = 10$, $k = 5$, $n = 100$, $S_b/S_e = 0.2$. Then the different conservative significances are:

Multiple comparison	0.048
Bonferroni-monotonic	0.196
Bonferroni-floating	0.630
Bonferroni-free	65.865

Here, the multiple comparison bound is much tighter than any of the three Bonferroni bounds.

If however $k = 2$, then the three significances are:

Multiple comparison	0.048	
Bonferroni-monotonic	0.0018	
Bonferroni-floating	0.0036	
Bonferroni-free	0.013	
Actual significance-free	≑0.008	(from Kass 1975)

Note that the multiple comparison bound is the same for all types of predictor, and is valid simultaneously for all values of k. It is sometimes tighter than the Bonferroni bound (when k is large) and some= times not (when k is small).

We would stress that the Bonferroni and multiple comparison pro= vide two mathematical bounds on the overall significance of a predic= tor. It is thus legitimate and desirable to compute *both* bounds in all cases, and then elect to use whichever of them yields the smaller bound on the tail area.

3.3 Many predictors

Finally, we take cognizance of the fact that in the AID there are generally a number of predictors that could be used for splitting at any stage, and it is necessary to select one of them. Early AID programs selected that predictor which explained the most model mis= fit. This criterion, however, favours free variables over monotonic, and those with more categories over those with fewer. Preferring par= simonious models, we aim to select that predictor which leads to the most significant split, and in practice, estimate this by the predic= tor whose conservative significance is greatest.

To control the stage-wise significance level, one could apply another level of the Bonferroni inequality by multiplying the conser= vative tail area of this most significant predictor by the number of predictors. In view of the conservatism built into the earlier deter= mination of significance, however, this is seldom necessary unless one has very many predictors, most having few categories.

The question of spurious significance due to the multiplicity of groups analyzed is very closely related. If due (e.g. Bonferroni) cognisance is taken of the number of predictors in analyzing any single group, then the probability of wrongly splitting that group can be controlled to be $\leq \alpha$. The probability of any descendant group's being wrongly split will be a further $\leq \alpha$, so that, ignoring the relatively weak effect of nonindependence in the splitting of a subset and its successors, there is a probability of only α^2 of creating two unjusti= fied levels in the dendrogram. The multiplicity-of-stages problem in the AID is thus confined effectively to the several different subsets at the same level of the dendrogram.

This number too is generally small in relation to the degree of conservatism built into the earlier stages of inference in the AID, and we consider that provided the overall α is in the usual range of 0.1 or less, one is unlikely to get spurious splits. One could however eliminate the effect altogether and get an analysis-wise overall size of $\leq \alpha$ by applying a further Bonferroni multiplier of the maximum number of subsets at the same level ever produced in the analysis.

3.4 The reduction to k categories

Armed with the known results on testing the overall significance of a grouped predictor, we may now return to the problem of the reduc=

tion to k categories.

We have already sketched the computational difficulties of the reduction from c categories to k, and our approach for resolving them. The inferential difficulties are considerably more severe: they are, very simply, the difficulties of determining into how many signifi= cant clusters the c categories fall, and we have already noted that this problem is by no means fully resolved yet. Our procedure, as sketched, takes no explicit account of the multiple comparison nature of the merging and (very much more infrequent) splitting, and so it is all but impossible to ensure in any generality that the groupings with which one ends are really significantly different and also internally homogeneous.

The naïve approach that suggests itself is to use the desired α significance level on each of the two-sample tests that arises in the merging and (much less common) splitting of the categories. However this approach will lead to an excessively high value of k with a proba= bility much greater than α. As an illustration of this, we carried out a repeated simulation of c category predictors, each category con= tains 10 independent N(0,1) variates and noted: (i) what significance level α^*, say, in the two-sample tests led to reduction to k=1 in 95% of simulations, and (ii) what the overall type II error rate was if a 5% test was used at each stage. This gave the following figures for α^*:

	c			
	2	5	10	20
Monotonic	0.05	0.01	0.008	0.005
Free	0.05	0.006	0.0^38	0.0^42

Using an α^* value of 0.05 for the merge at each stage, the percentage of samples failing to merge to a single grouping was

	c		
	5	10	20
Monotonic	17	28	33
Free	30	78	99

It is by no means clear how to carry out the individual two-sample tests in such a way that the "correct" value of k is exceeded with a controlled probability of (or approximating) the user-specified α. In the case of an XAID, however, one can go some distance in this direction. The critical value for a binary split may be obtained asymptotically from the theory of section 3.1. By applying this cri= tical value to all two-sample split/merge tests, one can ensure that the probability of merging down to k=1 when this is appropriate is ap= proximately 1-α. If however the true value of k exceeds 1, this ap= proach is likely to be highly conservative.

An alternative approach suitable for a free predictor is to recognize that the reduction to k categories may be made, in principle, using a multiple comparison or multiple range procedure (Miller 1966). This approach is interesting also because of its connection with the multiple comparison method of assessing the overall significance le= vel of the final grouping. It does however have the defect that it is not easy to see how to extend it to monotonic and floating predictors without incurring a great loss of power.

For criteria other than XAID, even these rather tenuous guide= lines are not available, and it would seem that at best one could use a table of simulation based α^* like that listed above to obtain some control over the ultimate probability of type I error.

4. OUTPUT FROM AN AID

It should be apparent already that an AID run involves a large number of computations and produces volumes of statistics. Many of these statistics are valuable in modelling and diagnosis, as we shall attempt to show in the following section, and so it is of interest to mention the statistics that are available and useful. We shall use XAID and CHAID to illustrate them.

First, for each subset analyzed one wishes to see:

(i) what grouping of the categories occurred, and
(ii) the significance of the grouped predictor;
(iii) some details (e.g. sufficient statistics) of the subsets actually created by the splitting at that stage.

This essential information can generally be printed as a compact halfpage summary of that stage. Figure 2 shows sample outputs from XAID and CHAID.

The other information that may be useful is a detailed accounting of all the splits and merges that took place when analyzing a predic= tor. This accounting should include the number and preferably also the sufficient statistics of the observations in each category of the pre= dictor and the value and significance of the split/merge test statis= tic. The overall test statistic and significance prior to any collap= sing is also valuable.

This latter, and much more voluminous report is usually needed only if one wishes to use the AID for diagnosis of a suitable parametric model to fit. If the AID is carried out as an end in itself, then the question of whether the AID might not have split on some other predic= tor at a particular stage is not likely to arise.

5. USE OF AID FOR MODELLING

We have already noted that while an AID analysis is often an end in itself, it may also be used to direct a parametric modelling exercise. We shall now discuss practical examples to illustrate these uses.

The first example is that summarized in the dendrogram in Figure 1. The dependent variable was the result obtained by a group of first year students in a term-end examination of a mathematical subject during 1979. This was categorized into 3 classes: clear pass, border= line fail and fail. The available predictors were:

(i) whether high school was completed locally or abroad;
(ii) sex;
(iii) year of study in the University;
(iv) high school English grade;
(v) course for which the student was registered;
(vi) whether high school mathematics was taken at Higher or Standard level;
(vii) year of completion of high school;
(viii) previous grade in this course (defined only for students repeating the course - otherwise missing);
(ix) grade on a voluntary pretest taken in the second week of the academic term;
(x) high school mathematics grade;
(xi) mid-term grade in the course.

The types and overall conservative significances of these pre= dictors at stage 1 are as follows:

(i)	Local matriculant	Mono	100%
(ii)	Sex	Mono	100%
(iii)	Year of study	Free	100%
(iv)	Matric English	Float	40%
(v)	Course	Free	100%
(vi)	Matric Maths type	Mono	100%
(vii)	Year of Matric	Free	0.09%
(viii)	Previous grade	Float	1.4×10^{-6}%

```
     ANALYSIS OF GROUP NO.  4
     PREVIOUS GROUP NO.  1
NO.        NAME                      MC-SIG(%)    BON-F-SIG(%) GROUPING
  2 GRADIENT (I)                      .100E+03      .100E+03   1  2
  3 SAND CLASS (S)                    .560E+00      .816E+00   2  /  1  3  /  4
  4 FLOW RATE (Q)                     .100E+03      .100E+03   7  3
  5 HEIGHT (H)                        .738E+00      .125E+00   1  2  /  3  4


    BEST PREDICTOR
  5 HEIGHT (H)                        .738E+00      .125E+00   1  2  /  3  4


STATISTICS FOR GROUPING:
GROUP NO.    MEAN   SAMPLEVARIANCE  NO.OF CASES    WEIGHT   WEIGHTED MEAN   VARIANCE(WEIGHTED MEAN)
    10     2.9793         .0524          16       16.0000          2.9793                .0033

    11     2.6370         .0622          15       15.0000          2.6370                .0041


    ANALYSIS OF VARIANCE:
                SUM OF SQUARES      MEAN SQUARE      DEGREES OF FREEDOM
    GROUPING             .9070          .9070                  1
    ERROR               1.6558          .0571                 29

    F-VALUE      = 15.886
    SIGNIFICANCE=    .416E-01
    BON-F-SIG.  =    .125E+00
    MC-SIG.     =    .738E+00
    CONSERV.SIG.=    .125E+00
    GROUPING IS SIGNIFICANT AT THE    .125E+00%- LEVEL(CONSERVATIVE)
```

Figure 2a XAID output

```
SUMMARY OF RESULTS
ANALYSIS OF SUBGROUP MATRICULATION MATHEMATICS (A,B)

 NO.       NAME                                            SIGNIF (%)   BONF-SIG (%) NUMBER OF GROUPS.
  1 MATRICULATION MATHEMATICS . . . . . . . . . . . . .    100.0000     100.0000   1    AB
  2 SOUTH AFRICAN MATRICULANT? . . . . . . . . . . .       2.7915869    2.7915869   2    Y  N
  3 MATRICULATION ENGLISH . . . . . . . . . . . . . .      100.0000     100.0000   1    12345
  4 YEAR OF MATRICULATION (1974-1978) . . . . . . . .      100.0000     100.0000   1    87564
  5 SEX (MALE,FEMALE) . . . . . . . . . . . . . . . .      100.0000     100.0000   1    MF
  6 COURSE(BCOM-FG,FL,PG;BACC-76,COM,CTA,M/A;OTHER)        100.0000     100.0000   1    ?7GLO
  7 YEAR OF STUDY . . . . . . . . . . . . . . . . . .      0.3623038    1.0869102   2    12  3
  8 TYPE OF MATRIC MATHS (HIGHER,STANDARD) . . . . .       0.0102052    0.0306155   2    H  ?S
  9 PREV M&S1B MARK IF REPEAT(1=FIRST,..,4=FAIL,..)        100.0000     100.0000   1    ?23456
 10 PRE-TEST (FAILED,BORDERLINE,PASSED) . . . . . .        0.1736363    0.8681811   2    ?F  BP
 11 FIRST MID-TERM(APRIL) TEST (FAIL,BORDER,PASS) .        6.6040E-08   1.3208E-07   2    FB  P

CHARACTERISTICS OF THE BEST PREDICTOR
  8 TYPE OF MATRIC MATHS (HIGHER,STANDARD) . . . .         0.0102052    0.0306155   2    H  ?S

****************************************************************************************************
PREDICTOR    8:  TYPE OF MATRIC MATHS (HIGHER,STANDARD) * ROW PERCENTAGES *
                     H       ?,S      TOTAL
          FAILED    5.2     40.0      12.4
      BORDERLINE    7.8     10.0       8.2
          PASSED   87.0     50.0      79.4
    TOTALS (100%)   77       20        97
                                   TABLE IS SIGNIFICANT AT    0.0102052% LEVEL
```

Figure 2b CHAID output

(ix) Pre-test	Float	4.3×10^{-8}%
(x) Matric Maths	Float	8.2×10^{-14}%
(xi) Mid-term test	Float	3.0×10^{-47}%

In the floating variables, the floating class corresponds to mis= sing information.

An obvious (and anticipated) use for the results of this analy= sis is in screening students on admission. For this purpose, the most significant predictor by far, namely mid-term test result, is of no value. It was however carried along in the analysis and reduced and tested for significance at each stage, but was not permitted to be used for splitting.

The predictors whose significance is given as 100% were reduced in the merging phase to a single category.

The order of significance of the predictors contains relatively few surprises. The lack of significance of Matric Maths type is one (on which we shall say more later), and the significance of Year of matric is another. Taking the latter effect first, summary statistics from the CHAID reveal the following results after merging of catego= ries.

	Year	
	1978	Pre 1978
% Fail	37	24
%Borderline	23	24
%Pass	40	52
n	330	458

The striking feature of this is the good performance of the stu= dents who completed high school a year or more previously, and the poor performance of those who had just completed. However this factor is confounded to a very large extent with the even more significant one of whether the student is repeating the course (which is impossible for a student who wrote the matriculation in 1978). It is thus not clear whether the significant effect of year of matriculation is an indirect consequence of the effect of repeating the course, or is an indication of some salutary effect of maturing. Later, we shall show how the subsequent detailed AID analysis resolves this question too.

From either the dendrogram or the printouts, it can be seen that the 6-point Matric maths grade scale was grouped as {A,B}, {C,D}, {E,?}. The grouping of missing (?) with E shows that not having a Matric maths grade recorded is a very bad omen. The grouping also suggests that, in a later parametric analysis, one could substitute the predictor E in any case where the Matric maths grade was not observed - an often use= ful corollary of using the floating option for missing information.

The Matric maths grade and the dependent variable, are both ob= tained by grouping of an ostensibly interval percentage mark. Thus it would be only natural if a parametric analyst were to analyze their interconnection by fitting linear regressions of these percentage marks. However, while not conclusive, the CHAID split casts some doubt on whether linear homoscedastic regressions can in fact be fitted: a salutary warning to the user to proceed with caution in trying to fit any such model.

At the next level of the tree, a model non-additivity appears. In the best group of students, the Matric maths type, whose lack of significance in the full set we found surprising, is highly signifi= cant and results in a split. In the other two groups of students, the predictor used is Previous grade, and it essentially separates those who are repeating the course from those who are not.

Thus the Matric maths type is an important predictor amongst the students obtaining an A or B in Matric maths, but not for those ob= taining more mediocre marks. This observation is presumably yet an= other piece of evidence that examination grades do not behave like interval variables near the edge of their ranges.

The lack of significance of Previous grade among the students with a Matric maths of A or B may be established from the CHAID summary printout exemplified in Figure 2b. Of course, as Figure 1 shows, rela= tively few of these students fail and repeat the course, so that this lack of significance might on the face of it be a sample size artefact. As we shall show below, however, this is not the case.

Further down the dendrogram, we find Matric maths type also en= tering as a splitting variable in what is essentially the set of new entrants obtaining a C in Matric maths. The pre-test enters twice, but (as must be very disheartening to the teachers who set it) it seems that what matters most is whether the students exercised their option to write the pre-test: the actual mark was largely immaterial. Those who wrote the pre-test (and who by inference are better motivated than those who did not), had a much higher final pass rate.

Sometimes, a dendrogram is able to confirm that a set of data is well approximated by an additive model in two predictors. This is the case if, after the first split occurs on one of the predictors, each of the subsets splits in the same way on the other predictor and exhi=

bits a constant effect of the change in levels of the predictor.

The converse is not true - it is possible for data which are in fact close to an additive model, to give rise to an unbalanced dendro= gram. In Figure 1 we see that while one group split on Maths type at level 2, the other two split on Previous grade. Before concluding that there is indeed some non-additivity involving the simultaneous effect of Matric maths grade, Maths type and Previous grade, we must ensure that the same predictor could not have been used equally well on all three subsets at level 2. From the more detailed output we obtain the following summary statistics:

Matric	A,B	C,D	E,?
Previous grade			
Merged significance	100%	0.05%	2×10^{-5}%
Unmerged MSCC	0.03	0.04	0.15
Maths type			
Merged significance	0.03%	9.6%	100%
Unmerged MSCC	0.17	0.02	0.01

By way of explanation, the MSCC (mean square contingency coeffi= cient) is defined as the χ^2 for the full, uncollapsed contingency table divided by N.

It is the basis of the Pearson, Cramer and Chuprov coefficients of contingency. It is analogous to a squared multiple correlation and since (unlike the significance) it is not affected by the value of N, it is a natural measure of the strength of association between the pre= dictors and Y in the three different-sized subsets.

As both the significance and the MSCC make clear, neither pre= dictor is equally valuable at different Matric maths grades. The pre= vious grade is very predictive in the worst group of students and poor in the better ones. Maths type is very good in the best group of stu= dents, and poor elsewhere.

The presence of this interaction has an important consequence for a more parametric follow-up analysis. It shows that if one were to fit a log-linear regression-type model to predict Y from A = Matric maths, B = Matric maths type and C = Previous grade, the model would, at a minimum, be (YAB, YAC), and so the CHAID results would probably be preferred for their better parsimony.

In an XAID using an interval dependent variable, two natural summary statistics corresponding to the MSCC would be the error mean square, and the squared multiple correlation coefficient. Both statis=

tics have their disadvantages:- the former in that a comparison be= tween the groups is dependent on an assumption of homoscedasticity which is otherwise little needed in AID. The defect in the latter is that R^2 is dependent on the "spread" in the design matrix, which need not be the same in the different groups being compared.

If a choice between these measures must be made, we prefer R^2, but feel that to use both is preferable.

5.1 Following a predictor through the analysis

Another very valuable exercise consists of following through the analysis predictors which are not significant enough to cause splitting. By monitoring how their predictive power varies as splits are made on other variables, one can gain a better insight into the interactions between the predictors and Y.

We will illustrate this point using the predictor Year of matric, which has a significance of 0.09% in the total group, but which inter= acts with the more significant predictor Previous grade. We ran another CHAID suppressing all predictors but Previous grade and obtained a split into: Previous grade ≥ 20% versus Previous grade < 20% or missing. For practical purposes, this is a split between new and repeat students, and we shall discuss it in these terms.

The grouping details on Year of matric after split/merge reduc= tion are as follows:

Total group

	1978	Pre 1978
% Fail	37	24
% Borderline fail	23	24
% Pass	40	52
	330	468

$\chi^2 = 33.31$, MSCC = 0.04.

After the split on Previous grade, we obtain the following figures:

	New students	Repeat students	
	All years	1977	< 1977
% Fail	34	6	14
% Borderline fail	24	12	31
% Pass	42	82	55
	653	72	73

New students: $\chi^2 = 8.02$, MSCC = 0.01, significance after merging 100%

Repeat students: χ^2=22.86,MSCC=0.16,significance after merging 0.2%

What do these figures say? They show that the significance of Year of matric is not solely a result of the fact that the repeat stu= dents fare better than the new students: the significance persists in the repeat students group. The strength of the association may be measured by summing the χ^2 values in the two subgroups and dividing by their total sample size to get a MSCC. This gives the value 0.04 - the same as that in the total group. It is thus fair to say that most if not all of the predictive power of Year of matric is not an arte= fact caused by confounding with Previous grade, but a genuine associa= tion in its own right. The analysis after splitting however shows that Year of matric has no significance in new entrants - it is thus imma= terial whether the student enrolls for the course fresh from high school or does something else first - but has a high significance in the repeat students. In this group, while the overall pass rate is greater than in new entrants, it is much greater amongst those matri= culating in 1977. Educationists would no doubt explain this by saying that the latter group had the potential to pass on their first attempt at the course, but failed because of problems in adjusting from the high school environment to the university one. To summarize: on admis= sion, whether the student is fresh from high school or not has no ef= fect on his likely performance. If he should fail, however, and repeat the course, then his chances of passing are much better if he enrolled fresh from high school than if he did not.

This is a rather intricate sort of interaction which is easy to isolate using AID-type procedures, but difficult to obtain in so con= cise a form using more conventional general linear model techniques.

As an example of similar calculations in a problem with an inter= val dependent variable, we now consider a set of data processed by XAID. There were 88 values, and it was intended to predict the interval variable Y from four interval predictors, I,S,Q and H, which had been observed at 2,4,7 and 4 levels respectively. Attempts to fit multiple regressions on polynomials in I,S,Q and H and on monotonic functions of them had yielded regressions with multiple correlations as high as 0.99, but all such functions showed a highly significant lack of fit. The data were therefore analyzed using XAID, which produced the dendro= gram in Figure 3.

The quite complicated structure of the dendrogram makes it clear why it was so difficult to set up a parametric regression function. At the first stage, Q is very highly significant. The next two levels of the dendrogram consist of splits on the predictors H and S, though in

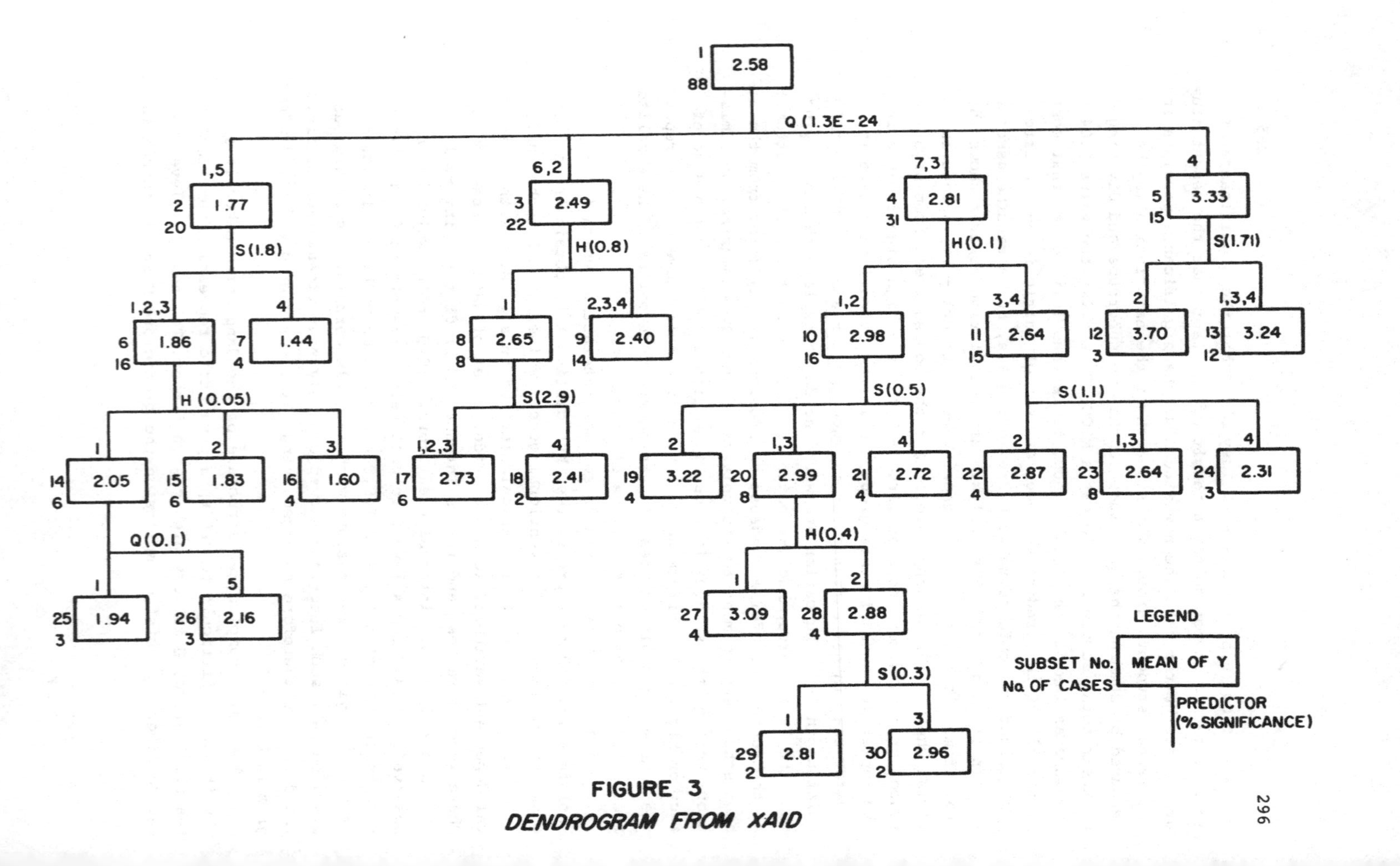

FIGURE 3

DENDROGRAM FROM XAID

the order S,H in groups 2 and 5, and H,S in groups 3 and 4. The pooled error standard deviation at level 4, when these splits on Q, H and S have been made, is 0.13, and comparing the means of the observations within the nodes at this level, one can see that an additive model in Q,S and H, while explaining most of the variability, cannot provide an acceptable fit. This impression is confirmed by the odd and non-matching splits in the next two levels.

In the event, AID was applied to this set of data only after more conventional exploratory methods had failed. Applying it first would have circumvented the effort wasted on trying to obtain a well-fitting regression model.

AID reduces the data down to a set of terminal nodes which are internally homogeneous but differ from one another. Thus the original data set may be reduced to these terminal nodes, and the modelling effort concentrated on finding a regression function which will ex= plain these nodes adequately. A subsequent attempt at finding a regres= sion function using these nodes rather than the original data fared much better.

6. SAMPLE CONSIDERATIONS

6.1 Sample size

The price paid for the extreme flexibility and modest modelling assumptions of AID methods is a profligate use of sample size. Each stage of an AID consists of further partitioning of the sample and an analysis of each subset of the partition in isolation from all others. It is a natural consequence of this that AID procedures cannot be used at all on small data sets, and work best on very large ones.

The most severe sample size requirements arise when one has a categorical dependent variable. As the splitting proceeds and the sample size diminishes, not only is there a progressive loss of power, but the distribution theory of the test statistic (which is only valid asymptotically) becomes more and more inaccurate. In any AID procedure for categorical data, it is prudent to incorporate a check on the sample size in any subset reached and to terminate the analysis of that subset if the sample size is not large enough to justify the distributional assumptions made. The actual sample size per subset needed depends on the number of categories in both the predictor and the dependent variables, and so it is difficult to specify an absolute limit. In CHAID, if one uses the normal rule of thumb of requiring cell expectations to exceed 5, then the minimal subset size is $5rc$, r and c denoting the maximum number of categories in the dependent and predictor variables.

The minimum size of the whole sample depends on a further and rather imponderable factor: into how many groups one anticipates partitioning the sample. We have carried out successful CHAID analyses on samples of size ranging from 500 upwards, but would be reluctant to use sample sizes much below 500.

In the other, more parametric members of the AID family, if the data are normally distributed as assumed, then there is no restriction on the sample size needed for inferential validity. Thus samples of any size may be analyzed validly, but small samples will give very truncated dendrograms because of the low power of even optimal tests in very small samples, and here we would regard 50 as a minimal sample size for analysis by XAID to be considered. If the data are not in fact normal, but one justifies the use of XAID by the central limit theorem, then once again, as with categorical dependent variables, it may become necessary to put in a restriction on the minimum size necessary for a subset to be analyzed.

There are some tricks that can be used to "stretch" a limited sample. One such is the AID analysis of residuals. Initially, one carries out an AID analysis of the data, which identifies a number of significant splits, but which fails to find all significant predictors because the limited number of observations is dissipated into too many terminal nodes. By regarding the mean value in each terminal node as a model prediction and subtracting it from its corresponding Y values, a new set of n residuals is obtained. These residuals may then be themselves subjected to an AID, and further significant splits may be found. Of course, at this second stage, the data being analyzed have induced intercorrelations so that the inferential theory is no longer strictly valid. In general, however, it will provide useful guidelines. If it seems advisable (i.e. if this analysis of residuals produces several subsets), then this process can be repeated.

Another useful modification of the basic procedure arises when one believes that the data are normal with a more-or-less known variance σ_o^2. This value of σ_o^2 and a corresponding number of degrees of freedom ν reflecting the strength of belief in the appropriateness of the σ_o^2 value may be specified to the AID routine. In the analysis of each group, a pooled estimate of variance, combining both the prior and the sample information on σ^2, is used in the testing.

This approach has the effect of smoothing out irregularities in the estimates of σ^2 at the nodes, and is especially useful if the total sample was small, giving rise to even smaller descendant sub= sets. The prior σ_o^2 could come from a preliminary XAID, or from some completely independent source.

Both these procedures adapt easily to multivariate normal data, but not to nominal dependent variables.

It is obvious, but worth noting, that AID works better and bet= ter for larger and larger samples. Thus while AID implementations may impose necessary restrictions on the maximum number of predictors or the number of categories per predictor, they should never constrain the maximum number of cases.

Computing time per group analyzed depends a little less than linearly on the number of cases analyzed and is not excessive even for problems with several thousand cases.

6.2 Sampling design

The preference of AID methods for large samples means that many of its applications are in the social sciences, in which one samples from finite and structured populations, and sampling may have been carried out by such methods as stratification and cluster sampling. The question then arises of the applicability of our earlier inferen= tial theory, which assumes simple random sampling.

For a continuous (scalar or vector) dependent variable, AID is a particular case of the general linear model, and ordinary, unweighted analysis will lead to a bias in both estimated effects and their stan= dard errors. Nathan and Holt (1980) give a method of eliminating both biases provided the design selection variable is also measured. If however it is not, then a standard weighted analysis may be used and this is asymptotically unbiased.

For a nominal dependent variable, Holt et al. (1980) and Brier (1980) show that by applying a simple correction factor to the χ^2 of a contingency table, one may bring many common and useful designs within the ambit of the inferential theory used in CHAID.

6.3 Outliers and robustness

The parametric AID procedures are non-robust, and all can be affected severely by outliers. The nonparametric variant based on the Kruskal-Wallis statistic is reasonably robust *provided* one carries out all analyses in terms of the ranks of observations within the imme= diate subset being analyzed. Thus for example when carrying out a two-group split or merge test, one must sort and rank only the observations within those two groups.

Fortunately, outliers usually make themselves glaringly apparent in the AID. In an attempt to achieve the high significance attainable by "predicting" an outlier, AID will carry out a sequence of splits until the outlier is stripped out into a subset of its own. This oc= currence of an isolated case or small set of cases, together with a visibly anomalous Y is the diagnostic for outliers.

In a sense, the outliers do not matter, as their contaminating influence is nullified by their being stripped off from the rest of the data set. However, there is a danger that they influenced earlier splits unduly, and so it is a good precaution to rerun the AID with such outliers eliminated.

Incidentally, the occurrence of outliers is not limited to in= terval dependent variables, but can occur with a nominal dependent variable as well - for example:

		X	
		1	2
Y =	1	1	11
	2	0	10 000

is highly significant, even by Fisher's exact test.

Thus while AID techniques can be affected by outliers, they will normally also diagnose the presence of these outliers automatically, and a rerun with the outliers omitted will provide a suitably robust analysis.

7. POSSIBLE DEVELOPMENTS IN AID

AID is like factor analysis in that its use as an operational if heuristic tool long preceded any proper understanding of its formal inferential properties. Today, most of the inferential problems have been resolved to the point at which at least some control over type I error can be applied. It can be expected that a full resolution of type I errors will not be long in appearing.

We have already mentioned an extension of AID to include covari= ates treated by normal linear modelling. This extension should enhance the power of the approach considerably.

Another needed extension is a facility to exert some greater de= gree of control over the variables used for splitting. One useful op= tion - adapted directly from stepwise regression - would be the use of a level indicator for each predictor, with no predictor being used for splitting if any predictor at a lower level was significant. In this way, the user could ensure preferential treatment for preferred pre= dictors. A different highly desirable option would be a facility for simultaneous investigation of subsets at the same level of the tree to test whether a common predictor, and common splits on that predic= tor, could be used to split all subgroups. This option could be used to find out in a single AID run whether an additive model is appro= priate.

A final development which should be mentioned is one that has in fact already arrived, but is not yet wide-spread. This is an interac= tive AID in which the computer routine reports interactively on the split/merge details and significance of each predictor in a group, and then allows the user to select which of the possible splits he would like to be made.

This interactive approach has many advantages over batch-proces= sing in that it is a great help in seeing to what extent an additive model can fit the data. Its main drawback is the difficulty of not overwhelming the user with valuable summary statistics.

AID is already an extremely potent and useful method of analysis for large data sets. It is to be hoped that the latest, and the future practical implementations will lead to its taking a proper place in the applied statistician's armoury of exploratory and confirmatory tools.

REFERENCES

BRIER, S.S. (1980). Analysis of contingency tables under cluster samp= ling. *Biometrika*, 67, 591-6.

ENGELMAN, L. and HARTIGAN, J.A. (1969). Percentage points of a test for clusters. *J. Amer. Statist. Assn.*, 64, 1647-8.

FISHER, W.D. (1958). On grouping for maximum homogeneity. *J. Amer. Statist. Assn.*, 53, 789-98.

GILLO, M.W. (1972). MAID, a Honeywell 600 program for an automatic survey analysis. *Behav. Sc.*, 17, 251-2.

GILLO, M.W. and SHELLY, M.W. (1974). Predictive modeling in multivariable and multivariate data. *J. Amer. Statist. Assn.*, 69, 646-53.

HARTIGAN, J.A. (1975). *Clustering Algorithms*. Wiley, New York.

HAWKINS, D.M. (1976). Point estimation of the parameters of piecewise regression models. *Appl. Statist.*, 25, 51-7.

HEYMANN, C. (1981). *XAID - an extended automatic interaction detector.* Internal Report, CSIR.

HOLT, D., SCOTT, A.J. and EWINGS, P.D. (1980). Chi-squared tests with survey data. *J.R.Statist.Soc.*, A, 143, 303-20.

KASS, G.V. (1975). Significance testing in automatic interaction de= tection (AID). *Appl.Statist.*, 24, 178-89.

KASS, G.V. (1980). An exploratory technique for investigating large quantities of categorical data. *Appl. Statist.*, 29, 119-27.

KENDALL, M.G. and STUART, A. (1973). *The Advanced Theory of Statistics, Vol.II*, Griffin, London.

KISH, L. (1967). *Survey Sampling*. Wiley, New York.

MILLER, R.G. (1966). *Simultaneous Statistical Inference*. McGraw-Hill, New York.

MORGAN, J.A. and MESSENGER, R.C. (1973). *THAID - a sequential analysis program for the analysis of nominal scale dependent variables.* SRC, Inst. Soc. Res., Univ. of Michigan.

MORGAN, J.A. and SONQUIST, J.N. (1963). Problems in the analysis of survey data: and a proposal. *J.Amer.Statist.Assn.*, 58, 415-34.

NATHAN, G. and HOLT, D. (1980). The effect of survey design on regres= sion analysis. *J.R.Statist.Soc.*, B, 42, 377-86.

PAYNE, R.W. and PREECE, D.A. (1980): Identification keys and diagnostic tables: a review (with Discussion). *Journal of the Royal Statis= tical Association, Series A*, 143, 253-292.

PETTITT, A.N. (1979). A non-parametric approach to the change-point problem. *Appl.Statist.*, 28, 126-35.

SCOTT, A.J. and KNOTT, M. (1976). An approximate test for use with AID. *Appl.Statist.*, 25, 103-6.

SCOTT, A.J. and SYMONS, M.J. (1971). Clustering methods based on the likelihood ratio. *Biometrics* 27, 387-98.

SONQUIST, J.N., BAKER, E.L. and MORGAN, J.A. (1971). *Searching for Structure (Alias-AID-III)*. Inst.Soc.Res., Univ. of Michigan.

WORSLEY, K.J. (1977). A non-parametric extension of a cluster analysis method by Scott and Knott. *Biometrics* 33, 532-5.

WORSLEY, K.J. (1979). On the likelihood ratio test for a shift in lo= cation of normal populations. *J.Amer.Statist.Assn.*, 74, 365-7.

CLUSTER ANALYSIS

DOUGLAS M. HAWKINS, MICHAEL W. MULLER AND J. ADRI TEN KROODEN
COUNCIL FOR SCIENTIFIC AND INDUSTRIAL RESEARCH

1. INTRODUCTION

The urge to classify objects must be recognized both as a basic human attribute, and as one of the cornerstones of the scientific method. A sorting and classification of a set of objects is the neces= sary prerequisite to an investigation into why that particular classi= fication works to the extent that it does.

The most striking example of this process is in biology. The Linnaean system of classification predated Darwin by a century, but the very success of a scheme which was able to describe biological species as if they were the leaves of a tree invited a model to explain such an effect. The theory of evolution provided just such a model.

Much later, the same evolutionary model and its consequent tree structure have been used in linguistics to study the evolution of languages. These two applications illustrate two possible uses for cluster analysis. The first is to take a set of objects with particu= lar interobject similarities and classify them "blindly". As a result of this, one can study the resultant typology, and use it to build up models to explain the typology. The second use is to take a set of ob= jects with a known form of typology (e.g. an evolutionary tree) and use an appropriate method to classify the objects.

There are other uses of cluster analysis in which the existence of a "real" typology is not presumed, but the analysis provides a con= venient summary of a large body of data. In market research for example, it may be convenient to group a large number of respondents according to their needs in a particular product area. In this operation, it is not necessary to assume that these groups correspond to different statis= tical populations, nor that the identical groups could be recovered af= ter an innocuous modification of the original data (for example permu= ting the cases); the analysis can stand or fall by the compactness or separation of the clusters found. Another, perhaps more familiar, example of this type arises when one measures an individual's IQ, say, and then classifies it into high, medium or low.

Ball (1971) lists these uses of cluster analysis, together with a further four:

Prediction based on groups

Hypothesis testing

Data exploration

Hypothesis generating

These latter applications tend more to the realm of Automatic Interaction Detection, and are adumbrated under that heading in Chapter 5. Here we will concentrate on methods for reducing data to a typology, and to a lesser extent, on how to determine the minimal number of clusters needed to represent the data adequately.

The term "cluster analysis" refers to a whole battery of quite divergent statistical methods which are used to obtain so-called "clusters". The term "cluster" is intuitively quite easily defined, but unfortunately it is not easy to turn intuition into an operational definition. For example, faced with a set of points looking like Figure 1 , few people would have difficulty in describing them as comprising two clusters:- the inner circle, and the outer annulus.

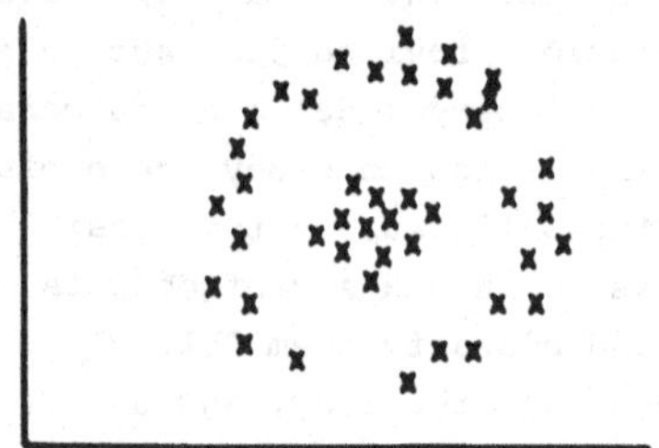

FIGURE 1

Such a description, however, would not necessarily be returned by a cluster analysis program, many of which would connect the central observations with the lower right arc of the annulus to form one cluster, with the remaining points of the annulus forming the other.

This difficulty permeates practical applications of cluster analysis, and any user should be well aware of it:- the statistical specification or, equivalently, the criterion used in the algorithm, imposes a particular geometry on the data and ensures that clusters corresponding to that geometry will be found. This difficulty creates little problem if the user already knows the form of typology; he must simply use a definition of a cluster which is compatible with that typology. Thus if he knows that an evolutionary model underlies his data then he should restrict himself to hierarchic approaches. When the form of the typology is not known, then there is a real difficul= ty and the user should be prepared to adopt an eclectic approach of analysing the data using several different models and comparing the results, as well as carrying out what checks of model fit he can.

It is convenient to recognize two broad classes of cluster ana= lysis techniques which are distinguished by the type of data they request.

The first class requires an ordinal measure of the similarity or dis-similarity between all $\frac{1}{2}n(n-1)$ pairs of objects, and the ana= lysis is then usually carried out using the linkage methods. This class, rightly or wrongly, accounts for the majority of all cluster analysis applications. It is particularly prevalent and appropriate in biological problems where a mixture of binary, categorical and in= terval observations on objects may be reduced to an acceptable "taxo= nomic distance" between objects.

The second class requires a profile of interval measurements made on each object. The analysis tends to make distributional assumptions about the profiles, and includes the mixtures of distributions models, and a number of likelihood maximizing iterative reallocation procedures.

In a purist sense, the distinction between these two methods is somewhat artificial, but in practice it is very real. On the face of it, if one has profiles of all objects on a number of variables then a suitable measure of similarity or dissimilarity between objects may be found. If, for example, the profiles are believed to follow multi= variate normal distributions, then the dissimilarity between objects should be measured by the Mahalanobis distance between them; under some circumstances this can be approximated adequately by a Euclidean distance. In many biological settings, the profile will consist of presence/absence observations; in this case it is more natural to use one of the many matching coefficients that have been defined. The lat= ter use of matching coefficients is common in biological problems, and has been shown by experience to work well. The application to normally distributed profile data via a suitable distance measure has been shown repeatedly to work badly. Despite this, it is used often:- to the ex= tent, for example, that a linkage method applied to Euclidean distances is the only method available in some program packages purporting to carry out cluster analysis.

On theoretical grounds, if one has a set of profiles together with an underlying normal model, then one would normally wish to para= metrize the clusters by their mean vectors and covariance matrices and express object to cluster dissimilarities in terms of these parameters. Depending on the particular distance metric appropriate to the normal model used, this parametrization may be irrecoverable after the data have been reduced to interobject distances. On practical grounds, if n, the number of objects, is large, then a profile-based method may be very much faster than a (dis)similarity-based one.

In the reverse direction, one might suppose that a matrix of in= terobject distances could be reduced by nonmetric multidimensional scaling to a set of profiles in a Euclidean space, following which a profile-based method of analysis could be used. This procedure, how= ever, has also been found to work poorly. A folk theorem holds (Kruskal, 1977) that multidimensional scaling is concerned with the large inter= object distances, and cluster analysis with the small ones. In other words, with the usual definitions of stress, it is the large interob= ject distances that have the greatest effect on the profiles found by the MDS.

Cluster analysis, however, is concerned primarily with a compari= son of the profiles of nearby points which are likely to be allocated to the same cluster and so it is important for cluster analysis that the profiles correctly depict the relative positions of mutually close points, distant points being unimportant. Only in the rather unlikely event of the distances being exactly reproducible in a low dimensional Euclidean space, will the MDS give profiles usable for cluster analysis.

The problem suggests its own solution, more details of which may be seen in Chapter 4:- a MDS in which the stress function pays close attention to small distances and less to large ones; however there does not yet appear to be any practical experience with this modified approach.

It is generally the case that when a cluster analysis is carried out, it is not known a priori how many clusters are present in the data, and the analysis will be carried out clustering the data into several typologies of different numbers of clusters. An extreme example is produced by the linkage methods which provide solutions for 1,2,...,n clusters. If this fact is taken into account, we recognize another di= mension along which cluster analysis algorithms may be classified. In some methods, any cluster in any $k > k^*$ cluster typology is contained completely in some cluster of the k^* typology. Such methods are called *hierarchic*. In other methods, termed *non-hierarchic*, this is not the case.

Some problems compel solution by hierarchic methods. In plant taxonomy, for example, every natural grouping of the objects can be expected to be an element of another grouping at a higher level, and so a hierarchic approach is both natural and meaningful.

In others, there is no apparent reason to insist on hierarchy. To see the effect of requiring a hierarchic solution, consider the following 8 objects, each with a single measurement:

-2.2, -2, -1.8, -0.1, 0.1, 1.8, 2, 2.2.

which quite clearly display 3 clusters. If however, we were to proceed hierarchically by first splitting into two clusters by maximizing the t value between the two "samples" so formed, the central cluster would be split yielding the clusters -2.2,-2,-1.8,-0.1, and 0.1,1.8,2,2.2. Further hierarchic splitting could not recover the clearly visible, compact, 3 clusters. To recover them using this hierarchic method, it would be necessary to continue to a four cluster solution - isolating -0.1 and 0.1 as individual clusters - and then fuse these two, which would run counter to the hierarchic approach.

For the sake of completeness, we might mention that a merging hierarchic solution which begins with each point constituting a cluster and then merges the clusters down would correctly recognize the set of three natural clusters, but would in turn be incapable of producing the optimal two-cluster division.

This trivial example illustrates a danger of using a hierarchic method when it is not appropriate for the data:- a poor allocation made at one stage of the analysis cannot be rectified later. Thus users should be very wary of using hierarchic methods if they are not clearly necessary.

A final categorization of cluster analysis problems is by whether there are any restrictions on the membership of clusters. Problems under the general headings of segmentation and change-point detection are examples in which there are such restrictions. If for example the objects represent the geochemistry of different points in a field, then it is obviously sensible to cluster together only points which are contiguous, so that the resulting clusters are geographically connected.

Cluster analysis models of this type are of importance when dealing with time series and with spatial data, and their discussion constitutes the final section of this chapter.

Later, we shall be discussing at some length multivariate normal mixture-based models. For these models it is fairly clear what criteria should be optimized by a proposed clustering, though the actual optimization involves approximations and major computational considerations.

For data other than normal profiles, the question of criteria is less clear on prior theoretical grounds. Gower (1974) has proposed the principle of "maximal predictive classification" in which the desired clustering is one which maximizes the number of binary attributes of each object correctly predicted by its cluster membership. The exten= sion of this principle to interval profile data would presumably yield similar, but not identical, results to the likelihood approaches.

2. SPECIFIC CLUSTERING METHODS

Clustering methods have been developed as the need arose in many unrelated fields. Psychologists, anthropologists, botanists and zoolo= gists were among the first to work in this area. Articles on clustering have appeared in the journals of such diverse disciplines as archaeology, computer science, criminology, economics, geology, market research and psychiatry.

Groups of workers in these different fields have worked in isola= tion, proceeding from differing views on what is meant by a cluster and describing their methods using their own specialized terminology. There has been a proliferation of methods, often of an ad hoc nature with inade= quate theoretical justification.

Several attempts have been made to produce general surveys and to place the various methods in a consistent framework. Among these are the books by Anderberg (1973) and Bijnen (1973) and the survey article by Cormack (1971). Between them these surveys have provided good coverage of the literature on clustering but questions regarding the relative value of the methods have remained largely unanswered. In the following sec= tions some of the more widely used methods are introduced and some con= clusions based on practical experience with using the methods are given.

2.1 Hierarchic methods

2.1.0 *Definition of hierarchic methods*

An hierarchical clustering scheme (Johnson, 1967) can be defined as follows. Let $\{C_k, k = 0,1,2,\ldots,m\}$, be a sequence of partitions of n objects into disjoint clusters. The sequence forms a hierarchical clustering scheme if the following conditions hold.

1.) C_o is the partition formed by placing each distinct object in a separate cluster.
2.) For each level i, $i = 1,2,\ldots,m$, every cluster in C_i is either in C_{i-1} as well, or is the union of some subset of clusters in C_{i-1}.

3.) C_m is the partition which groups all n objects together in one cluster.

4.) With each C_i there is associated a value α_i such that $f(i) = \alpha_i$ is a monotonic function of i.

The value α_i may be treated as a measure of the resemblance be= tween objects in clusters which combine at level i of the hierarchy.

A theoretical framework for considering the construction of hier= archic classifications from data in the form of dissimilarity coefficients has been given by Johnson (1967) and Jardine, Jardine and Sibson (1967). Several specific hierarchical clustering methods are now presented fol= lowed by a discussion of the choice of a method in practice.

2.1.1 *Single linkage cluster analysis*

Single linkage cluster analysis is one of the simplest of all clus= tering methods. This method has been widely used since it was first pro= posed by Sneath (1957). The method operates on a matrix of distance (or similarity) coefficients between groups which is revised as each succes= sive level of the hierarchy is generated. Initially each object to be classified forms a group by itself and the distance matrix between groups at this stage is given by the original distance matrix between objects. The distance between any two groups at each subsequent stage is taken to be the smallest distance between a pair of objects with one member in each group. The classification at the next level of the hierarchy is then formed by uniting the two closest groups and updating the distance matrix by deleting the rows and columns corresponding to each of these groups, and replacing them by a single row and column of inter-group distances between the combined group and the remaining groups.

Gower and Ross (1969) have given an efficient method of computing a single linkage cluster analysis. Given a matrix of distances or simi= larities between objects, the first step is to find the minimum spanning tree of the graph representing the linkage between objects. A tree span= ning n points is a set of straight line segments joining pairs of points, such that

(i) no closed loops occur,
(ii) each point is included in at least one line segment, and
(iii) the tree is connected.

The length of each segment is proportional to the distance or in= verse similarity between the objects represented by the end-points of the segment. The length of the tree is the total length of the segments com= prising it. A minimum spanning tree (MST) is a spanning tree of minimum length, and can be found iteratively as follows. Let A be the set of segments already forming part of the MST at any stage, and B the set of unallocated segments. Initially A is empty and at each stage the short= est segment in B which does not form a closed loop with any of the seg= ments in A is added to A. The tree is complete when A contains (n-1) segments. It is clear from this algorithm that if all segments of length $\geq d_o$ are removed from the MST, the connected sub-trees will cor= respond to the groups at level d_o in the hierarchy of a single linkage cluster analysis. By varying the value d_o the entire hierarchy can be obtained. It may be noted that starting with the minimum value of d_o and increasing it will yield precisely the same results as starting with the maximum value of d_o and decreasing it. Thus the same hierarchy is obtained in a single linkage cluster analysis starting with all objects in the same group and at each stage dividing a group into two, as would be obtained starting with each element in a separate group.

In practice it is sufficiently accurate to alter d_o by discrete steps, thus obtaining an approximation to the hierarchy. Gower and Ross (1969) recommend an alternative algorithm given by Prim (1957) as the most efficient way of finding the MST. Initially any point is chosen and the shortest segment including that point is assigned to A. Then, at each stage, the shortest segment from B which connects with at least one point of A but does not form a closed loop is added to A. Iteration stops when A contains (n-1) segments. This algorithm produces the same results as the previous one unless several equal segments of minimum length occur, but in any event the same hierarchy is obtained from the MST in each case.

The single linkage method can detect clusters of arbitrary shape and is in this sense a very general method. The fact that the presence of scattered intermediate points lying between denser clusters of points tends to cause these clusters to link together prematurely has been noted as a defect in the method (Lance and Williams, 1967). It could be expec= ted that this 'chaining' tendency, as it is called, would be particularly apparent when the data consist of a large number of objects forming a random sample from a mixed distribution with the components not widely separated. Practical experience with this method has shown that this chaining tendency is a major problem in many cases and very often pre= vents useful results being obtained, especially when a classification at one particular level is required and the hierarchy as a whole is not of particular interest.

One way of overcoming this problem is to modify the single linkage method in such a way as to avoid chaining. Jardine and Sibson (1968) considered the logical basis for such modifications by laying down a set of mathematical or logical criteria which they felt should be satis= fied by any hierarchical clustering method. After evaluating the common hierarchical methods in terms of these criteria they came to the conclu= sion that the only satisfactory hierarchical clustering method was single linkage clustering and that the limitation of this method were the limi= tations of hierarchical clustering itself. They also concluded that any attempt to overcome these limitations within the framework of hierarchi= cal clustering would merely substitute other and perhaps even more un= desirable characteristics for the defect of chaining. Their solution to the problem is to extend the definition of hierarchical clustering by allowing the clusters at any level of the hierarchy to overlap. They de= fine a family of clustering methods indexed by a parameter k, where k-1 specifies the maximum number of objects in the overlaps between classes at the same level of the hierarchy. When k=1 this method reduces to single linkage and when k=2 they refer to it as the double-link method in which clusters may overlap to the extent of one object.

While this proposal helps alleviate the chaining problem, it does not help the situation in which a mutually exclusive set of clusters is needed. Thus users who accept Jardine and Sibson's desiderata but cannot accept long straggly clusters caused by chaining, should turn their atten= tion to other non-hierarchic methods.

There is another practical use (Gower 1974) to which single link= age clustering, or indeed any other unequivocal and computationally easy method, may be used, and that is to provide an initial approximate clus= tering for an iterative solution by some other algorithm. This option is attractive if n is small (since single linkage involves time greater than order n), and the second algorithm is computationally tedious, but is expected to give a final result similar to that of single linkage. In general, however, we have found this two-stage approach to be more trouble than it is worth.

It is perhaps worth reminding the reader that single linkage uses distances at the ordinal level of measurement, and so is unaffected by arbitrary strictly monotonic transformations of these distances.

2.1.2 *Mode analysis*

The single linkage method is conceptually simple, without imposing any arbitrary restrictions on cluster shape and it is relatively efficient from a computational standpoint. It would therefore be useful if a way could be found to overcome the tendency to chain distinct groups of ob=

jects together. Mode analysis is a method developed by Wishart (1969b) in an attempt to overcome the chaining problem by first removing rela= tively isolated objects and then clustering the remaining objects using a single linkage technique.

The method also operates on a matrix of ordinal distances (d_{ij}) between objects, although similarity coefficients can also be used by treating them as inverse measures of distance. Imagine the objects re= presented as points X_i, $i=1,2,\ldots,n$, in a space in which inter-point distances are given by (d_{ij}). Let $N_r(X_i)$ be the closed neighbourhood of X_i such that $X_j \in N_r(X_i)$ iff $d_{ij} \leq r$. Let $r_k = r_k(X_i)$ be the mini= mum value of r for which k other points $X_j (j \nmid i)$ belong to $N_{r_k}(X_i)$. For example, if (d_{ij}) is a matrix of Euclidean distances, then r_k is the radius of the smallest hypersphere centered at X_i which contains k neighbouring points. If k is chosen to have an appropriate value, r_k will be high for isolated points and low for points in a densely occu= pied region. The points are first sorted into increasing order on $r_k(X_i)$ and then the points are introduced one at a time in this order. At each cycle, when point X_i is introduced, a distance threshold r is set equal to $r_k(X_i)$. As the second and each subsequent point is con= sidered one of three conditions can arise:

1) the new point does not lie within a distance r of any previous point, in which case it initiates a new cluster,
2) the point lies within a distance r of points belonging to one cluster only, in which case the point is added to that cluster,or
3) the point lies within a distance r of points from several clus= ters, in which case the clusters are fused and the new point added to the combined cluster.

If at any stage the nearest neighbour distance between two clus= ters falls below the current value of r, the clusters are combined. Just prior to any fusion of two clusters the membership of the clusters at that point is noted, as these levels summarise the useful informa= tion about the formation of clusters during the analysis. Each level is characterised by the current distance threshold r. The proportion of points included in the clusters at any stage is known as the "en= closure ratio". A high enclosure ratio is, according to Wishart, asso= ciated with a stable clustering of the data. When a specified minimum number of clusters nc has been reached (usually nc = 1) and when a spe= cified enclosure ratio p_o has been obtained (usually $p_o = 0.8$), the process is terminated. The analysis normally starts with one cluster and ends with one cluster. At some intermediate level the number of clusters will be a maximum and Wishart has suggested that this group= ing should be chosen as the most natural clustering of the data. When there is more than one grouping with this number of clusters, then the one with the highest enclosure ratio would be the first choice. If it is desired to cluster all the data points at this level the remaining points may be added to the clusters to which they are closest. When k=1 the method is equivalent to single linkage clustering, except for the fact that each point which has not yet been included in the analy= sis is allocated to the cluster closest to it instead of forming an ad= ditional single-point cluster. This method may be used with any ordinal distance or similarity coefficient.

A comparison of several clustering methods on seven sets of data consisting of samples from trivariate normal distributions was made by Muller (1975). It was found that mode analysis produced much better results than the unmodified single linkage method, and it is presumably generally true that this approach, which deals with the problem of chaining by deferring the allocation of the points which cause the chaining, will be superior to a direct single linkage analysis of the data.

Muller's experiments also show, however, that mode analysis is inferior to both Ward's method (discussed later), and methods making explicit use of assumptions that the clusters are multivariate normally distributed.

2.1.3 *Group average linkage*

This method operates on a matrix of distance or similarity coeffi= cients between objects. Product moment correlation coefficients were used in the original definition of the method, but any other similarity coefficient can also be used as can squared Euclidean distance. For the purpose of describing the algorithm a matrix of distance coefficients $D = (d_{ij}^2)$ will be assumed, but the modification for a similarity coeffi= cient merely involves a change of notation. Let $\Sigma_{pq}\ c_{pq}^2$ denote the sum of all distances c_{pq}^2 between an object in group g_p and an object in group g_q. Then the distance d_{pq}^2 between group p and group q is defined to be

$$d_{pq}^2 = 1/(n_p p_q) \sum_{p,q} c_{pq}^2 \quad , \qquad (2.1)$$

where n_i is the number of objects in group g_i.

Initially each object is in a group by itself, and at each stage the groups g_p and g_q for which d_{pq}^2 is a minimum are combined and the row and column of D corresponding to g_q is deleted. Group g_p is replaced by $g_r = g_p \cup g_q$ and the new distances d^2_{ri} are given by

$$d^2_{ri} = \frac{1}{n_r . n_i} \sum_{r,i} c^2_{ri} \quad ,$$

from which it follows that

$$d^2_{ri} = \frac{n_p}{n_r} d^2_{pi} + \frac{n_q}{n_r} d^2_{qi} \tag{2.2}$$

This procedure is iterated with successively updated distance matrices until all the objects are joined together in a single group.

It is easy to show that if the distance coefficients are Eucli= dean distances (or more generally Mahalanobis distances) then d^2_{pq} re= duces to the Euclidean distance between the centroids of the two groups plus the within group scatter. It thus follows that group average link= age will tend to produce compact, spherical clusters.

Some results of Digby and Gower (1981) give reason to believe that with a suitable definition of the group centroid, this property is true of more general similarity and distance measures.

This method was one of the earliest hierarchic clustering methods as the basic idea was first proposed by Sokal and Michener (1958). It has been recommended by several users and in an empirical comparison by Cunningham and Ogilvie (1972) it was rated as the best from a group of seven hierarchic methods including all those discussed here with the exception of mode analysis. The method was found to perform poorly on samples drawn from a mixture of normal distributions in the study by Muller (1975) mentioned in the previous section. An examination of the data used by Cunningham and Ogilvie shows that the method outperformed the other methods mainly on data sets which exhibited a fair degree of inherent chaining and thus could not be regarded as coming from a mixture of a small number of normal parent distributions. The samples used by Cunningham and Ogilvie were also small with only twenty ob= jects in all divided into four clusters of five objects each.

While both single linkage and mode analysis (along with some of the methods discussed later) are invariant under an arbitrary strictly monotonic transformation of the distances, group average linkage is not. Thus it is not advisable to use it unless one is satisfied that the distance function lends itself to averaging.

2.1.4 *Median linkage*

In this method the procedure for calculating the new distances d^2_{ri} between any group g_i and the group g_r which is formed by fusing groups g_p and g_q is independent of the relative sizes of g_p and g_q. Thus if extra objects identical to objects already in the sample are added the clustering procedure is unaffected. For this method

$$d_{ri}^2 = \tfrac{1}{2}(d_{rp}^2 + d_{rq}^2) - \tfrac{1}{4} d_{pq}^2 \ . \qquad (2.3)$$

If d_{ri}^2 is squared Euclidean distance then geometrically g_r is located at the midpoint of the line joining the points representing g_p and g_q. The value of d_{ri}^2 is the square of the length of the median of the triangle $g_i\ g_p\ g_q$ drawn from g_i. This is illustrated in Figure 2.1.

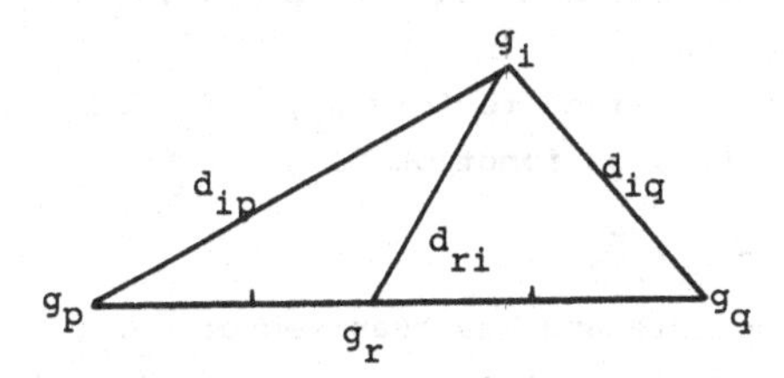

Fig.2.1 Geometrical illustration of median linkage

This method is also appropriate for use where $\{d_{ij}^2\}$ are non-metric distance measures. More details of median linkage are given by Gower (1967) and Lance and Williams (1967).

2.1.5 *Ward's method*

A method in which the loss of information which results when ob= jects are grouped together is quantified in terms of some appropriate functional relationship was suggested by Ward (1963).

The object of the method is to produce a hierarchy in which the loss of information in passing from each level of the hierarchy to the next is minimized. At the first level each object is in a separate subset, and at each stage two subsets are united to form the next level. If the previous level contains k subsets there are k(k-1) possible unions and that one is selected which minimizes the increase in the loss function.

If two or more such unions each yield the minimum value, an arbi= trary choice is made between them. The process is terminated when all objects are in the same group.

Although Ward considered the use of an arbitrary loss function, the name "Ward's method" is usually associated with the analysis of profile data using as a loss function the sum of squared deviations of all observations from their cluster means. This criterion is the tr(W) criterion discussed more fully below. An efficient algorithm for this loss function is due to Wishart (1969a):-

Suppose without loss of generality that the sample mean vector $\bar{X}$ taken over all groups is zero. The loss function may be written as the Analysis of Variance decomposition

$$tr(W) = tr(T) - tr(B). \tag{2.4}$$

where T, B and W are the total, between and within cluster dispersion matrices.

If groups g_p and g_q are combined to form g_r, tr(T) remains un= changed and the increase in the loss function is given by

$$I_{pq} = n_p\bar{X}_p^T\bar{X}_p + n_q\bar{X}_q^T\bar{X}_q - n_r\bar{X}_r^T\bar{X}_r \quad , \tag{2.5}$$

where group g_i contains n_i objects and has mean vector $\bar{X}_i$.

Let $d_{pq}^2 = (\bar{X}_p-\bar{X}_q)^T(\bar{X}_p-\bar{X}_q)$, be the squared Euclidean distance be= tween the means of groups g_p and g_q. It follows that

$$I_{pq} = (n_pn_q/n_r)d_{pq}^2. \tag{2.6}$$

It can be shown that

$$d_{ir}^2 = (n_p/n_r)d_{ip}^2 + (n_q/n_r)d_{iq}^2 - (n_pn_q/n_r^2)d_{pq}^2. \tag{2.7}$$

After g_r has been formed by combining groups g_p and g_q, the fur= ther increase in the loss function at the proposed fusion of g_r and any other group g_i is

$$I_{ir} = [n_in_r/(n_i+n_r)]d_{ir}^2$$

$$= \frac{n_i\ n_r}{(n_i+n_r)}\left[\frac{n_p}{n_r}d_{ip}^2 + \frac{n_q}{n_r}d_{iq}^2 - \frac{n_pn_q}{n_r{}^2}d_{pq}^2\right]$$

i.e.

$$I_{ir} = \frac{1}{(n_i+n_r)}\left[(n_i+n_p)I_{ip} + (n_i+n_q)I_{iq} - n_iI_{pq}\right]. \tag{2.8}$$

The process starts with the matrix D of squared Euclidean distan= ces between objects, and each object in a separate group. Initially, those two objects p and q are combined for which

$$I_{pq} = \tfrac{1}{2}d_{pq}^2 \tag{2.9}$$

is a minimum. The row and column corresponding to q are deleted from the matrix and the elements d_{ip}^2 are replaced by $d_{ir}^2 = 2I_{ir}$. Thus a new value for d_{ip}^2 is given by

$$(d_{ip}^*)^2 = \frac{2}{(n_i+n_r)} [(n_i+n_p)I_{ip} + (n_i+n_q)I_{iq} - n_i I_{pq}],$$

i.e.,

$$(d_{ip}^*)^2 = \frac{1}{(n_i+n_r)} [(n_i+n_p)d_{ip}^2 + (n_i+n_q)d_{iq}^2 - n_i d_{pq}^2]. \quad (2.10)$$

This process may be repeated with the modified matrix D, although D is no longer an Euclidean distance matrix.

Muller (1975) found that Ward's method performed better than the other hierarchic methods discussed here on samples drawn from a mixture of trivariate normal distributions, but not quite as well as the relo= cation and mixture methods described in Section 3.

2.1.6 *Lance and Williams' flexible method*

Many of the most common hierarchic strategies can be regarded as special cases of a single general method which uses a particular type of combination of the inter-group distances before fusion of clusters to calculate the updated distances after fusion. Lance and Williams (1967) have used this fact as the basis for a flexible clustering method. As before we consider the group g_r formed by combining groups g_p and g_q and the distances between group g_r and any other group g_i. Lance and Williams start by considering the relation

$$d_{ir}^2 = \alpha_p d_{ip}^2 + \alpha_q d_{iq}^2 + \beta d_{pq}^2 + \gamma |d_{ip}^2 - d_{iq}^2|, \quad (2.11)$$

where d_{ij}^2 is the measure of distance between groups g_i and g_j. By vary= ing the values of α_p, α_q, β and γ all the methods discussed above with the exception of mode analysis can be obtained. The values corresponding to these methods as well as to furthest neighbour (Johnson 1967) are given in Table 2.1.

In this table n_p is the number of objects in group g_p, n_q the number in g_q and $n_r = n_p + n_q$.

Lance and Williams' flexible strategy is based on a constrained version of this relation where $\gamma = 0$ and the values α_p, α_q and β satisfy

$$\alpha_p + \alpha_q + \beta = 1,$$

$$\alpha_p = \alpha_q$$

and

$$\beta < 1.$$

Table 2.1 *Values of parameters in flexible clustering method corresponding to some well-known techniques*

Method	α_p	α_q	β	γ
Single linkage	½	½	0	-½
Group average	$\frac{n_p}{n_r}$	$\frac{n_q}{n_r}$	0	0
Ward's method	$\frac{n_p+n_i}{n_r+n_i}$	$\frac{n_q+n_i}{n_r+n_i}$	$\frac{-n_i}{n_r+n_i}$	0
Furthest neigh= bour (Complete linkage)	½	½	0	½
Median	½	½	-¼	0

This strategy is designed to produce a family of methods exhibi= ting varying degrees of chaining. When $\beta = 1$ complete chaining will occur, while as β drops to 0 and becomes negative more and more "intense" groupings are produced.

2.2 Non-hierarchic methods

2.2.0 *Non-hierarchic methods in general*

These methods usually contain a parameter which can be set to the desired number of clusters. A solution obtained by such a method cor= responds to a single level within a hierarchy rather than to the hier= archy as a whole. If a second solution is obtained by increasing the number of clusters the new clusters will not necessarily be subsets of clusters in the first solution as would be the case with a hierarchic method. Most of the methods of this type operate on a matrix of pro= file data although as mentioned in the introduction this need not be the case.

2.2.1 *Mixture analysis*

Let $\{f(X,\theta_i)\}$ be a family of probability density functions defined on an p-component random column vector X and indexed by a m-component parameter vector θ_i. A finite mixture of these density functions is de= fined to be

$$h(X) = \sum_{i=1}^{r} c_i \, f(X,\theta_i)$$

subject to $c_i > 0$ and $\Sigma_1^r c_i = 1$ where r is the number of components in the mixture and c_i is the proportion of component i in the mixture. If it can be assumed that all the within - cluster distributions have density functions belonging to the family then cluster analysis can be formulated as the problem of estimating the parameters r, c and θ_i $(i = 1,2,\ldots,r)$ from the sample mixture density function h(X) or the sample cumulative distribution function H(X). This approach is particularly appropriate when cluster analysis is used as an inferential method since it provides direct estimates of the population parameters. If it is required to assign each object to a cluster, then standard classification techniques may be used in a second stage when population estimates of the parameters have been obtained.

A basic theoretical question which arises in connection with the mixture model is whether the decomposition of the function h(X) into components is unique. This has been investigated for several families of univariate distributions, but the result most relevant to the clustering problem was obtained by Yakowitz and Spragins (1968). They have shown that the family of all p-dimensional normal distributions is identifiable in the sense that the equation

$$\sum_{i=1}^{r} c_i \, N(X,\theta_i) = \sum_{i=1}^{r^*} c_i^* \, N(X,\theta_i^*),$$

where $N(X,\theta_i)$, $i = 1,2,\ldots,r$ are the cumulative distribution functions of r distinguishable p-dimensional normal distributions, implies that

$$r = r^*,$$

and that $c^* = (c_1^*, c_2^*, \ldots, c_r^*)$ is a permutation of c, while θ^* is the same permutation of the columns of θ, where θ and θ^* are matrices formed from the column vectors θ_i and θ_i^* respectively.

Yakowitz (1969) has shown further that it is possible to construct consistent estimators for the parameters in the above model, for any family of cumulative distribution functions $\{F(X,\theta)\}$ which generates identifiable finite mixtures and for which F is continuous with respect to θ. Although this result does not provide a suitable numerical method for estimating the parameters, it gives a firm theoretical justification for attempts to find such methods.

Stanat (1968) has derived a numerical method for the multivariate normal family which is based on the decomposition of the sample cumulative distribution function. The Fourier transform of this function is found numerically and an operator is defined which can be used sequen= tially to isolate the components of the mixture. Stanat noted, however, that this procedure is severely affected by sampling error. It does not appear to provide a practicable estimation procedure, particularly in the multivariate case.

Day (1969) considered several estimation procedures for multi= variate normal mixtures. He concluded that moment estimators behaved poorly, except in the univariate case, and that the computation invol= ved in minimum chi-square and Bayes estimation was prohibitive. Day then considered maximum likelihood estimation and presented a numeri= cal procedure for the special case of a mixture of two multivariate normal distributions with equal covariance matrices.

In the case of unequal covariance matrices he showed that the likelihood function is not bounded above and thus the method of maxi= mum likelihood may break down.

Wolfe (1970) gave the following general approach to the solution of the likelihood equations for families of distributions where the probability density function is a twice differentiable function of its parameters. Each iteration consists of two steps. In the expectation step the sufficient statistics for the complete data are estimated and in the maximization step the maximum likelihood estimates are obtained as though these values were obtained from observed data. In this context the term data can include what are usually regarded as parameters. Be= fore discussing EM algorithms any further, however, Wolfe's solution for the mixture problem will be outlined.

Initially it is assumed that r, the number of components in the mixture, is known. For a given column vector of observations X, the posterior probability of membership of group g_s which corresponds to the mixture component distribution with density function $f_s(X) = f(X,\theta_s)$ is defined as

$$p(s|X) = \frac{p(s).p(X|s)}{p(X)} = \frac{c_s f_s(X)}{h(X)}, \quad s = 1,2,\ldots,r \quad . \tag{2.12}$$

The log likelihood function of a sample of n random vectors drawn from the mixture h is given by

$$L = \sum_{k=1}^{n} \ln h(X_k). \tag{2.13}$$

To maximise this function subject to the constraint

$$\sum_{s=1}^{r} c_s = 1,$$

a modified log likelihood function L* is formed, given by

$$L^* = \sum_{k=1}^{n} \ell n\, h(X_k) - \omega(\sum_{s=1}^{r} c_s - 1),$$

where ω is a Lagrange multiplier. The maximum likelihood estimates obtained by maximizing this function when $h(X)$ is a mixture of multivariate normal distributions differing in both mean and covariance matrix are

$$\hat{c}_s = \frac{1}{n} \sum_{k=1}^{n} \hat{P}(s|X_k), \quad s = 1,2,\ldots,r, \tag{2.14}$$

$$\hat{\mu}_s^T = \frac{1}{n\hat{c}_s} \sum_{k=1}^{n} \hat{P}(s|X_k)X_k^T, \quad s = 1,2,\ldots,p, \tag{2.15}$$

and

$$\hat{\Sigma}_s = \frac{1}{n\hat{c}_s} \sum_{k=1}^{n} \hat{P}(s|X_k)(X_k-\mu_s)(X_k-\mu_s)^T, \quad s = 1,2,\ldots,r; \tag{2.16}$$

where μ_s is the mean vector of distribution s and Σ_s is the corresponding covariance matrix. These estimates are obtained on the assumption that there are no functional relationships between the parameters. If the distributions have a common covariance matrix then the previous result must be replaced by

$$\hat{\Sigma}_s = \frac{1}{n} \sum_{k=1}^{n} X_k X_k^T - \sum_{s=1}^{r} \hat{c}_s\, \hat{\mu}_s \hat{\mu}_s^T . \tag{2.17}$$

Formal maximum likelihood estimates of all the parameters may be found by solving these sets of equations. In the case of unequal covariance matrices, however, the method of maximum likelihood breaks down owing to the presence of singularities in the likelihood function. Nevertheless, since most numerical methods for obtaining maximum likelihood estimates search for a local maximum of the likelihood function, reasonable estimates of the parameters might be found in practice using these results. This question is investigated further below.

Wolfe proposed the following method of successive approximation for the solution of these equations. Initial estimates are provided for the parameters c_s, μ_s and Σ_s, $(s=1,2,\ldots,r)$. These estimates are substituted to give $\hat{f}_s(X_k)$ and hence $\hat{h}(X_k)$. Estimates of $p(s|X_k)$ are then calculated using

$$\hat{p}(s|X_k) = \hat{c}_s\, \hat{f}_s(X_k)/\hat{h}(X_k),$$

These estimates $\hat{p}(s|X_k)$ are then substituted to obtain new estimates of c_s, μ_s and Σ_s. This procedure is iterated until convergence is obtained. It is easily seen that this is an example of an EM algorithm where the expectation step is the calculation of the $\hat{p}(s|X_k)$ and the maximization step is the calculation of the new parameter estimates. The general theory of EM algorithms given by Dempster et al. (1977) shows that the value of the likelihood function must be strictly in= creasing on each iteration and that a maximum likelihood solution is a fixed point of the EM algorithm.

Once the parameters in the model have been estimated, the objects can be classified into groups corresponding to the component distribu= tions using any standard classification procedure. For example, the estimated likelihood $\hat{p}(s|X_k)$ that observation X_k arose from distribution f_s may be used to assign X_k to that distribution f_{s_o} for which $\hat{p}(s_o|X_k)$ is a maximum. This procedure is the analogue of the minimum risk deci= sion procedure for classification with known population parameters with these parameters replaced by their maximum likelihood estimates.

In some trials of Wolfe's algorithm it was found to perform very effectively in separating mixtures of three trivariate normal distri= butions with a slight to moderate degree of overlap between objects from the parent distributions. Details of these runs are given in Muller (1975).

In this approach, Bayes' theorem is used to get the posterior probability $p(s|X_k)$. An alternative approach is to note that each X_k follows only one of the distributions in the mixture; and so it seems reasonable to restrict $p(s|X_k)$ to the values 0 and 1.

This latter formulation is obviously equivalent to an allocation of each X_k to some cluster s, and so the reallocation approaches dis= cussed in section 2.2.3 may be applied. An alternative maximum likeli= hood procedure is that due to John (1970).

Despite the fact that X_k can, in fact, only come from one of the components of the mixture, the restriction of the model to an unequivo= cal allocation of X_k to a cluster introduces inconsistency in the esti= mates of the components of the mixture, which is not present in the proper maximum likelihood procedure of Wolfe. We shall comment further on this defect in later sections.

2.2.2 *Latent profile analysis*

A model which is also expressed in terms of mixtures of distribu= tions is the latent profile model.

Instead of assuming a specific functional form for the distribu= tions making up the mixture, this model postulates that within each distribution $f_s(X)$, $s = 1,2,\ldots,r$, the variates $X_1,X_2,\ldots,X_p$ are statis=

tically independent, that is $f_s(X)$ can be factorised as

$$f_s(X) = \prod_{i=1}^{m} f_{si}(X_i) , \quad s = 1,2,\ldots,r. \tag{2.18}$$

Let c_s be the probability that an observation drawn at random from the mixture originates from distribution f_s. Then

$$\sum_{s=1}^{r} c_s = 1.$$

The expected value of X_i is given by

$$E(X_i) = \sum_{s=1}^{r} \int\ldots\int x_i c_s \prod_{j=1}^{m} f_{sj}(x_j)dx_1,\ldots,dx_p \quad i = 1,2,\ldots,p$$

i.e.

$$E(X_i) = \sum_{s=1}^{r} c_s E(X_i|s). \tag{2.19}$$

Similarly, for any continued product $\Pi_{(t)}X_i$, where (t) denotes the subset of variates X_i included in the product,

$$E(\Pi_{(t)}X_i) = \sum_{s=1}^{r} \Pi_{(t)} E(X_i|s). \tag{2.20}$$

These equations are known as the accounting equations. Lazarsfeld and Henry (1968) discuss procedures for obtaining parameter estimates by replacing the quantities on the left hand side of these equations by sample estimates obtained from the mixed sample and solving the equations. This approach is not entirely satisfactory since the avail= able moments are not all used and are not used in a symmetrical manner. Thus a variety of solutions can be obtained according to the particular way in which the moments are used. The effect of sampling error on this procedure has not been determined, but might prove serious in practice, particularly as higher order moments are involved.

If the additional assumption of multivariate normality is made for each distribution then the condition of statistical independence between variates within a distribution holds if and only if all corre= lations between variables are zero within each distribution.[See, for example Anderson (1958), Theorem 2.4.2]. Thus the model becomes a spe= cial case of Wolfe's mixture analysis model with unequal, diagonal co= variance matrices.

When the observed variates X_i can only take on the values zero or one a model corresponding precisely to latent profile analysis is defined, known as latent class analysis. Methods of solution for this case are given by Lazarsfeld and Henry (1968), but these methods are not very satisfactory. Goodman (1974) gives a better method of solution using maximum likelihood. This solution can be computed using a general log-linear model program such as GLIM (Baker and Nelder, 1978).

2.2.3 *Allocation methods*

As we hinted briefly in section 2.2.1, the mixture model may be viewed in two complementary ways. Prior to sampling, the distribution for an arbitrary observation is

$$h(X) = \sum_{s=1}^{r} c_s f_s(X)$$

and by working directly with this mixture distribution and estimating its parameters one may find the components of the mixture.

Sampling itself may also be regarded as consisting of a multinomial random sampling in which S, a random one of the r components of the mixture is selected, followed by sampling X at random from $f_S(X)$. Thus the sample drawn may be regarded as consisting of N_1 random observations from $f_1(\cdot)$, N_2 from $f_2(\cdot),\ldots,N_r$ from $f_r(\cdot)$. Thus another possible approach is to attempt to partition the sample into those members coming from $f_1(\cdot)$, those from $f_2(\cdot),\ldots$ and so on. From this partitioning, one could deduce estimates for the c_s and any unknown parameters in the $f_s(\cdot)$.

This is the approach followed by the allocation methods. In likelihood terms, we may regard the allocation approach as defining parameters

$$\gamma_{ks} = \begin{cases} 1 & \text{if } X_k \text{ is in cluster S} \\ 0 & \text{otherwise} \end{cases} ,$$

the likelihood of the allocation being defined by

$$\prod_{k=1}^{n} \{ \sum_{s=1}^{r} \gamma_{ks} f_s(X_k) \} \tag{2.21}$$

An optimal allocation is one which maximizes this likelihood, and one normally attempts to find it by trying different allocations, and stepping from one trial allocation to another.

As an historical note, we might mention that this process actually evolved backwards in that criterion functions for the optimality of a particular allocation were defined first on intuitive grounds, and only later was it shown that some criteria correspond to particular and more or less plausible assumptions of particular forms for the $f_s(\cdot)$.

In theory, one could carry out an allocation optimization by generating all possible partitionings of the n data points into k clusters, computing L for each allocation, and selecting the largest such L.

In practice, however, the number of possible partitions of n ob= jects into k clusters is the very rapidly increasing Stirling number $S_{n,k}$ tabulated in Table 2.2

Table 2.2 *No. of Distinct Partitions* $S_{n,k}$ *of n objects into k Groups*

n \ k	2	3	4	5	$\sum_{i=1}^{n} S_{n,i}$
6	31	90	65	15	203
8	127	966	1701	1050	4140
10	511	9330	34105	42525	115975
12	2047	86526	611501	1379400	4213597
14	8191	788970	10391745	40075035	190899322
16	32767	7141686	171798901	1096190550	10480142147
18	131071	64439010	2798806985	28958095545	682076806159

It may be seen from Table 2.2 that even with n as low as fourteen, the number of possible partitions is far too large for the examination of all possibilities to be feasible, even with the aid of a fast digi= tal computer.

Iterative allocation techniques for finding the optimum partition have been proposed by several authors. The basic method used is to start with some initial partition and then at each stage follow some predetermined strategy in an attempt to form a modified partition which gives an improvement in the value of the criterion function. When the strategy fails to yield an improved partition the procedure is termina= ted. Optionally, the improvement obtainable by moving pairs of objects may then be investigated, and this operation, while time-consuming if n is large, provides some protection against local optima.

There are many variants and options in this broad approach (Ball and Hall 1967, Friedman and Rubin 1967, Freeman 1968, Banfield and Bassill 1977). For example one may use a random initial allocation, or one may first do a discriminant analysis to cluster the data around some initial "seeds" (which themselves may or may not be optimized). Then one may reallocate the first object to give an improvement in the criterion function, or the object giving the greatest improvement - we have found the former approach to be much faster.

In all the variants, however, there is no guarantee of reaching the global optimum. If the optimality of the final clustering is im= portant, then it is necessary to repeat the clustering several times altering such factors as the initial allocation and the ordering of the data vectors (the procedure is in general not invariant under per= mutation of the data vectors).

Since the criterion function is changing monotonically and the number of possible partitions is finite this procedure must terminate. A partition at which the procedure ends is called a 1-move local op= timum. The whole procedure may be repeated for several different ini= tial partitions. The partition corresponding to the best criterion value obtained is then selected. When a 1-move local optimum is reached, secondary strategies such as pairwise reallocation may be used in an attempt to improve the criterion value further.

A limitation of the procedure as described above is the fact that the number of groups k must be specified beforehand. In practice it is usual to repeat the analysis for each of a range of values of k. When this is done a combination of the allocation technique and Ward's method (see Section 2.1.5) may conveniently be used. The allocation method is first applied specifying the number of groups equal to a value slightly larger than the maximum number of groups anticipated in the data. At each subsequent stage an initial partition with one less group is found by considering each partition formed by uniting two of the groups from the previous stage. That union is performed which yields the best value for the criterion function and the relocation method is again used in an attempt to improve the criterion value further. This procedure is then repeated until the desired range of values of k has been covered. This technique was suggested by Howard (1966).

In order to define the criterion functions most commonly used with the allocation technique some further notation is now needed. Let $X = (X_{ij})$ of order nxm be the new raw data matrix with columns repre= senting variables and rows the n objects to be grouped. Without loss of generality it may be assumed that the variables are standardized to have zero means, i.e. that $\Sigma_i X_{ij} = 0$, $j = 1,2,\ldots,m$.

Suppose the objects have been partitioned into r groups g_s, $s = 1,2,\ldots,r$, each containing n_s objects, where $\Sigma_1^r n_s = 1$.

Let X_k^T be the kth row of X, and $\Sigma^s X_k$ denote summation over those values of k for which the observation X_k is assigned to group g_s. If the mean vector for the group g_s is $\bar{X}_s$, the within-group scatter ma= trix W_s is defined as

$$W_s = \Sigma^s (X_k-\bar{X}_s)(X_k-\bar{X}_s)^T. \tag{2.22}$$

The pooled within-group scatter matrix is

$$W = \sum_{s=1}^{r} W_s , \tag{2.23}$$

the total scatter matrix is

$$T = X^T X,$$

and the between-groups scatter matrix is

$$B = \sum_{s=1}^{r} n_s \bar{X}_s \bar{X}_s^T. \tag{2.24}$$

These matrices satisfy the analysis of variance identity

$$T = W + B. \tag{2.25}$$

Some criteria which have been proposed can now be defined in terms of these matrices. A simple and widely used criterion is given by

$$\min \operatorname{tr}(W) = \min \sum_{s=1}^{r} \Sigma^s (X_k-\bar{X}_s)^T (X_k-\bar{X}_s). \tag{2.26}$$

This criterion is simple to compute and has the advantage of being clearly interpretable geometrically, since minimizing tr(W) is equiva= lent to maximizing tr(B). Thus the criterion may also be expressed as

$$\max \operatorname{tr}(B) = \max \sum_{s=1}^{r} n_s \bar{X}_s^T \bar{X}_s.$$

Geometrically, this is equivalent to maximizing the weighted sum of squared Euclidean distances of the group mean vectors $\bar{X}_k$ from the grand mean vector, with weights proportional to the number of objects in each group.

This criterion was originally proposed, and is still often used, merely as a computationally simple and reasonable-looking criterion of the compactness of the clusters. In fact, it also has a likelihood in= terpretation in that tr W is the likelihood criterion if it is assumed that

$$f_s(\cdot) = N(\xi_s, \sigma^2 I)$$

with σ^2 arbitrary. The superficially more general situation

$$f_s(\cdot) = N(\xi_s, \sigma^2 \Gamma), \qquad \Gamma \text{ known}$$

may be handled by a slight generalization of the criterion, but is more sensibly treated by a preliminary transformation of X to $\Gamma^{-\frac{1}{2}} X$.

In practice, the user is seldom so fortunate as to know the co=variance matrix of his clusters up to an arbitrary constant multiple. Nevertheless, the tr(W) criterion is computationally so simple that it is often used whether statistically appropriate or not. Frequently, the sphericity assumption can be made at least a reasonable approxima=tion by a preliminary transformation of the data. One possible trans=formation is to the principal components of the total sample, scaled to unit variance:- an additional bonus if this is done is that by dis=carding all but the leading components, one can also reduce dimension=ality. We have found this approach to work well, especially in the difficult cases in which the clusters are not widely separated.

It is a well-known practical feature of the tr(W) criterion that, having assumed that the clusters are spherical, it tends to find clus=ters that are spherical. Thus if the covariance matrices of the compo=nent distributions are far from having the form $\sigma^2 I$, tr(W) will pro=duce poor clustering.

Another commonly used criterion is $|W|$. This is a likelihood criterion on the assumption that

$$f_s(\cdot) = N(\xi_s, \Sigma), \ \Sigma \text{ unspecified}$$

i.e. that the different components of the normal mixture differ in mean, but are homoscedastic. From the computational point of view, it is at best of order p times as difficult to optimize as is tr(W), but may be desirable simply on the grounds of its weaker assumption about the com=mon covariance matrix assumed for the clusters.

Another criterion is

$$GH = \min \prod_{s=1}^{r} |W_s/n_s|^{n_s} \tag{2.27}$$

which is the likelihood criterion if

$$f_s(\cdot) = N(\xi_s, \Sigma_s)$$

i.e. a general heteroscedastic normal model. A variant of this crite=rion is

$$IH = \min \prod_{s=1}^{r} |\text{diag } W_s/n_s|^{n_s}$$

which involves the additional assumption that the off-diagonal elements of all Σ_s are zero. This assumption is commonly made in the latent pro=file model with an assumed normal distribution within clusters.

Allocation models, especially those using the tr(W) criterion, are used very widely. There are two practical problems with their use, however. One is the question of convergence. While convergence will always occur, it is not necessarily to the global optimum. In fact, with the two heteroscedastic criteria, if any cluster is allocated too few members, we immediately get a value $-\infty$ for the criterion which, while certainly minimal, is not a helpful solution. Thus if one needs the global optimum and not some local one, repeated restarts with ran= dom initial allocations must be used.

The other problem is that unlike the mixture methods, the allo= cation methods usually produce inconsistent estimates of the parameters of the $f_s(\cdot)$. As a simple illustration of this, suppose one has a sample of m N(0,1) data and m N(1,1).

An optimal tr W (equivalently $|W|$) allocation would put all ob= servations less than 0.5 in one cluster, and all greater than 0.5 in the other. For large m, the means of the data in the two clusters will tend to -0.55 and 1.55, while the pooled variance will tend to 0.7. The Mahalanobis distance between the clusters will tend to 6 instead of the correct value 1.

This illustrates the general point that the objects allocated to the clusters will always underestimate the within cluster variability and overestimate the separation between the clusters.

There is some evidence that the discriminant functions for classi= fying future observations to one of the clusters are affected only mildly, so if one is concerned with estimating cluster parameters only for the purpose of setting up such discriminant functions, then the bias in the parameters may not matter. Otherwise however, one should use the cluster statistics to provide initial values for a few EM itera= tions of the mixture model.

It is interesting to note a parallel here with Jardine and Sib= son's extension of single linkage cluster analysis mentioned in Section 2.1.1. In both cases a method which uses discrete allocation of objects to clusters is shown to have undesirable characteristics and these are overcome by allowing objects to be associated with more than one clus= ter. In the mixture model this is done by giving each object a vector of probabilities of membership of each cluster and in Jardine and Sibson's method an object can be simultaneously included in up to k clusters where k is a parameter indexing a family of such methods.

2.2.4 *Q-mode factor analysis*

This is another method that may be used with profile data. A normal, or R-mode factor analysis expresses the *variables* measured in terms of a smaller number of factors (see Chapter 2)

$$X_{pxn} = L_{pxk} F_{kxn} + U_{pxn} \tag{2.28}$$

where L is the loading matrix, and F the factor score matrix. A Q-mode analysis attempts a similar explanation of the *objects* as

$$X^T_{nxp} = G_{nxk} M_{kxp} + U_{nxp} \tag{2.29}$$

Here G is again a loading matrix while M is a matrix of scores on $k < n$ "archetypes". As in an R-mode factor analysis, these archetypes may be interpreted by noting the cases on which they load most heavily. For ease of interpretation, it is vital that G contain a minimum of appreciably negative elements.

Q-mode factor analysis as a descriptive method has been used to good effect in a number of areas, notably geology, where it is sensible to regard all possible types of rock as being mixtures (i.e. linear combinations) of a small number of distinct "pure" forms - the archetypes. See for example Jaquet et al. (1975).

The analysis may also be used to cluster the objects by associating into a cluster all objects which load most strongly on the same archetype.

As the technique is most commonly applied, the term "factor analysis" is a misnomer, and should be replaced by "component analysis". The R-mode analysis involves finding principal components by solving the equation

$$(X^T X - \lambda I)a = 0 \qquad \text{(R-mode)}$$
$$\Rightarrow X(X^T X - \lambda I)a = 0$$
$$\Rightarrow (XX^T - \lambda I)Xa = 0 \qquad \text{(Q-mode)} \tag{2.30}$$

There is thus a simple and extremely useful correspondence between the R-mode and Q-mode principal components. This correspondence can be used not only to reduce computations (Gower 1966), but also to aid interpretation.

One vexed question in Q-mode component analysis is the scaling of the variables since, unlike the situation in R-mode analysis, it is necessary to add and multiply the different variables to one another. It is conventional to avoid this difficulty by scaling all variables to unit standard deviation, but this should not be done if the variables are known a priori to be on a common scale.

As this discussion makes clear, Q-mode component analysis does not rest on very sound statistical foundations, and its major benefits are the ease and speed with which it can be carried out.

3. PRACTICAL CONSIDERATIONS

The first practical problem facing a potential user of cluster analysis is which of the bewildering array of methods he should use. Part of the decision may be made for him:- if the data available con=sist only of inter-object similarities or dissimilarities, then, bar=ring some application of nonmetric multidimensional scaling using a suitable stress function, he must use a distance-based approach. We regard the principal competitors here as the single linkage and mode analysis techniques - the former largely on the grounds of its desirable theoretical properties, and the latter because of its ability to deal with odd-shaped clusters and straggling points that lead to chaining in single linkage. Between the two methods, we have had the better re=sults with mode analysis, and would use single linkage only if there are good prior reasons for doing so.

If the data consist of profiles, then one has the option of using a profile-based method or of converting the data to similarities or dissimilarities and using a suitable technique for the latter. If the user has reason to believe that a normal mixture model fits his data, or that an easily accessible (eg logarithmic) transformation would bring his data to a form in which a normal mixture would fit reasonably well, then a normal profile based method is preferable. If however the user does not see good prospects for fitting a normal, then he is ef=fectively compelled to use the distance approach. A common situation where this occurs is with biological material where the profile may consist of the presence or absence of various characteristics. Such data are commonly, and very successfully, treated by turning the bi=nary variables into a suitable matching coefficient and then using single linkage analysis or by using Gower's maximal predictive classi=fication approach. With profiles of non-normal interval data, some ingenuity may be necessary to define a suitable similarity or dissimi=larity coefficient.

Turning finally to the case of potentially normal profile data, the user must decide between using the mixture or reallocation algo=rithms and must decide what assumptions to make about the covariance matrices of the clusters. We can summarize the results of these deci=

sions as follows:

	Reallocation	Mixture
$\Sigma_i = \sigma^2 I$	tr(W)	-
Σ	$\|W\|$	Homoscedastic
Σ_i	GH	Heteroscedastic

(Note: the mixture model with Σ_i all assumed to be of the form $\sigma^2 I$ is easily formulated, but is not used commonly).

As already noted, the same basic statistical model underlies the mixture and reallocation approaches. Thus the choice between them is dependent on the ultimate use to which the results will be put, and the computational considerations.

Taking the computational considerations first in the reallocation methods, one must, at each stage, investigate the effect of reclassify= ing the object currently under consideration to each of the other clus= ters. This involves computing the effect on the criterion of each po= tential allocation, and actually implementing the change and updating the relevant summary statistics if a change is indicated. The checking is an order kp operation for tr(W) and an order kp^2 operation for both $|W|$ and GH (though the latter criterion takes longer to evaluate be= cause of the evaluation of k logarithms:- this is a significant part of the computation only if p is small). The updating following a reallo= cation is of order p for tr(W) and p^2 for the other two criteria.

It should be noted that in this, we are assuming that efficient updating formulae for W and the determinants are used. (Plackett 1950)

The mixture methods effectively involve a discriminant analysis for each observation (this is an order kp^2 computation) to get $Pr[s|X_i]$, followed by a recomputation of the summary statistics at the end of each cycle - an order $np^2 + p^3$ operation.

There is thus relatively little to choose between the computa= tions of one full cycle of the reallocation and the mixture procedures. The main difference, however, comes in the number of iterations through the full sample needed for convergence, where the reallocation methods generally require far fewer iterations than the mixture models. The reason for this is that in the reallocation models, $Pr[s|X_i]$ can only take the values 0 or 1, and this limitation on the possibilities leads to much faster convergence.

Turning now to objectives, if the desired end result of the ana= lysis is a classification of the objects into clusters, then realloca= tion is clearly the method of choice:- it is both faster than a mixture analysis, and requires no post-processing to give the needed results. Even if one wants both an allocation and discriminant function to classify future observations, reallocation is attractive, as the bias in the estimated discriminant functions is relatively small.

If one needs estimates of the parameters of the clusters, then a reallocation alone is inadequate, as it produces badly biased esti= mates of the cluster parameters. Thus at least some EM cycles of the mixture model are necessary to remove the inconsistency in the estima= tes. Even in this case, however, at least some preliminary reallocation cycles can be of value in speeding up convergence, though at the cost of more complicated programming.

The other choice to be made is between the three possible assump= tions about Σ_i. As tr(W) is the criterion most easily optimized, there is an incentive to use it. However this should not be done blindly; rather an attempt should be made to check the assumption $\Sigma_i = \sigma^2 I$, and if it is not appropriate, to achieve it at least approximately by a preliminary transformation. (This is not even theoretically possible unless the data are in fact homoscedastic).

The general homoscedastic model using $|W|$ should be used if there is reason to expect quite markedly non-diagonal Σ.

The general heteroscedastic criterion is very much the last re= sort. It is more sensitive to departures from normality than the other two criteria, requires more computer time and storage, and tends to fall into degenerate solutions in which some n_s falls below p making the corresponding W_s singular.

We have had some success in reducing this last defect. If the ob= servations in cluster s form a random sample from $N(\xi_s, \Sigma_s)$, then

$$E[|W_s/n_s|^{n_s}] = |\Sigma_s|^{n_s}.b(n_s)$$

the factor $b(n_s)$ being easily computed from Anderson (1958) p.171. Changing the criterion to minimizing $\Pi[W_s/n_s|^{n_s}/b(n_s)]$ reduces, but does not eliminate, the tendency to degenerate solutions.

These comments have been phrased in terms of the reallocation ap= proach, but they hold essentially without modification for the mix= ture approach as well.

4. INFERENCE ON MODEL FIT AND NUMBER OF CLUSTERS

The problems of inference on the number of clusters "actually" present in a set of data, and of testing for model fit, have not yet received much successful attention (Sokal, 1977) but are increasingly recognized as important.

As in the discussion of techniques, we shall distinguish between methods based on similarities or distances, and those based on pro= files. We shall simplify the situation slightly by assuming, in line with our earlier comments and recommendations, that the profile data discussed come from underlying normal models, while the distance-based methods are hierarchical.

Taking the latter type of situation first, we consider it the major virtue of the widely used single and complete linkage and mode analysis that their classifications are not affected by an arbitrary strictly monotonic transformation of the distance function, and that the consequent mildness of necessary assumptions on the distance func= tion gives the methods a broader applicability than have methods with more parametric assumptions. Thus, with these models, there are really no assumptions to verify. Neither is the question of the actual number of clusters a well-posed one, since a complete dendrogram of the ob= jects is produced, a k-cluster solution being obtained by cutting the dendrogram at some threshold distance. To specify a suitable threshold distance however, would require additional and undesirable assumptions about the distance function.

It is however, entirely legitimate and desirable to check on how well the dendrogram agrees with the original interobject distances. Letting d^*_{ij} denote the minimum distance at which objects i and j are connected in the dendrogram one may check the agreement between d_{ij} and d^*_{ij} using one of the many coefficients devised for the purpose (eg. Cormack, 1971). Such coefficients serve a useful descriptive purpose, but in the absence of stronger assumptions about the nature of the d_{ij}, one cannot as a rule attach any probabilistic interpretation to them.

We now turn to the test of model fit for profile data believed to follow, and modelled by, normal distributions.

4.1 Tests of normal fit of profile data

As we have pointed out, the normal mixture may be viewed margin= ally - i.e. each X_i comes from the mixture of distributions - or condi= tionally - i.e. each X_i comes from the J th component of the mixture, where J is a random variable. Corresponding to these two viewpoints and the corresponding algorithms, we have two approaches to testing model fit.

There are close analogies between the tests of model fit we pro= pose here and those of discriminant analysis. As the latter are set out in some detail in Chapter 1 of this volume we shall set out our tests here with a minimum of detail.

As in discriminant analysis, we shall concentrate on tests based on quadratic forms, as we believe these reflect accurately the key dis= tributional requirements of the model.

Consider first the mixture model

$$X_i \sim N(\xi_j, \Sigma_j) \text{ w.p. } c_j.$$

and suppose that the parameters c_j, ξ_j and Σ_j are known, or have already been estimated consistently from a large sample. Initially, we will as= sume $\Sigma_j = \Sigma$ - i.e. a homoscedastic model is fitted.

Let $\Delta_{mj} = (\xi_m-\xi_j)^T\Sigma^{-1}(\xi_m-\xi_j)$ denote the Mahalanobis distance be= tween the m th and j th distributions in the mixture, and let

$$V_{im} = (X_i-\xi_m)^T \Sigma^{-1}(X_i-\xi_m). \tag{4.1}$$

Then it follows at once that V_{im} has a mixture of noncentral chi-squared distributions

$$V_{im} \sim \chi^2_{p,\Delta_{jm}} \text{ w.p. } c_j \tag{4.2}$$

This immediately supplies the basis for a test:- Compute $V_{im}, i=1,\ldots,n$, form their empiric distribution function, and test it for compatibility with the mixture of χ^2 distributions supposed by the model. Use of a tail-emphasizing statistic such as the Anderson-Darling statistic will provide good power against potentially harmful deviations from model.

The value of m, the population from which the deviation was mea= sured, has not been specified. In practice, one would usually pick either that m for which c_m was a maximum or that for which the Δ_{jm} col= lectively were a minimum. The reason for either of these choices is to make the mixture distribution close to a central χ^2_p.

There is no difficulty other than computational tedium associated with this approach. In well-separated populations, however, one could introduce an approximation by selecting for each i, $\min_m V_{im}$, and checking these n values for compatibility with a χ^2_p distribution. To the extent that all Δ_{jm} are large, this approximation of the actual situ= ation will be valid.

An even cruder approximation involves classifying the data to their most probable source (this is most easily done by retaining for this purpose the $\Pr[s|X_i]$ from the last iteration of the EM algorithm). Then taking these allocations as the true source of the data, one may pass the data through a routine which checks discriminant analysis data for compatibility with model.

Unlike the two more exact approaches, this extends essentially without modification to heteroscedastic models.

An advantage of this last and roughest approach is that it allows the use of routines for the already well-established problem of testing discriminant analysis data for model fit, while the two sounder approaches do not. In fact, as we shall see later, the crude approach works fairly well, and so leaves little incentive to develop the other and better approaches.

This approach of classifying the data into the components of the normal mixture and then checking for compatibility with the discriminant analysis model clearly lends itself even better to the reallocation methods than the mixture models.

In addition to the test for model fit applied to the data, one should as a matter of course, also list the means, standard deviations, and correlation matrices of each of the components of the mixture. If the analysis was performed using the reallocation method, then in addi= tion, one should preferably run through a few cycles of the EM algorithm to remove the worst of the bias in these estimates. These summary statis= tics are often useful in providing a bird's eye view of the end result of the cluster analysis, and can also help detect model misfit.

4.2 Tests on number of clusters

A considerably more difficult problem is testing for the number of "significant" clusters. Part of the problem is easy, namely the choice of a test statistic. Both the mixture and reallocation approaches are based on the maximization of a likelihood; thus a suitable test statis= tic is any convenient monotonic function of the generalized likelihood ratio

$$L = \frac{\text{maximized likelihood for } k \text{ clusters}}{\text{maximized likelihood for } \ell < k \text{ clusters}} \tag{4.3}$$

The question that arises, and that remains largely unresolved still, is what the null distribution of L is if the data actually contain ℓ clusters.

Before discussing this thorny problem, we shall briefly mention some other approaches to the problem. One is the "elbow" method, in which one plots against k the maximized likelihood or any convenient function of it. On the assumption that this will improve dramatically with increasing k until one reaches the correct k, and will increase more or less steadily thereafter, the plot will show an "elbow" at the correct value of k.

Two theoretical problems with this are that the appearance of the plot can be altered completely by an innocuous (eg log) transformation of the criterion function; and that there is no reason to suppose that the criterion will show the same sort of increase on going, say, from 3 clusters to 4 as from 4 to 5 when 3 or fewer clusters are present. Thus the "elbow" method rests on a rather shaky theoretical foundation, and in practice, often produces multiple elbows or none at all, even when other approaches are clear on the number of clusters.

A related suggestion for the tr(W) case (Everitt, 1977, p.60) is to plot

$$C = \frac{(N-k)\ \mathrm{tr}(B)}{(k-1)\,\mathrm{tr}(W)} \tag{4.4}$$

against k. A maximum at k in this plot suggests k clusters. However, it is easy to see that if the data consist of, say, two very widely sepa= rated clusters with a third which is close to one of the other two,then the plot of C will wrongly suggest two and not three clusters.

Another graphical plot proposed is of $k^2|W|$ (Marriott, 1971). On the assumption that the profiles form a *single* sample from a *uniform* distribution. Marriott argues that $k^2|W|$ should be constant for all k, and thus if the plot shows a maximum at any k, then that k is the indi= cated number of clusters.

As Marriott points out, this suggestion would fail to resolve an equal mixture of two univariate normals whose means were separated by the not-inconsiderable amount of 3σ. The proposal has the further weak= ness that its theory is predicated on the null assumption of a single cluster, and so it is not clear whether it is capable, even under the most favourable conditions, of correctly distinguishing between, say, four clusters and five.

Returning to the likelihood ratio statistic, we shall first con= sider the test, for two clusters versus one using the allocation method. Often one knows what the distribution of L would be if the sample were allocated *at random* into the two clusters, and in this case one may ob= tain a conservative test of size $\leq \alpha$ using Bonferroni's inequality: L is the maximum of $L_1, L_2, \ldots, L_M$ where L_i corresponds to the i th of M possible allocations of the n observations into two clusters. Thus

$$\begin{aligned} \Pr[L > c] &= \Pr[\max_{1 \leq j \leq m} L_j > c] \\ &\leq \sum_{j=1}^{M} \Pr[L_j > c] \end{aligned}$$

For the division into two clusters, $M = 2^{n-1} - 1$. By choosing c so that the latter probability is α, we obtain a conservative test based on L.

Let W denote, for the moment, the sum of squares and cross products matrix for the complete sample, and W_1 and W_2 the corresponding matrices for the first and second clusters. Then special cases of the Bonferroni approach are the following:-

(a) $\Sigma_i = \sigma^2 I$:- tr(W) used.

Here the likelihood ratio L is a monotonic function of

$$F = \frac{tr(W-W_1-W_2)}{tr(W_1 + W_2)} \cdot \frac{(n-2)p}{p} \qquad (4.5)$$

which, for a random allocation into two clusters, would follow an F distribution with p and p(n-2) degrees of freedom. The Bonferroni approach is to test F against the α/M fractile of this F distribution.

(b) $\Sigma_i = \Sigma$:- |W| used.

The likelihood ratio is

$$L = \frac{|W|}{|W_1+W_2|}$$
$$= 1 + p\,F/(n-p-1) \qquad (4.6)$$

where, under a random allocation F would follow an F distribution with p and n-p-1 degrees of freedom. It may thus be tested against the α/M fractile of this distribution.

(c) Σ_i general - GH used.

The likelihood ratio here is the familiar "identity of two normal populations" statistic: e.g. Anderson (1958) p.256. The user should be warned that, because of the tendency of the heteroscedastic criterion to split the data into unbalanced classes, the usual asymptotic approximations to the distribution of L_i are of rather dubious validity here, and so the test should be regarded as indicative rather than definitive.

The Bonferroni approach is well-known to be highly conservative in this context. Thus while a significant value certainly indicates the need for two clusters, a non-significant value does not necessarily deny this need.

For this reason, Engelman and Hartigan (1969), and Scott and Knott (1976) have derived some non-conservative asymptotic results for the case p=1 for which case the tr(W) and |W| approaches are identical.

We shall simply sketch an extension of these univariate results to $p > 1$. As the clustering given by the $|W|$ criterion is invariant under non-singular transformation, it makes no difference to the clus= tering using $|W|$ what the value of Σ is, and so we shall assume $\Sigma = I$.

The key to the $p > 1$ situation is an understanding of how tr(W) will divide a $N(0,I)$ sample into two clusters. A little thought sug= gests that effectively, though not exactly, it will use the sample's leading principal component as a discriminant function, and will divide the sample near the middle of the range of this component, producing two clusters of about equal size. The p-1 directions orthogonal to the principal component will contribute random χ^2 distributed components to tr(W), but no more.

Provided the sample is reasonably large, the sample values of the leading principal component will be distributed approximately as $N(0,\lambda\sigma^2)$ where λ is either the observed leading eigenvalue of W, or a suitable measure of central tendency, such as the mean or median, of the eigen= value of a Wishart (I,n-1) matrix. Defining

$$R = \ell n \left[\frac{\mathrm{tr}(W-W_1-W_2)}{\mathrm{tr}(W_1+W_2)}\right]$$

then asymptotic distribution calculations analogous to those of Hartigan (1975), p.98 suggest that asymptotically, R is approximately normal with mean

$$\mu \doteq \ell n \left[\frac{2\lambda n + \pi(p-1)}{n(\pi p - 2\lambda)}\right] \doteq \ell n \left[\frac{2\lambda}{\pi p - 2\lambda}\right]$$

and variance

$$\tau^2 \doteq \frac{2\pi^2 p(\pi p - 2\lambda - 2p + 1)}{n(\pi p - 2\lambda)^2}$$

and this result may be used to establish the approximate significance of a binary split using tr(W), when the underlying model $\Sigma_i = \sigma^2 I$ is appro= priate.

This approximation is rather rough and ready, but, especially in large samples, provide a reasonable guide to the significance of a bi= nary split.

The $|W|$ criterion has been studied by Lee (1979) who presents em= pirical evidence that the α fractile of $\ell n \{|W|/|W_1 + W_2|\}$ may be approxi= mated, at least for small p, by

$$- \ln(1-2/\pi) + Z_{\alpha^*}(n-2)^{-\frac{1}{2}} + \{2.4 + 5.2(p-1)\}(n-2)^{-1}$$

where Z_{α^*} denotes the p/α fractile of a $N(0,1)$ distribution.

Turning now to the more general decision of testing k clusters against k-1, we find that these binary split results still have some value. If in fact the sample consists of a mixture of k-1 well-separated normal populations, then to form the k cluster solution, the algorithm can do little better than to carry out a binary split of one of the k-1 clusters. This fact, incidentally, provides a useful quick visual diagnostic in a reallocation analysis:- if the allocation begins to depart radically from hierarchic form with the k cluster solution bearing little resemblance to the k+1 or k-1 solutions, then it is likely that overfitting has occurred and spurious clusters are being found.

It is thus commonly the case, if the k-1 cluster solution consists of "real" clusters, that the k cluster solution differs from it by a binary split, or a reallocation closely approximating a binary split. The overall significance of the k cluster solution as against the k-1 cluster one may thus be assessed by finding the significance of the split (or approximate split) actually made, and applying a Bonferroni multiplier of (k-1) to it.

Turning now to the mixture model, an unthinking application of the regular asymptotic theory of likelihood ratio tests might lead one to suppose that 2 log L would follow a χ^2 distribution with degrees of freedom obtained by the usual parameter-counting method. This however, would be wrong, since the null hypothesis tested is not an interior point of the model's parameter space.

For the homoscedastic case, Wolfe (1971) has carried out some simulations suggesting that if

$$L = \frac{\text{maximized likelihood for k clusters}}{\text{maximized likelihood for } \ell < \text{k clusters}}$$

then

$$2(n-1-p-\tfrac{1}{2}\ell)n^{-1} \ln L$$

follows an approximate χ^2 distribution with $2p(k-\ell)$ degrees of freedom. (Note that the erroneous parameter counting argument would suggest $(p+1)(k-\ell)$ degrees of freedom). Our own experience with simulated normal data disagrees sufficiently with Wolfe's suggestion to make us apprehensive about its validity for large values of n or p, however it seems unlikely to lead one seriously astray on moderate values of p.

4.3 Illustration

As an illustration of testing for model fit, we simulated some data from 3 populations with 40 cases per population. Six sets of data were simulated:

1: p=2, all 3 populations N(0,I)

2: p=20, all 3 populations N(0,I)

3: p=2, the first component in the three populations having mean 0, 1.5 and 3 respectively.

4: As in case 3, but with p=20

5: As in case 3, but with the second component in the three populations having variances 1, 2.9 and 5.8 respectively.

6: As in case 5, but with p=20.

Cases 1 to 4 correspond to $\Sigma_i = I$, while cases 5 and 6 are hetero=scedastic. Cases 1 and 2 contain only a single cluster, and so are in=formative about null hypothesis behaviour of the analysis.

The covariance matrices of all cases are the identity I, except for the two heteroscedastic runs. The appropriate criterion is thus tr(W); using this in all 6 runs also gave us the opportunity to measure the performance of tr(W) on heteroscedastic data.

The minimized values of tr(W) for the 6 cases and different num=bers of clusters are given in Table 4.1

Table 4.1

	Set					
k	1	3	5	2	4	6
10	40	63	80	1666	1758	1990
9	42	69	85	1682	1747	2019
8	45	79	96	1732	1778	2061
7	54	88	111	1768	1826	2109
6	63	103	132	1800	1880	2169
5	76	113	157	1867	1945	2241
4	96	134	195	1939	2015	2304
3	117	169	240	2022	2095	2417
2	172	226	370	2126	2214	2600
1	251	438	544	2255	2473	2842

For the case of 4 or fewer clusters, the solutions for all six sets were reasonably near to hierarchic. The asymptotic theory requires the value of λ - the leading eigenvalue of the sample covariance matrix. This was not computed; we used the expected value of λ for a sample of size 120, which is $\lambda \doteqdot 1.12$ for p=2 and $\lambda \doteqdot 1.85$ for p=20.

The significance as assessed by the Bonferroni and the asymptotic approaches are given in Table 4.2.

Table 4.2

	Set					
	1	2	3	4	5	6
2 vs 1						
Bonferroni	$2\text{x}10^{16}$	$4\text{x}10^{16}$	82	$2\text{x}10^{-5}$	$1\text{x}10^{16}$	533
Asymptotic	.9	.6	$1\text{x}10^{-4}$	$3\text{x}10^{-6}$	.9	$2\text{x}10^{-3}$
3 vs 2						
Bonferroni	62	$1\text{x}10^{6}$	$2\text{x}10^{5}$	$8\text{x}10^{4}$	.01	.9
Asymptotic	.8	.9	1	.7	.6	.1
4 vs 3						
Bonferroni	$3\text{x}10^{8}$	71	54	6000	$3\text{x}10^{4}$	$6\text{x}10^{5}$
Asymptotic	1	1	1	1	1	1

While the Bonferroni approach is certainly shown in this table to be extremely conservative, the asymptotic theory also appears to be con= servative for testing more than two clusters. For n=120 and p=2,20, the asymptotic mean and variance of R are given by the theory as -0.59, 0.14 for p=2, and -2.77, 0.14 for p=20. We ran an independent test with 45 sets of 120 $N(0,I_2)$ and 33 sets of 120 $N(0,I_{20})$ vectors, fitting one, two and three clusters, and got the following means and standard devia= tions for R.

	Simulated		Asymptotic	
	mean	s.d.	mean	s.d.
p=2				
2 vs 1	-0.57	0.11	-0.59	0.14
3 vs 2	-0.75	0.17		
p=20				
2 vs 1	-2.74	0.09	-2.77	0.14
3 vs 2	-3.05	0.11		

In both cases, for testing 2 clusters versus 1, the approximate theory is astonishingly accurate, but for 3 clusters versus 2, it is conservative. Clearly much research still needs to be done on the distri= bution theory for distinguishing $k > 2$ clusters from k-1.

For testing the appropriateness of the homoscedastic normal model we used the classification determined by the 3 cluster tr(W) solution and submitted the data to a discriminant analysis routine incorporating the Anderson-Darling statistics for model fit discussed in Chapter 1.

Table 4.3

Set 1

Cluster	n	Anderson-Darling	Normal component 1	2
1	61	1.62	-1.34	1.05
2	35	0.77	-0.33	0.69
3	24	0.63	0.47	0.07
Overall	120	1.34	-0.92	1.15

Set 2

Cluster	n	Anderson-Darling	Normal component 1	2
1	47	0.74	-0.32	-1.58
2	35	0.69	0.82	-0.71
3	38	1.02	-0.64	-0.99
Overall	120	0.81	-0.11	-1.93

Set 3

Cluster	n	Anderson-Darling	Normal component 1	2
1	50	0.70	-0.32	-1.04
2	44	0.36	-0.35	-0.52
3	26	0.51	-0.78	0.06
Overall	120	0.69	-0.78	-0.33

Set 4

Cluster	n	Anderson-Darling	Normal component 1	2
1	32	0.63	0.51	-1.14
2	46	0.51	-0.42	-1.03
3	42	0.29	-0.33	-0.20
Overall	120	0.60	-0.19	-1.34

Set 5

Cluster	n	Anderson-Darling	Normal component 1	2
1	29	1.69	-0.29	2.23
2	49	7.66	-3.61	0.50
3	42	2.08	0.41	0.95
Overall	120	5.84	-2.21	1.98

Set 6

Cluster	n	Anderson-Darling	Normal component 1	2
1	24	2.75	-1.47	-2.42
2	60	1.40	1.10	-1.39
3	36	1.18	-0.72	-0.12
Overall	120	1.08	-0.27	-2.13

Several striking features of these results catch the eye. First, in sets 1 and 2, the clusters in fact consist of a trisection of a single normal population, and so cannot be truly normal. Yet the minimum A_{ij} have so nearly uniform a distribution as to give no hint of a model mis= fit. However much of a disadvantage this might be in other application areas, it means that the proposed test for multivariate normality will pass the data if they are normal and homoscedastic, *even if* the number of clusters is overspecified. Note that the model fit can only improve if any separation between the clusters is introduced; thus these two sets represent worst cases for genuinely spherical normal data.

There are no great surprises in sets 3 and 4:- the fit is good, and inspection of the cluster mean vectors shows that the analysis has recovered the actual components of the mixture.

Sets 5 and 6 show the unmistakeable signs of the heteroscedasticity in the data, fairly modest though this is. Actually plotting out the data and the discriminant functions of set 5 shows that one of the clus= ters recovered is the spherical component of the mixture, while the re= maining two divide the data, not along the axis of separation of their means, but along the axis of greater scedasticity. In the face of this clear evidence of heteroscedasticity of the clusters, we then analyzed the data using the heteroscedastic criterion. This proved to be very sus= ceptible to settling into local optima, but, given good starting values (or sufficient random restarts) was able to converge to the global opti= mum corresponding to the actual components of the mixture.

5. SEGMENTATION

5.1 Introduction

Segmentation is a specialized form of cluster analysis with restric= tions on cluster membership. Two broad classes of problems will be dis= cussed, namely the segmentation of a sequence and that of spatial data.

The segmentation of a sequence has received much attention in the literature recently and seems to be fairly well covered. This type of problem arises when dealing with time series or when there is some order that must be retained in the data. The problem of segmenting spatial data usually occurs in geographic setups where observations are taken at dif= ferent points in a plane and the resulting clusters must consist of points which are geographically connected.

We shall discuss the segmentation of profile data only. The segmen= tation of (dis)similarity data, may, however, be carried out by simple analogues of the techniques we mention for the more difficult problems of profile data.

5.2 Segmentation of a sequence

Let $X_1, X_2, \ldots, X_n$ denote a sequence of observation vectors each having p components. The segmentation of a sequence into k segments consists of finding the changepoints $0 = b_o < b_1 < \ldots < b_k = n$ such that the set of X_j, with $b_{i-1}+1 \le j \le b_i$, are in some sense homogeneous.

A suitable statistical model may be given by

$$X_j \sim N(\xi_i, \Sigma_i),\ b_{i-1}+1 \le j \le b_i, \quad 1 \le i \le k.$$

It is assumed that the ξ_i differ from segment to segment. Depending on whether the Σ_i also differ between segments, two classes of problems can be distinguished, namely the homoscedastic model where $\Sigma_i = \Sigma\ \forall i$, and the heteroscedastic model where the Σ_i may vary from one segment to another.

The problem of locating the changepoints relates closely to one of hypothesis testing, since the optimal changepoints are those which maximize the significance of the test for identity of the k segments defined by the changepoints, and the criteria which are used are similar to those in ordinary cluster analysis.

$$\text{Let } \bar{X}_{\ell,m} = \sum_{i=\ell+1}^{m} X_i/(m-\ell)$$

and

$$W_{\ell,m} = \sum_{i=\ell+1}^{m} (X_i - \bar{X}_{\ell,m})(X_i - \bar{X}_{\ell,m})^T$$

The pooled within-group scatter matrix is then given by

$$W = \sum_{i=1}^{k} W_{b_i-1,b_i}\ ,$$

and the total scatter matrix is given by

$$T = \sum_{i=1}^{N} (X_i - \bar{X})(X_i - \bar{X})^T \equiv W_{0,N}.$$

For the homoscedastic model no single criterion exists which is uniformly better than all the others, but the following criteria are most commonly used:

(i) The largest characteristic root of $T\,W^{-1}$

(ii) The determinant $|TW^{-1}|$

(iii) The trace $tr(TW^{-1})$

If it is reasonable to suppose that the Σ_i are not only equal, but spherical, then the tr(W) criterion could be used. In segmentation, however, there is little computational need for this simplification, and so it is not commonly made.

For the heteroscedastic case only one standard criterion is used, namely the log likelihood ratio criterion which is given by

$$\min \sum_{i=1}^{k} n_i \, [\ell n|W_{b_{i-1},b_i}| - \ell n|T| - p \, \ell n \, n_i + p \, \ell n \, n]$$

where $n_i = b_i - b_{i-1}$.

Although the heteroscedastic model is more general and practical experience has indicated that fairly often the data are found to be heterogeneous, extra care must be taken when using this model since it is much more sensitive to departures from the assumed normality than are the criteria for the homoscedastic model.

To obtain an optimal solution a substantial amount of matrix computation is necessary for both the homoscedastic and heteroscedastic criteria, and for large values of p the computational effort becomes excessive. Two heuristic proposals have been proposed to oversome this difficulty, namely principal components and the canonical variates approach.

When using principal components the minor components are discarded and this will hopefully reduce the dimensionality of the problem so that an exact analysis of the retained components becomes feasible. The disadvantage however, is that there is no guarantee that the minor components can safely be discarded. It is for example possible that a very strong general factor exists along the sequence, but it might not be relevant for distinguishing the changepoints, for example it may be a "growth effect". This factor will dominate the principal components and the actual differences between the segments will only show up in the minor components. Thus retention of only the major components will be fatal. In our experience such irrelevant "growth factors" have been more common in segmentation than cluster analysis and militate against a principal component reduction of dimensionality.

For the canonical variate approach a specific statistical model is assumed, namely

$$X_j \sim N(\xi_i, \Sigma), \quad b_{i-1}+1 \leq j \leq b_i$$

and $\quad \xi_i \sim N(\xi, \Gamma) \; , \quad 1 \leq i \leq k,$

and is thus only applicable in the homoscedastic case. A discussion of this approach and of the estimation of the canonical variates from the data can be found in Hawkins and Merriam (1974) and Hawkins and Ten Krooden (1979b). The advantage of using the canonical variates is that under the assumptions of the statistical model it is guaranteed that the leading canonical variates are the only important variables and the minor canonical variates can thus be discarded with safety. If a strong general effect exists it will be de-emphasized in the canonical variate approach but not in the principal component analysis.

5.3 Methods for segmentation

5.3.1 *Global optimization*

In cluster analysis, global optimization is computationally in= feasible except for almost trivially small data sets, but it is possible in segmentation of quite big data sets.

Using the appropriate criterion for either the homoscedastic or heteroscedastic model, this method finds that set of changepoints $\{b_o, b_1, \ldots, b_k\}$ which maximizes the criterion over all possible sets of changepoints. This maximization may be performed by using the optimiza= tion method of dynamic programming, which is discussed in Hawkins and Merriam (1974) for the homoscedastic case and in Hawkins and Ten Krooden (1979a) for the heteroscedastic case.

From the theoretical point of view the global optimization method provides the best solution to determining the changepoints. Unfortunately both storage and time requirements are quadratically dependent on the length of the sequence, and the segmentation of very long sequences, say $N \geq 300$, might not be practically feasible. Two heuristics have been pro= posed to allow the segmentation of long sequences. The one is to allow only every ℓ th point as a potential changepoint, which will reduce the requirements for both storage and time by a factor of ℓ^2. Another heuris= tic is to divide the sequence into several manageable sections and to segment each section. Thereafter the sections may be spliced together and it is only necessary to check whether the first segment of any sec= tion needs to be pooled with the last segment of the previous section. Both heuristics remove not only the computational difficulties but also the guarantee that the optimal set $\{b_i\}$ is found.

5.3.2 *Split moving window*

This segmentation method was proposed by Webster (1978). In con= trast with the global optimization the split moving window attempts to locate only one changepoint at a time. First a window width 2h say has to be defined. This window is then moved along the sequence from one end to the other, one observation at a time (assuming that the sampling in= terval is constant). At each position the exposed portion of the sequence is divided about its mid point, and the effect of placing a changepoint at this position is measured by using any of the standard statistical criteria for testing the difference between the two groups. The test criterion is computed for every position of the window and plotted against the mid points. The changepoints will appear as peaks or troughs (depending on the criterion) in the resulting graph.

A major disadvantage of this method is the necessity of specify= ing the window width in advance. On the one hand one wishes to make h as large as possible so that small shifts can be detected reliably. On the other hand, if 2h is so large that more than one changepoint is contained in the window, the performance of the method can be inhibited badly.

The advantage of this method is that its computer requirements are minimal, and large sequences can be segmented with ease, so despite its theoretical difficulties, it is often used in practice.

5.3.3 *Hierarchic methods*

These methods include the hierarchic aggregative, the hierarchic disaggregative and the stepwise methods. In the disaggregative method one starts with only one segment, consisting of the whole sequence, and then starts adding changepoints as they are shown up by the data. Thus at stage i it is supposed that i changepoints have already been deter= mined. For each existing segment an extra changepoint is determined, using any of the standard criteria, and that segment for which the changepoint is most significant is then split into two new segments. This process is repeated until the desired number of segments is ob= tained.

The aggregative method involves the deletion of changepoints. At each stage of the process the least significant changepoint is deter= mined and the two corresponding segments are merged to form one new seg= ment. This process is repeated until all the segments have been merged into one, or until the desired number of segments has been obtained. The starting point of this method needs care. One might wish to start with all N points as changepoints, but when for example the heteroscedastic criterion is used, and a segment has fewer than p points, the determi= nant of the covariance matrix is zero, and its log infinite, giving a degenerate solution to the problem. An heuristic which has been found to work well, is to assign some positive value δ to all $|W_{i-1,i}|$ and let the merging proceed from there.

The aggregative method seems to be attractive, since at any stage when considering whether to merge two segments, one has as far as possible ensured that the segments are as wide as they can be, thus ensuring maxi=

mum power. On the other hand, the two segments have each already passed tests of homogeneity and so are not likely to consist of several dis= parate segments. Nevertheless, the method is not perfect. In the early stages of the segmentation when the segments are still short and the statistical tests are consequently not yet powerful, it is possible to merge segments that later on when additional segments have been merged in, appear incompatible. A final adaptation which helps to alleviate this difficulty is the stepwise method which is a combination of the aggregative and disaggregative methods.

At each stage of the stepwise method the most similar pair of ad= jacent segments are merged if they are not significantly different. Other= wise, the existing segments are tested for possible subdivision, and a segment is split if the appropriate criterion is a maximum over all seg= ments, and exceeds a prespecified cutoff. It is also possible to tighten and relax the cutoffs for merging and splitting alternately, which has the effect of allowing individual points or short segments to become detached from a segment and then reattached to the neighbouring segment if that should be more compatible.

Of the non-global methods it is felt that the stepwise method is the best. It avoids the difficulties of the need to specify a window width in the split moving window and of a choice of changepoint at one stage of the purely hierarchic methods which appears at a later stage to have been erroneous. Despite our lumping it with the two purely hier= archic methods, it need not necessarily produce a hierarchy.

5.4 Segmentation of spatial data

Consider a two-dimensional region and assume that a vector with p components has been observed at each of N points in this region. The problem now is to divide this region into k subregions such that the set of vectors within a subregion is homogeneous. The statistical model is similar to the model used in the segmentation of a sequence, namely

$$X_j \sim N(\xi_i, \Sigma_i), \text{ for all } X_j \text{ in subregion } i,$$

where the ξ_i may differ between regions. Again we have the homoscedastic model where $\Sigma_i = \Sigma$, for all i, and the heteroscedastic model where the Σ_i may also vary from one subregion to another.

5.4.1 *Methods for segmentation*

Theoretically a global optimization technique based on any of the standard criteria will give the best solution, but in the case of spatial data (unlike sequence data) this technique seems to be impossible to im= plement. Two main methods for the segmentation of spatial data have been proposed so far.

The first method is a disaggregative one, and for practical reasons the data points need to be evenly spaced on a grid. At each step of the method a straight line is used to divide a region into two parts. To make it more feasible practically, only horizontal, vertical and diagonal lines are used. The process starts with only one region. Hotelling's T^2 is then used to determine the best position to split this region into two subre= gions using a straight line. At stage i when i subregions have already been found the best split position for every subregion is considered and that region for which the split is the most significant is then divided into two. This process continues until the desired number of regions has been obtained.

This method relates closely to the hierarchic disaggregative method for segmenting a sequence and the same type of problem may be encountered here, for example a line that in the early stages of the process seemed to be the best split position may at a later stage appear to have been erroneous. Another criticism against this method is that in real life situations one will usually find irregularly shaped subregions and thus this straight line segmentation method will not be suitable.

Another possibility is to use the reallocation techniques of clus= ter analysis, but to restrict cluster membership so that the clusters al= ways consist of points which are geographically contiguous. Thus in the reallocation process only the border points of a cluster can be realloca= ted to another cluster and then only to the geographical adjacent cluster. This method however seems not to be very stable and is easily trapped in= to local optima. Several random restarts will thus be necessary to deter= mine the global optimum.

This technique may also be broadened to include single linkage on (dis)similarity data by setting to infinity the dissimilarity between non-contiguous points.

5.5 Illustrations

Example 1: Segmentation of the Sandford transect

This transect near Sandford St Martin in North Oxfordshire lies across a succession of gently dipping and folded lower Jurassic sediments. It is 3200 m long and the soil was described to 1 m from pits at 10 m in= tervals, by recording 63 properties at each pit. The data were transformed to principal components and only the first six components were retained for further analysis.

Global optimization and the split moving window method were ap= plied to the data and the results are given in Webster (1978) and Haw= kins and Ten Krooden (1979a).

The split moving window method used the Mahalanobis D^2 criterion with a window width of 140 m, i.e. 14 sampling intervals. Only four boundaries could reliably be determined.

Global optimization was performed using both the tr(W) criterion for the homoscedastic case on the first six canonical variates, and the heteroscedastic criterion on the first six principal components. Fair= ly similar results were obtained in these two analyses. Nearly all the boundaries which could be recognized in the field were found by the global optimization techniques. Some of the boundaries are slightly dis= placed, but this occurred where the changeover is either very gradual or small.

The reason why the global optimization methods performed much bet= ter on this particular set of data is that the changes are almost al= ways gradual and not very large. Since the optimization is done globally, boundaries will be placed in a gradual transition zone provided that the difference between the segments on either side is large compared with other changes along the transect. On the other hand the split moving win= dow method uses only local information to determine a changepoint and gradual changes cannot be determined reliably. In cases where the change is sharp the split moving window method will easily recognize a boundary.

Example 2: Spatial segmentation of Cae Ruel data

This set of data consists of 442 observations on a 15 m grid. At each grid point 7 variables were measured, 6 chemical variables and the stone content, describing the top soil.

First the data were transformed to canonical variates. For the straight line segmentation method Hotelling's T^2 was used as splitting criterion and the region was subdivided until 10 subregions were obtained. In the reallocation method we started with a random division of the data into 10 subregions and then used the tr(W) criterion to find the best allocation into 10 subregions. The two most similar regions were then merged and again the tr(W) was used to find the 9 subregion solution. This process was repeated until all the regions were merged into one region.

Very similar results were obtained by both methods for the 2,3 and 4 subregion solutions. For more than 4 subregions, however, the results differ markedly.

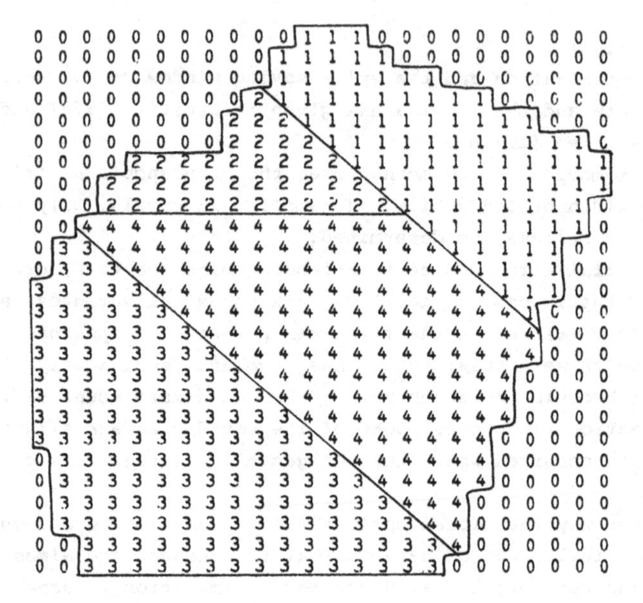

Figure 5.1.a

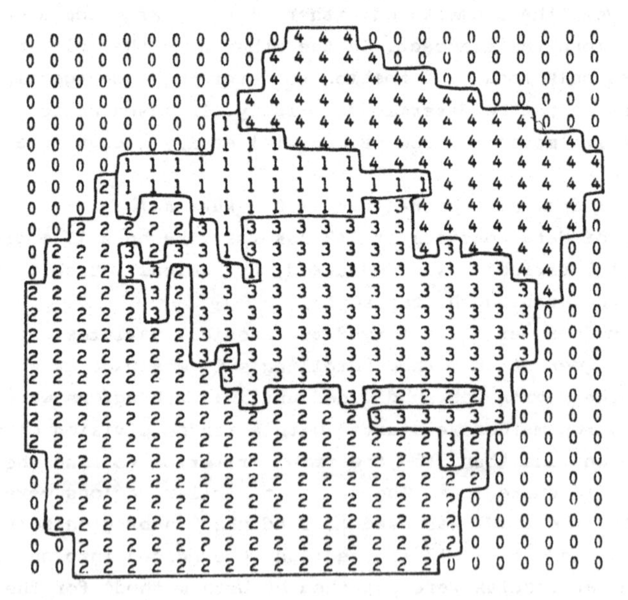

Figure 5.1.b

The results for 4 subregions are given in Fig. 5.1.a for the straight line segmentation and in Fig. 5.1.b for the reallocation method. The numbers indicate to which subregion the observation vector, measured at the grid point, was allocated. A zero indicates that no measurements were taken at the particular point. We are grateful to Dr. R. Webster for supplying the original data for both these examples.

6. FURTHER READING

Recently, several books on cluster analysis have appeared: notably Van Ryzin (1977), Hartigan (1975), Everitt (1977), Spaeth (1977), and serious users of cluster analysis methods would do well to read them. We have adopted an uncompromisingly model-based approach. This has as a natural consequence that, since only a few models are commonly used, only the few algorithms deriving from them are of interest. Most other writers (and all the above works, to varying degrees) lay more stress on algorithms and criteria in the belief that intuitively reasonable criteria should produce good results over a wide range of possible (and generally unstated) models.

For example, we regard the tr(W) criterion as predicated on a model of normal data with spherical within-cluster covariance matrices. In view of known robustness properties, we would have no hesitation in ap= plying it to other light-tailed data, but not to heavy-tailed data. We would be reluctant to apply it to any data whose within cluster covariance matrix we thought to be non-spherical, and would regard the results of such an analysis at best as providing a summary statistic of the data about which no inferential statements could be made. By contrast many other authors would apply tr(W) even in the face of evidence of non-spherical clusters, or would use Euclidean distance as a metric (which boils down to very much the same thing).

The reader should be aware of these differences in approach, and of the consequent differences in emphasis and content.

REFERENCES

ANDERBERG, M.R. (1973). *Cluster Analysis for Applications*. New York, Academic Press.

ANDERSON, T.W. (1958). *An Introduction to Multivariate Statistical Analysis*. New York, John Wiley.

BAKER, R.J. and NELDER, J.A. (1978). *The GLIM System Release 3*. Oxford, Numerical Algorithm Group.

BALL, G.H. (1971). *Classification Analysis*. Stanford Research Institute, Stanford.

BALL, G.H. and HALL, D.J. (1967). A Clustering Technique for Summari= zing Multivariate Data. *Behavioral Science*, 12, 153-155.

BANFIELD, C.F. and BASSILL, L.C. (1977). A Transfer Algorithm for Non-Hierarchical Classification. *Appl. Statist.*, 26, 206-210.

BIJNEN, E.J. (1973). *Cluster Analysis*. The Netherlands, Tilburg Univer= sity Press.

CORMACK, R.M. (1971). A Review of Classification. *J. Roy. Statist. Soc. A*, 134, 321-353.

CUNNINGHAM, K.M. and OGILVIE, J.C. (1972). Evaluation of Hierarchical Grouping Techniques: A Preliminary Study. *Computer Journal*, 15, 209-213.

DAY, N.E. (1969). Estimating the Components of a Mixture of Normal Distributions. *Biometrika*, 56, 463-474.

DEMPSTER, A.P., LAIRD, N.M. and RUBIN, D.B. (1977). Maximum Likelihood from Incomplete Data via the EM Algorithm. *J. Roy. Statist. Soc. B*, 39, 1-22.

DIGBY, P.G.N. and GOWER, J.C. (1981). Ordination between- and within-groups applied to soil classification. *Syracuse University Geo= logy Contributions*, 8.

ENGELMAN, L. and HARTIGAN, J.A. (1969). Percentage points of a test for clusters. *J.Amer.Statist.Assn.*, 64, 1647-8.

EVERITT, B. (1977). *Cluster Analysis*. Heinemann Educational Books, London.

FREEMAN, J.J. (1968). *Experiments in Discrimination and Classification*. NRL report 6742, Naval Research Laboratory, Washington, D.C.

FRIEDMAN, H.P. and RUBIN, J. (1967). On Some Invariant Criteria for Grouping Data. *J. Amer. Statist. Assoc.*, 62, 1159-1178.

GOODMAN, L.A. (1974). The Analysis of Systems of Qualitative Variables when Some of the Variables are Unobservable. I. A Modified Latent Structure Approach. *Amer. J. Sociology*, 79, 1179-1259.

GOWER, J.C. (1966): Some distance properties of latent root and vec= tor methods used in multivariate analysis, *Biometrika*, 53, 325-338.

GOWER, J.C. (1967). A Comparison of some Methods of Cluster Analysis. *Biometrics*, 23, 623-637.

GOWER, J.C. (1974). Maximal predictive classification. *Biometrics*,30, 643-654.

GOWER, J.C. and ROSS, G.J.S. (1969). Minimum Spanning Trees and Single Linkage Cluster Analysis. *Appl. Statist.*, 18, 54-64.

HARTIGAN, J.A. (1975). *Clustering Algorithms*. Wiley, New York.

HAWKINS, D.M. and MERRIAM, D.F. (1974). Zonation of multivariate sequen= ces of digitized geologic data. *J. Math. Geol.*, 6, pp. 263-269.

HAWKINS, D.M. and TEN KROODEN, J.A. (1979a). Zonation of sequences of heteroscedastic multivariate data. *Computers and Geosciences*, 5, pp. 189-194.

HAWKINS, D.M. and TEN KROODEN, J.A. (1979b). A review of several methods of segmentation. In D.GILL and D.F. MERRIAM (ed.), *Geomathematical and Petrophysical Studies in Sedimentology*. Pergamon Press.

HOWARD, R.N. (1966). Classifying a Population into Homogeneous Groups, pp. 585-594. In: J.L.LAWRENCE, ed., *Operational Research and Social Sciences*. London, Tavistock Publications.

JAQUET, J.M., FROIDEVAUX, R. and VERNET, J.-P. (1975). Comparison of automatic classification methods applied to take geochemical samples. *J.Math.Geol.*, 7, 237-266.

JARDINE, C.J., JARDINE, N. and SIBSON, R. (1967). The Structure and Con= struction of Taxonomic Hierarchies. *Mathematical Biosciences*, 1, 173-179.

JARDINE, N. and SIBSON, R. (1968). The Construction of Hierarchic and Non-hierarchic Classifications. *Computer Journal*, 11, 177-184.

JOHN, S. (1970). On Identifying the Population of Origin of Each Obser= vation in a Mixture of Observations from Two Normal Populations. *Technometrics*, 12, 553-563.

JOHNSON, S.C. (1967). Hierarchical Clustering Schemes. *Psychometrika*, 32, 241-254.

KRUSKAL, J. (1977). The relationship between multi-dimensional scaling and clustering. In: *Classification and Clustering*, J.van Ryzin, ed., Academic Press, New York.

LANCE, G.N. and WILLIAMS, W.T. (1967). A General Theory of Classificatory Sorting Strategies 1. Hierarchical Systems. *Computer Journal*, 9, 373-380.

LAZARSFELD, P.F. and HENRY, N.W. (1968). *Latent Structure Analysis*. Boston, Houghton Mifflin.

LEE, K.L. (1979). Multivariate tests for clusters. *J.Amer.Statist.Assn.*, 74, 708-714.

MARRIOTT, F.H.C. (1971). Practical problems in a method of cluster ana= lysis. *Biometrics*, 27, 501-14.

MULLER, M.W. (1975). *A Comparison of some Clustering Techniques*. CSIR Special Report PERS 224, National Institute for Personnel Research, Johannesburg.

PLACKETT, R.L. (1950). Some theorems in least squares. *Biometrika*, 37, 149-157.

PRIM, R.C. (1957). Shortest Connection Matrix Networks and some Generali= zations. *Bell. System Technical Journal*, 36, 1389-1401.

SCOTT, A.J. and KNOTT, M. (1976). An approximate test for use with AID. *Appl.Statist.*, 25, 103-6.

SNEATH, P.H.A. (1957). The Application of Computers to Taxonomy. *J. Gen. Microbiology*, 17, 201-226.

SOKAL, R.R. (1977). Clustering and classification: background and current directions. In: *Classification and Clustering*, J.van Ryzin,ed., Academic Press, New York.

SOKAL, R.R. and MICHENER, C.D. (1958). A Statistical Method for Evalua= ting Systematic Relationships. *University of Kansas Science Bulletin*, 38, 1409-1438.

SPAETH, H. (1977). Cluster-Analyse-Algorithmen zur Objektklassifizierung und Datenreduktion. R.Oldenbourg Verlag, Munich.

STANAT, D.F. (1968). Unsupervised Learning of Mixtures of Probability Functions, pp. 357-388. In: KANAL, L.N., ed., *Pattern Recognition*. Washington D.C., Thompson Book Company.

VAN RYZIN, J. (1977) ed. *Classification and Clustering*. Academic Press, New York.

WARD, J.H. (1963). Hierarchical Grouping to Optimize an Objective Func= tion. *J. Amer. Statist. Assoc.*, 58, 236-244.

WEBSTER, R. (1978). Optimally partitioning soil transects. *J. Soil Sci.*, 29, 388-402.

WISHART, D. (1969a). An Algorithm for Hierarchical Classifications. *Biometrics*, 25, 165-170.

WISHART, D. (1969b). Mode Analysis: A Generalization of Nearest Neigh= bour which Reduces Chaining Effects, pp. 282-308. In: COLE, A.J., ed., *Numerical Taxonomy*, London, Academic Press.

WOLFE, J.H. (1970). Pattern Clustering by Multivariate Mixture Analysis. *Multivariate Behavioural Research*, 5, 329-350.

WOLFE, J.H. (1971). A Monte Carlo Study of the Sampling Distribution of the Likelihood Ratio for Mixtures of Multinormal Distributions. *Technical Bulletin STB72-2*, Naval Personnel and Training Research Laboratory, San Diego.

YAKOWITZ, S. (1969). A Consistent Estimator for the Identification of Finite Mixtures. *Ann. Math. Statist.*, 40, 1728-1735.

YAKOWITZ, S.J. and SPRAGINS, J.D. (1968). On the Identifiability of Finite Mixtures. *Ann. Math. Statist.*, 39, 209-214.

INDEX